Ecological Genetics
and Evolution

Buhler 1970.

Ecological Genetics
and Evolution

�֎

ESSAYS IN HONOUR OF

E.B.FORD

EDITED BY

ROBERT CREED

NEW YORK

APPLETON-CENTURY-CROFTS

EDUCATIONAL DIVISION

MEREDITH CORPORATION

BLACKWELL SCIENTIFIC PUBLICATIONS

OXFORD AND EDINBURGH

© 1971 by Blackwell Scientific Publications
5 Alfred Street, Oxford, England and
9 Forrest Road, Edinburgh, Scotland

ISBN 0 632 08360 3

First published 1971

Printed and bound in Great Britain by
William Clowes & Sons Limited
London, Colchester and Beccles

Dedication

This book describes some examples of the experimental study of evolution and adaptation, carried out by means of field-work and laboratory genetics. Much of the work reported here has been inspired and encouraged by Professor E.B.Ford, and we hope that, in some small way, this volume will be complementary to his own book, *Ecological Genetics*. It is therefore with the greatest of pleasure that his friends dedicate this collection of essays to him on the occasion of his seventieth birthday.

In 1963 the Nuffield Foundation gave to the University of Liverpool the extremely generous grant of £350,000 to establish a unit of medical genetics. It is an open secret that in the background of this munificence was Professor E.B.Ford, and the four papers from the Nuffield Unit of Medical Genetics are offered as a small tribute to his constant inspiration.

Foreword

One of the privileges of appointment to a Chair at another University is that it gives one the right to talk to many distinguished people about their work and ideas. E. B. Ford was known to me before I came to Oxford as the author of a book on butterflies and as somewhat of an eccentric, but I was quite unprepared for the welcome he gave me into the Department of Zoology and for the enormous interest of the subject which he gradually revealed to me. My contact with the Genetics Laboratory was made easier by one of the first things I had to do. Within a few weeks of my arrival, it came to light that a new building for another department was to be erected on a piece of land, known to us as 'Henry's weed garden' but generally regarded as being derelict. Even my, at that time, elementary, knowledge of ecological genetics made it easy to realize that the population of caterpillars that had been under continuous observation there for eleven years put it in a rather special category of wilderness; although I did not succeed in saving it, I was able to persuade the university to substitute another experimental plot and this may have helped the geneticists to appreciate that the new professor was not only interested in electrical apparatus.

Later, in All Souls, in a study full of art treasures, I began to savour the full beauty of the subject and to appreciate the intellectual power and depth of understanding of the man who helped to create this new branch of biology. I had not previously realized that one could experiment with evolution as precisely as with a laboratory preparation: that a 'stimulus' can be delivered to a natural population by the release of a genetically aberrant stock and the 'response' recorded as accurately as in any experiment in neurophysiology. I learned of the new measurements of the force of natural selection and of the delicate balance between conformity and diversity that it maintains in the natural world. I could not help feeling on some of these occasions that I was back in the world of Charles Darwin, with the shield of material comfort and security again making it possible to embark on a journey of the mind

into some of the most beautiful and challenging labyrinths of understanding. I am glad to have the opportunity to join with the contributors
to this volume in expressing my gratitude to Henry Ford for this experience that he has given us.

Dr Creed has suggested that I might write something about the
history of genetics in Oxford and its plans for the future. The various
branches of biology are like the individual animals of a species. As
a multicellular animal grows from a single germ cell, so each can trace
its origin to a single event. With genetics, there was a diapause, but
then development started, with much differential growth and first one
and then another facet becoming the focus of attention. In the scientific
world in general, the growth of genetics has recently been lop-sided,
until to many people the organism as a whole appears almost to be
composed of a single part, molecular biology. I hesitate to diagnose
malignancy, but the invasiveness is undoubtedly there, with a noticeable
lack of contact inhibition. In Oxford this particular phenomenon did
not occur; under Ford in Zoology and Darlington in Botany, two other
parts developed into limbs and we had, until recently, a very active
creature, but one lacking a certain cohesion. It is no insult to ecological
genetics to say that, though extremely important, it is incomplete in
itself; biological departments other than zoology and botany made it
clear that they would welcome a more comprehensive basis of teaching
and research.

E. B. Ford has now (officially) retired and, by the effluxion of time,
Darlington must follow soon. When these unfortunate events forced
themselves on our attention, we other biologists in Oxford decided that
we would try to give Genetics a proper focus and cease to rely on the
accident that zoologists and botanists with an interest in the subject
would be appointed to permanent posts. By pooling resources from
several departments we have now created a statutory Chair of Genetics
and have established the nucleus of a new department. We are fortunate
that Walter Bodmer has come to begin the task of building it up, for
his researches and interests cover a very wide field. In such a situation,
it is inevitable that ecological genetics will cease to be a distinguishing
feature of the subject in Oxford. The addresses of contributors to this
volume show that we shall have a serious rival in a university of the
north, but work will continue here in the department of zoology as
well as in the new laboratory of genetics. Our own students will see that
the ecological approach survives. And in any case, the very success of

the outlook which Henry Ford has fostered, coupled with the resurgence of interest among biologists and others in the natural environment in which we live, makes it certain that ecological genetics will now be recognized as an integral part of the subject in most universities of the world. This is the cultural heritage that he has given us and it will long outlast even the continuing career of the great man to whom this volume is dedicated.

Department of Zoology,   J. W. S. PRINGLE
Oxford

Contents

List of Contributors

ERRATUM

Add to list of contributors (p. xiii)

A.D. BRADSHAW, Department of Botany, University of Liverpool, Liverpool, England

Ecological Genetics and Evolution

Department of Psychology, University of Washington, Seattle, U.S.A.

L. BROWER Department of Biology, Amherst College, ssachusetts, U.S.A.

ER Department of Biology, Amherst College, Amherst, s, U.S.A.

JRY The Royal Society for the Protection of Birds, Sandy, , England

Department of Zoology, University of Liverpool, Liverpool,

E Department of Zoology, University of Edinburgh, Edinburgh, esent address: Genetics Laboratory, School of Botany and Zoology, rk, Nottingham, England)

KE Nuffield Unit of Medical Genetics, University of Liverpool, ngland

J.B.CLEGG Nuffield Unit of Medical Genetics, University of Liverpool, Liverpool, England

L.M.COOK Department of Zoology, University of Manchester, Manchester, England

E.R.CREED Department of Zoology, University College, Cardiff, Wales

C.D.DARLINGTON Botany School, University of Oxford, Oxford, England

Th. DOBZHANSKY Rockefeller University, New York, U.S.A.

W.H.DOWDESWELL School of Education, Bath University, Bath, England

D.A.P.EVANS Nuffield Unit of Medical Genetics, University of Liverpool, Liverpool, England

P.T.HANDFORD Genetics Laboratory, Department of Zoology, University of Oxford, Oxford, England

xiii

H.B.D.KETTLEWELL Genetics Laboratory, Department of Zoology, University of Oxford, Oxford, England

D.R.LEES Genetics Laboratory, Department of Zoology, University of Oxford, Oxford, England

K.G.McWHIRTER Department of Genetics, University of Alberta, Edmonton, Canada

J.J.MURRAY Department of Biology, University of Virginia, Charlottesville, Virginia, U.S.A.

P.O'DONALD Department of Genetics, University of Cambridge, Cambridge, England

The Hon. MIRIAM ROTHSCHILD Ashton, Peterborough, England

J.R.G.TURNER Department of Biology, University of York, Heslington, York, England

D.J.WEATHERALL Nuffield Unit of Medical Genetics, University of Liverpool, Liverpool, England

E.B.Ford: bibliography

It is usual for a bibliographer to pick out a few papers of outstanding quality and high-light them. This is an impossible task in the case of E.B.Ford's work, because of its almost relentless excellence. There are, however, a number of his papers which introduce original ideas of general significance and these must be mentioned. Thus for example both the 1925 and 1927 papers, in collaboration with Julian Huxley (working with *Gammarus chevreuxi*) demonstrated for the first time that genes control the time of onset and rate of development of processes within the body. The paper on the variability of species in the Lepidoptera (1928b) was published jointly with R.A.Fisher. The collaboration between these two men was extraordinarily felicitous for E.B.Ford took Fisher's brilliant but purely theoretical and mathematical predictions into the fields and the woods and demonstrated that they actually 'worked' in nature. The paper on *Melitaea aurinia* written in collaboration with his father and published in the *Transactions of the Entomological Society* (1930b), pointed out and described the relationship between fluctuation in numbers and rapid evolution, stressing the fact that variability increases as the organisms become more numerous. The 1937 paper shows that successful industrial melanics are characterized by physiological advantages as well as colour changes. The investigation of the pigments of Lepidoptera (1938, 1941 & 1942) was one of the earliest and most successful attempts at relating chemistry to classification. *Butterflies* (1945), number one of the New Naturalist series, is probably the best popular book of its kind ever written. It provides the ordinary reader with an entirely fresh interest in his hobby, opening up areas of research within the grasp of any intelligent collector, and skilfully blending pure natural history with genetics, ecology and chemistry. This book and its companion volume *Moths* (1955) has brought great pleasure and added interest to thousands of readers.

Between 1930 and 1940 E.B.Ford 'invented' polymorphism. The term is defined in his 1940 paper in *The New Systematics* and two years later in his book *Genetics for Medical Students* he treated the human blood groups, for the first time, as polymorphisms. This was a revolutionary step forward. Three years later (1945), with almost prophetic insight, he predicted the association of specific diseases with the human blood groups.

One of the notable qualities of gifted men is their versatility, and therefore it comes as no surprise to find that E.B.Ford has made a distinguished contribution in quite another field. He is one of the eminent band of amateur archaeologists who have added to the high quality of the work of the Prehistoric Society. His special interest is in the Iron Age (La Tène) buildings of ultimate Cornwall, their souterrains or *fogous*, on which he and his collaborators have contributed two scholarly papers (1953e, 1957e).

Perhaps the most outstanding feature of E.B.Ford's bibliography is its unity and balance. The author has set out to place the study of evolution on an experimental basis, by a combination of laboratory genetics and field ecology. He has not been diverted or sidetracked but he devises new techniques when they are required. Thus, marking, release and recapture was used to demonstrate selection and estimate its force in a wild population (1947c), and again in 1949 (b) to show average differential survival between populations of known size. A gift of paramount importance—that of selecting the right type of material for a certain type of investigation—is well illustrated in the long series of papers (with various collaborators) on *Maniola jurtina* (1949 onwards).

The bibliographer perceives this work proceeding by a series of logical steps, touched with intuitive insight and sustained by arduous field work. With the publication of *Ecological Genetics* (1964) a synopsis of the achievements and future aims of this central theme is presented with stimulating and triumphant clarity. It should be noted that the first edition of this book generated a remarkable surge of interest in the subject generally, so that before the third edition has been printed E.B.Ford has been invited to organize laboratories designed for the study of ecological genetics in many parts of the world as far afield as Finland, Jordan, Canada and France.

M.R.

H.B.D.K.

BOOKS

1931   *Mendelism and Evolution.* Methuen, London
      8th edition 1965
      Spanish translation by Ovidio Nunez, 1950, Buenos Aires

1933   *Mimicry*, with G.D.H.Carpenter. Methuen, London
       Spanish translation by M.M.Grassi & A.P.L.Digilio, 1949,
       Buenos Aires
1938   *The Study of Heredity*, Oxford University Press, London
       2nd edition 1950
1942   *Genetics for Medical Students*. Methuen, London
       6th edition 1967
       Italian translation by L. Cavalli Sforza, 1948, Longanesi, Milan
1945   *Butterflies*. New Naturalist Series, Collins, London
       48 colour plates, 24 black-and-white plates, 32 maps
       2nd reprint of 3rd edition 1967
1951   *British Butterflies*. Penguin Books, London
       16 colour plates
1955   *Moths*. New Naturalist Series, Collins, London
       32 colour plates, 24 black-and-white plates, 12 maps
       2nd edition 1967
1964   *Ecological Genetics*. Methuen, London
       3rd edition 1971
       Polish translation by K.Zacwilichowska, 1967, Warsaw
1965   *Genetic Polymorphism*. All Souls' Studies, Faber & Faber,
       London

                              PAPERS

1924   The geographical races of *Heodes phlaeas* L. *Trans. ent. Soc.
       Lond.* (1924) 692–743.
1925   Mendelian genes and rates of development. *Nature, Lond.* **116**,
       861–3 (with Julian Huxley)
1926   Variability of species. *Nature, Lond.* **118**, 515–16 (with R.A.Fisher)
1927   Mendelian genes and rates of development in *Gammarus chevreuxi*.
       Brit. J. exp. Biol. **5**, 112–34 (with Julian Huxley)
1928a The inheritance of dwarfing in *Gammarus chevreuxi*. *J. Genet.*
       20, 93–102.
1928b The variability of species in the Lepidoptera with reference to
       abundance and sex. *Trans. ent. Soc. Lond.* (1928) 367–84 (with
       R.A.Fisher)
1929   Genetic rate-factors in *Gammarus*. *Wilhelm Roux Arch.
       EntwMech. Org.* **117**, 67–79 (with Julian Huxley)

1930a The theory of dominance. *Amer. Nat.* **64**, 560–6

1930b Fluctuation in numbers, and its influence on variation in *Melitaea aurinia. Trans. ent. Soc. Lond.* **78**, 345–51 (with H.D.Ford)

1931 The health and parasites of a wild mouse population. *Proc. zool. Soc. Lond.* (1931) 657–721 (with C.Elton and J.R.Baker)

1936 The genetics of *Papilio dardanus. Trans. R. ent. Soc. Lond.* **85**, 435–55

1937 Problems of heredity in the Lepidoptera. *Biol. Rev.* **12**, 461–503

1938 The genetic basis of adaptation, from *Evolution* (pp. 43–55), Oxford University Press

1939 Taste-testing the anthropoid apes. *Nature, Lond.* **144**, 750 (with R.A.Fisher and Julian Huxley)

1940a Polymorphism and taxonomy, from *The New Systematics* (pp. 493–513), Oxford University Press

1940b Genetic research in the Lepidoptera (being the Galton Lecture of London University, delivered August 1939 to the Seventh International Congress of Genetics). *Ann. Eugen.* **10**, 227–52.

1940c The quantitative study of populations in the Lepidoptera. 1. *Polyommatus icarus. Ann. Eugen.* **10**, 123–36 (with W.H. Dowdeswell and R.A.Fisher)

1941 Studies on the chemistry of pigments in the Lepidoptera, with reference to their bearing on systematics. 1. The anthoxanthins. *Proc. Roy. ent. Soc. Lond.* A **16**, 65–90

1942a Ibid. 2. Red pigments in the genus *Delias. Proc. Roy. ent. Soc. Lond.* A **17**, 87–92

1942b The proportion of the pale form of the female in *Colias croceus* Fourcroy. *Entomologist* **75**, 1–6

1944a Studies on the chemistry of pigments in the Lepidoptera, with reference to their bearing on systematics. 3. The red pigments of the Papilionidae. *Proc. Roy. ent. Soc. Lond.* A **19**, 92–106

1944b Ibid. 4. The classification of the Papilionidae. *Trans. Roy. ent. Soc. Lond.* **94**, 201–23

1945 Polymorphism. *Biol. Rev.* **20**, 73–88

1947a A murexide test for the recognition of Pterins in intact insects. *Proc. Roy. ent Soc. Lond.* A **22**, 72–6

1947b Studies on the chemistry of pigments in the Lepidoptera, with reference to their bearing on systematics. 5. *Pseudopontia paradoxa* Felder. *Proc. Roy. ent. Soc. Lond.* A **22**, 77–8

1947c The spread of a gene in natural conditions in a colony of the moth *Panaxia dominula* L. *Heredity* **1**, 143–74 (with R.A.Fisher)

1948 The genetics of habit in the genus *Colias*. *Entomologist* **81**, 209–11 (with W.H.Dowdeswell)

1949a Early stages in allopatric speciation. *Princeton Bicentennial Volumes: Symposium of Genetics, Palaeontology and Evolution*

1949b The quantitative study of populations in the Lepidoptera. 2. *Maniola jurtina. Heredity* **3**, 67–84 (with W.H.Dowdeswell and R.A.Fisher)

1949c Genetics and cancer. *Heredity* **3**, 249–52

1950 The Sewall Wright effect. *Heredity* **4**, 117–19 (with R.A.Fisher)

1951a Organic evolution, from *A Century of Science*, 1851–1951 (ed. Herbert Dingle), Hutchinson, London

1951b Genetics, from *Scientific Thought in the Twentieth Century* (ed. Prof. A.E.Heath), Watts & Co., London

1952a The distribution of spot-numbers as an index of geographical variation in the butterfly *Maniola jurtina. Heredity* **6**, 99–109 (with W.H.Dowdeswell)

1952b The influence of radiation in the human genotype, from *Biological Hazards of Atomic Energy* (ed. A.Haddow), Oxford University Press, pp. 67–71

1953a The experimental study of evolution. *Aust. & N.Z. Assn. Adv. Sci.* **28**, 143–54

1953b Polymorphism and taxonomy. *Rept. 7th Sci. Congress, Roy. Soc. N.Z.* (1951), 245–53

1953c The genetics of polymorphism in the Lepidoptera. *Advanc. Genet.* **5**, 43–87

1953d The influence of isolation on variability in the butterfly *Maniola jurtina. Symp. Soc. exp. Biol.* **7**, 254–73 (with W.H.Dowdeswell)

1953e An above-ground storage pit of the La Tène Period. *Proc. Prehist. Soc.* **19**, 121–6 (with E.V.Clark)

1954 Problems in the evolution of geographical races, from *Evolution as a Process*, Allen & Unwin, London, pp. 99–108

1955a Evolution in polymorphic forms, *Proc. 9th Internat. Congr. Genetics. Caryologia* (Firenze), supp. to vol. **9**, 463–8

1955b A uniform notation for the human blood groups. *Heredity* **9**, 135–42

1955c Ecological genetics of *Maniola jurtina* on the Isles of Scilly. *Heredity* **9**, 265–72 (with W.H.Dowdeswell)

1956a Rapid evolution and the conditions which make it possible. *Cold Spr. Harb. Sym. quant. Biol.* **20** (1955), 230–8

1956b Evolutionary processes in animals. Appendix to *Chromosome Botany* by C.D.Darlington, Allen & Unwin, London, pp. 169–173.

1957a *Polymorphism.* Chambers's Encyclopaedia. World Survey **4**, 83–5

1957b Further studies on isolation in the butterfly *Maniola jurtina*, L. *Heredity* **11**, 51–65 (with W.H.Dowdeswell and K.G.McWhirter)

1957c The study of evolution by observation and experiment, being the 1st Woodhull Lecture delivered 15.3.57; *Proc. Roy. Instn.* **36**, 1–9

1957d Polymorphism in plants, animals and man. *Nature, Lond.* **180**, 1315–9

1957e The Fogou of Lower Boscaswell, Cornwall. *Proc. Prehist. Soc.* **23**, 213–19 (with E.V.Clark and C.Thomas)

1958 Darwinism and the study of evolution in natural populations. *J. Linn. Soc. (Zool.)* **44**, 41–8

1959a Evolutionary studies on *Maniola jurtina* on the English mainland 1956–1957. *Heredity* **13**, 363–91 (with W.H.Dowdeswell, K.G.McWhirter and E.R.Creed)

1959b An evolutionary study of the butterfly *Maniola jurtina* in the north of Scotland. *Heredity* **13**, 353–61 (with Bruce Forman and K.G.McWhirter)

1960a Evolution in progress, from *Evolution after Darwin* (ed. Sol Tax) **2**, 181–96. Chicago Press

1960b Further studies on the evolution of *Maniola jurtina* in the Isles of Scilly. *Heredity* **14**, 333–64 (with W.H.Dowdeswell and K.G.McWhirter)

1961a Butterflies and moths in scientific research. *Proc. Somerset Arch. & Nat. Hist. Soc.* **104**, 28–42

1961b The theory of genetic polymorphism. *Symp. R. ent. Soc.* **1**, 11–19

1962 Evolutionary studies on *Maniola jurtina*: The English mainland 1958–60. *Heredity* **17**, 237–65 (with E.R.Creed, W.H.Dowdeswell and K.G.McWhirter)

1964 Evolutionary studies on *Maniola jurtina*: The Isles of Scilly, 1958–59. *Heredity* **19**, 471–88 (with E.R.Creed and K.G.McWhirter)

1966a Genetic polymorphism. *Proc. Roy. Soc.* B **164**, 350–61

1966b Natural selection and the evolution of dominance. *Heredity* **21**, 139–47 (with P.M.Sheppard)

1968 Ecological genetics. *Advanc. Sci.* **25**, 1–9

1969a Ecological genetics. Chapter 8 in *Aspects of Scientific Thought, 1900–1960* (ed. R.Harré), Oxford University Press

1969b Le Mélanisme industriel. *Bull. Soc. zool. Fr.* **94**, 119–28

1969c The *medionigra* polymorphism of *Panaxia dominula*. *Heredity* **24**, 561–9 (with P.M.Sheppard)

1970 Evolutionary studies on *Maniola jurtina*: The 'Boundary Phenomenon' in Southern England 1961–8, from *Essays in Evolution and Genetics in Honor of Theodosius Dobzhansky* (eds. M.K.Hecht and W.C.Steere), Appleton-Century-Crofts, pp. 263–87 (with E.R.Creed, W.H.Dowdeswell and K.G.McWhirter)

1 ❋ The Evolution of Polymorphic Systems

C.D.DARLINGTON

I. POINTS OF VIEW

Genetics is a framework which we are engaged in fitting into our understanding of life. This understanding has been built up from what are academically described as biology, medicine, and agriculture, anthropology and history. The framework is being built up by a variety of methods which are inherent, not so much in the variety of living things as in a variety of points of view.

What are these points of view? They were established already in the last century by Galton, Mendel and Weismann. It is in this order that we must put them for Galton wanted to understand the character of whole organisms and to separate the internal and external causes of their differences; Mendel wanted to discover and separate the different internal causes; and Weismann, considering the causes established, wanted to separate them from their consequences in their proper evolutionary sequence. For him sexual reproduction, by recombining differences, conserved variation and provided the materials for natural selection to work upon. Natural selection thus favoured the origin and development of sexual reproduction itself. Together, Weismann came near to saying, they initiated the evolution of genetic systems.

This variety of approach and of opinions is still with us. We have the original contrast of aim and method. We have the integral or individual point of view which sees whole and visible organisms as the reality to be understood by comparison and in relation to the communities to which they belong. We have the analytical or differential point of view which sees the separation of invisible factors or elements, genes or nucleotides, as the reality to be discovered by experiment. But we have also a microscopic point of view which grudges to allow any reality to these determinants until they have been seen and recorded as concrete and chemical microscopic objects; thus offering us a mirror image of both the earlier

I

views of our science. And beyond these we have the evolutionary point of view which adds a new dimension to all the others. These several ways of looking at life and heredity and variation have given their exponents their own frames of reference, their own languages, their own real worlds of genetics.

These various worlds are each apparently capable of growing without limit. And their separate development is characteristic of modern education and research which provide separate niches for activities which will suit the individual preferences of every one of us. Yet on a long view such separation is out of place in this field. A grave error lies at its root. For the integral approach takes account of individual organisms and even real people. It appraises properties such as intelligence and fertility which are beyond analysis. That is bad enough. But worse follows. It carries with it the dreadful suggestion, which some time ago paralysed the science of botany, that it may be useful. It may be trembling on the brink of application. This misplaced fear underlies the decay of the integral and the prestige of the analytical approach to heredity.

Another grave error grows out of this separation. For the business of genetics is to seek out connections, connections between organisms, between determinants, between processes, and between all these internal orders and the external world in which they have their being. Separation therefore will defeat our prime purpose unless we defeat it first.

2. FORDIAN POLYMORPHISM

It is just here, in this business of seeing connections that we have to look at the work of Edmund Ford. Here we have a naturalist with a discerning eye for the visible forms of animals and plants, an eye practised above all in knowing the Lepidoptera; and seeing beyond what genetics could do for the Lepidoptera to the larger question of what the Lepidoptera could do for genetics. What they did do in Ford's hands however proved to be something out of the ordinary and for reasons which are worth our while examining.

Already before the publication of Morgan's *Theory of the Gene* the crucial question in evolutionary thought was whether, and if so how, mutations of the type which had been used to demonstrate the theory of heredity in *Drosophila* could also have been used as the materials of evolutionary change. How could the good have come out of the bad?

To this question several answers could be and were attempted. But the one which gave the most immediately intelligible result, since like *Drosophila* genetics it was based on the experimental breeding of insects, was given us by Ford. It was contained in his principle of balanced polymorphism.

Ford's statement of this principle by its austerity and abstraction was calculated and perhaps intended to deter any but a rigorous mind entering into the discussion. In the event it concealed some of its implications which only slowly became evident. What he had discovered was (to use one of the possible modes of expression) that mutant genes could exist in a species balanced against their alternatives by selective forces. And that when they did so they became clustered in adaptive aggregates. By these evolutionary steps they could become the basis of a stable polymorphism within a single breeding group or genetic species.

3. ACTIVITY AND NEUTRALITY

The question now arose as to why differences of a type which we (following Darwin) might expect to lead to the splitting of a species into two parts could remain balanced within it. The facile assumption was sometimes made that these were neutral differences as opposed to the active ones which split species. Such a notion is parallel to that applied to the supernumerary chromosomes of plants and animals, supposedly inert or even parasitic, to which I shall return later. It involves the absurdity of imagining that the most vital and selectable of living materials are unable to rid themselves of their own grossly defective parts. The argument in relation to polymorphism was however much stronger than for inert chromosomes. The clusters or complexes of polymorphic systems could only have been generated by powerful selective forces due to the general advantages of diversity or the specific advantage of the heterozygote or of a breeding system generating heterozygotes. This view led Ford to predict the activity of polymorphic genes or super-genes in relation to disease resistance.

When we turn from visible or phenotypically recognized polymorphisms such as that of *Panaxia dominula* and look at the concealed or genotypic polymorphisms revealed by inversion differences in *Drosophila* the argument of activity is overwhelmingly confirmed. The species *D. willistoni* in the aggregate is known to be heterogeneous and heterozygous for 50 different inversions. Particular populations average in their heterozygosity as many as 9 inversions. These as a rule are separate and able to recombine *inter se*. But within the inversion recombination is suppressed. The materials are inherited as a unit. This immense and highly specific heterozygosity seems to be having the same active effect as that of the visibly expressed heterozygosity of *Panaxia*. In both cases the differences are not concerned with splitting the species. Are they concerned rather with holding it together? Here the evidence of human evolution is most instructive.

4. MAN: AN EXPANDING SPECIES

The development of the brain in man was so dominating in effect that, like the development of warm-bloodedness in the original mammals, it might have given rise to a whole new class of animals. Instead it led to the appearance of numerous competing and presumably inter-sterile species. Repeatedly, it would seem, by their conflicts all but one were destroyed. By this fragmentation and extinction the diversification of the group was interrupted and its internal polymorphism contracted. At the time of the great human expansions leading to the colonization first of Eurasia and then later of America, Australia and Polynesia a new evolutionary pattern appeared. Mankind had made good these earlier losses and now in addition contained sufficient geographical diversity to provide for dozens of ordinary systematic species. Yet this time the unity of the species, the inter-fertility of its parts, was preserved. The main factor in this preservation was probably the development of man's outbreeding system resting on a now universal incest taboo which restrained inbreeding and so prevented fragmentation.

A subsidiary factor in preserving this unity was the development of a reservoir of balanced polymorphism in the central body of mankind. How valuable this reservoir could be was proved in the last great expansions. These were so rapid that the new peripheral colonizers shed a part of their polymorphism. Ford has illustrated this peripheral loss in many animals in terms of form and behaviour. In *D. robusta* a similar change is indicated by a loss of inversions. In man it is shown by the disappearance of certain blood groups. It is also indicated by the loss of resistance to the diseases which have continuously infested the central body of mankind. In the dense populations of this main mass, densely parasitized, polymorphism evidently preserved and developed resistance especially to virus diseases. But when, following the great navigations, the peripheral peoples were brought into contact with these central diseases they were often destroyed before they could acquire resistance by the only possible means which was hybridization.

Thus polymorphism and the incest taboo seem to have been the means of holding together an expanding species, of preserving unity-with-diversity. The success has been unexampled and until recent times, when man broke out of the old continents, the success has been complete.

5. THE CAPTURE OF THE BREEDING SYSTEM

What man's evolution has been able to show has been by the range of non-experimental studies, descriptive, historical and statistical, so far un-assisted by the third method, the use of the chromosomes. It was by the chromosome method however, as Ford early observed, that the mechan-isms underlying the evolution of polymorphism were put into relation with his own understanding of the problem (table 1.1).

The gain from this insight has been on both sides and much, perhaps most, of it still lies in the future. We may see this from considering the problems of *Oenothera* and of sex chromosomes in plants and animals. These have both been in the forefront of our minds during the last forty years. And they both depend on the possession of permanent unanalysable differences.

In both these types of situation heterozygosity was often conveniently marked for us by structural differences between the chromosomes pairing at meiosis. Sometimes they were differences of size, sometimes of inter-change or fragmentation, and sometimes they were made up of several differences. Where interchanges were concerned clearly linkage groups were being united and recombination was being correspondingly reduced. Where fragmentation was concerned the reverse was the case.

Here were situations where powerful genetic differences were taking effect but the differences could not be broken up into any plausibly elementary genetic units. They were complex differences. The complexes and the polymorphisms were however revealed in altogether new situa-tions. For example in the sex chromosome systems of the higher plants and animals there is generally, and in varying degrees, a heterozygous disadvantage—first pointed out by Haldane. In *Oenothera* the contrary

TABLE I.I. Polymorphic systems.

Type	Notation	Function	Evolution	Incidence	References
1. ALLELIC or MENDELIAN (floating)	AA, Aa, aa	maintains diversity or develops heterozygous advantage (HA) restricts recombination	single nucleotide to supergene with inversions, interchanges, duplications, fusions, heterochromatin	Man PANAXIA DROSOPHILA TRILLIUM etc.	E.B. Ford 1965 Darlington 1965 (Table 81) Dobzhansky 1951 Novitski et al. 1965 Darlington & Shaw 1959
2. BACK-CROSSING (unstable)	(i) ss : Ss (ii) XX : XY	enforces outbreeding - by heterostyle incompatibility (HA) - by diploid sex determination (HD)	increasing divergence of differential segments with suppressed recombination	PRIMULA LYTHRUM etc. all dioecious P. & A.	Mather 1950 Dowrick 1956 Darlington 1965
3. PERMANENT HETEROZYGOTE	(i) A+B : AB (ii) X+Y : XY	maintains complex hybridity with self fertilisation (HA) maintains haploid sex differentiation	(i) interchanges with balanced lethals (ii) differential segments give haploid sex dimorphism	OENOTHERA ISOTOMA BRYOPHYTA	Darlington 1958 1965 James 1970 Darlington 1965
4. MULTIPLE ALLELIC	$S_1 S_2 S_3 ...$	enforces outbreeding by incompatibility (HA)	accumulates differences by total heterozygosity near S locus	HIGHER & LOWER PLANTS	Crowe 1964
5. PARTHENO-GENETIC (frozen)	ABC, abc	stabilises élite heterozygote (HA)	heterozygosity for multiple inversions	DROSOPHILA mangabierai	Carson 1962
6. SUPER-NUMERARY	(i) $2x + 0, 1, 2, 3B..$ (ii) $2x \rightleftharpoons 2x - 1$	generates diversity by meiotic irregularity	(i) B chromosomes (ii) Iso-chromosomes	short-lived P. & A. NICANDRA	Darlington & Wylie 1956 Darlington 1965
7. CYTOPLASMIC or INFECTIOUS	:	controls sex-ratio. controls race relations	kills embryos of heterozygous sex kills non-carriers	DROSOPHILA PARAMECIUM ·Yeast	Sakaguchi 1962, 1965 Preer 1967 Somers & Bevan 1968

HA: Heterozygous advantage.   HD: Heterozygous disadvantage

is true; there is an extreme heterozygous advantage. It is so complete that homozygotes of both kinds are lethal and the polymorphism is effective only in the haploid state, $A + B$ alternating with AB. This situation is closely paralleled by the inheritance or determination of sex in the Bryophyta where $X + Y$ alternate with XY. But whereas in the Bryophyta it is the outbreeding mechanism itself which creates and maintains the complexes, complex hybridity is forced on its victims by a switch from outbreeding to inbreeding; whether in nature on *Oenothera* or *Isotoma* or in experiment on *Campanula* or *Periplaneta*.

We know in many such cases that the interchange of segments of chromosome which brings them to our notice has the effect of holding together whole chromosomes by preventing recombination between them. Indeed in the extreme case by causing a whole chromosome set to segregate in a block. We also know that in all plants and animals, when (as was first discovered in *Drosophila*) segments of chromosome are inverted, crossing over is inhibited or frustrated in the inversion hybrid. In these two situations we see the respective bases of origin of great complexes and of smaller units like those to which we have given the name of super-gene. Between the two we see many gradations. For what may be a super-gene in distinguishing the sex chromosomes of *Humulus* or *Rumex* may be attached by interchanges to form a chain of chromosomes incorporating (in the centipede *Otocryptops*) most of the complement. In other words we see evolutionary processes in which chromosome changes attach themselves in succession to a variety of balanced polymorphisms.

Which is captor and which is captive in these transactions it is useless to ask since the two roles alternate. But we can ask whether, on an evolutionary time scale, the capture is reversible or irreversible. The answer is that individual interchanges are irreversible. But the large complexes built up by successive interchanges are occasionally and partially reversible. For by crossing over between opposite complexes they can be recombined and the earlier evolution thereby reversed. The main mutations in *Oenothera*, in my view, happen by this kind of reversal.

Another kind of capture which is definitely irreversible happens with parthenogenesis. If a new gene combination appears allowing suppression of meiosis and the capacity to develop without fertilization, and if this comes about in a particularly well-favoured individual, then at one stroke a new parthenogenetic species appears. In both plants and animals such species are commonly triploid. But in the fly *D. mangabeirai* it happened to an individual heterozygous for three inversions. The whole species is uniformly heterozygous for these three inversions and is thus evidently an identical extension of this individual. In a word parthenogenesis has frozen or pegged the shifting apparatus of polymorphism at a point evidently of the highest heterozygous advantage. A rare and improbable combination of events has become the means of origin of a new species.

Reversible polymorphism is as important as the irreversible but it depends on a different sequence of events, a different evolutionary strategy. How does it come about?

Reversible polymorphism is best illustrated by the behaviour of the heterostyle super-gene and reversibility we may take to be the distinguishing mark of this class of super-gene. Heterostyly is known in 51 genera of 18 families of Angiosperms but it has been most thoroughly studied in a few species of primroses. Over a hundred years of experiment have succeeded in showing how the super-gene differences between pin and thrum manipulate the style as a pollen-catcher and pollen-filter and the pollen as a style penetrator so as to favour outbreeding. In the genus *Primula*, over millions of generations, 120 heterostyle species have evolved with many still-to-be-explored variations on the same theme. Today, between the component parts of the pin-thrum heterozygote, crossing over can still take place. It gives reciprocal long and short 'homostyle' plants of various types according to the place where the sequence of components is broken. The breakdown of the super-gene thus reveals the linear order of these components (Fig. 1.1).

This crossing over takes place both in nature and in experiment. And it takes place most often in new tetraploids. Why? Because in the tetraploids the positions of crossing over are shifted. The process is analogous to the crossing over which breaks down the complexes in a new species of *Oenothera* and gives mutations. But in this case the super-gene has been protected, not by interchanges or inversions, but by lying outside the normal range of crossing over.

The need for protection from crossing over is one of the two or three great paradoxes of genetics and it is the one which lies at the root of the evolution of discontinuity, both within and between species.

All the properties of genetic systems, notably their polymorphisms and their crossing over, can be discussed in opposed terms which are the reciprocal of one another. These are the terms which make, as we saw, the several dialects of the experimental breeder and the microscopist, of the social and the molecular geneticist. We need not suppose that one dialect is superior to another. But when Fisher tells us of the *advantage* of intensified linkage we may easily forget that this means the *disadvantage* of too much crossing over. We may then fail to ask the question, the crucial question, of why crossing over cannot be reduced to any specifiable frequency, in fact to an optimum frequency.

The answer is twofold. In the first place chromosome observations show that there is no unconditional optimum frequency. In the second

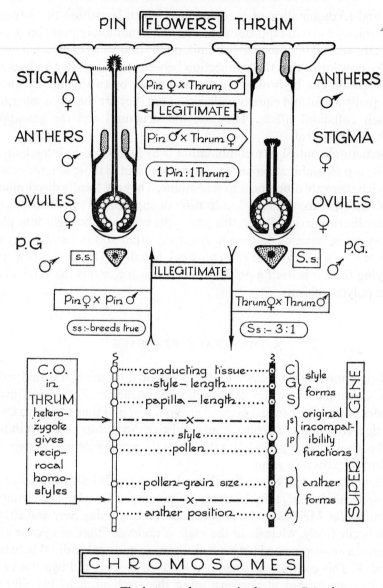

FIGURE I.I. The heterostyle system in the genus *Primula*.

place (in my view) the crossing over serves two quite different ends. It is necessary for the recombination of genes within each pair of chromosomes. But it is also necessary, by its effect of chiasma formation, for the main sequence of events in normal meiosis: that is for the pairing, reduction and recombination of chromosomes which constitute the internal mechanism of sexual reproduction. The first condition imposes an upper limit; the second imposes a lower limit, on the frequency of crossing over.

We now know that this connection between crossing over and sexual reproduction can be broken. It can be broken by two methods. These are equally useful and equally widespread but they are sharply contrasted in their collateral effects. They are the structural and the genotypic methods of control.

Structural control of recombination is by inversions or interchanges which stop recombination just in those places and in those heterozygotes in which there are differences to recombine. They are locally discriminating. Genotypic control is by localization or suppression of crossing over. It is undiscriminating. When this genotypic control is brought into play by any sexual species or genetic system it often allows itself a way of escape: a way which I will call two-track heredity. This device is worth studying since it is itself a polymorphism which controls the survival of other polymorphisms.

7. TWO-TRACK HEREDITY

How does the two-track system work? The obvious method would seem to be sequential. For example by cyclical parthenogenesis the short-lived aphides extend their sexual cycle to a whole year and thus cut down their genetic recombination to once a year and give themselves genetic stability for the whole of that period. But this is not the genuine article: it gives no choice; it leaves no option.

Genuine two-track heredity depends on a parallel differentiation in the mechanism of heredity between the two sexes. The classical example is *Drosophila*. Meiosis in the female is normal: crossing over and chiasmata occur freely. Meiosis in the male is abnormal: crossing over and chiasmata are suppressed between autosomes, and are localized between X and Y. This contrast between two kinds of meiosis and two tracks in heredity was crucial in demonstrating the mechanism of heredity in experiment. In my opinion, it has been equally crucial in exploiting this mechanism in the course of evolution. It means that special gene sequences, including super-gene polymorphisms and inversions, if they are beneficial, can be handed down in the male line for ever. And without causing sterility in the heterozygote.

The incidence of this form of two-track heredity has been surveyed by John and Lewis. It is characteristic of many groups of short-lived animals, the heterozygous sex being always the one with suppressed recombination. It is found not only in the higher Diptera but also in the Orthoptera, Mecoptera, Lepidoptera and Homoptera, in the Acarina, Copepoda and scorpions. And with male haploidy in the Hymenoptera and other groups the same two-track effect is achieved.

Nor is two-track heredity confined to animals. For we now find just the same contrast as in *Drosophila* established as a difference between meiosis in the male and female lines of hermaphrodite plants, species of *Fritillaria*: the pollen mother cells have meiosis without chiasmata. Does this differentiation occur in other plants? Very likely. In the higher plants as a whole however a simpler device is always available and is perhaps universally applied. For in them crossing over is always higher in the embryo-sac than in the pollen mother cells and it is always distributed differently in the two types of meiosis. In addition it is always distributed differently in chromosomes of different shapes and sizes. Incipient super-genes of complexes can, and therefore will, be placed by ordinary processes of breakage and reunion in regions of low crossing over suitable for the character required of any particular combination.

We thus find ourselves with an instructive contrast between two primary methods of restricting recombination, both of them used in animals and in plants (table 1.2).

The question may well be asked at this point: do these contrasted systems, structural and genotypic, exclude one another or do they combine with one another? Are they competitive or complementary? The answer is that they combine with one another and we must now consider how they do so.

TABLE 1.2. Modes of restriction of recombination which are capable of favouring the development of polymorphism.

Control	Reversibility	Effectively discriminating between		
		Sexes	Chromosomes	Het/Homozygotes
Structural	−	−	+	+
Genotypic (Two-track)	+	+	−	−

8. HIERARCHICAL SYSTEMS

Breeding systems, like anatomical structures, develop by a hierarchy of processes which follow one another in evolutionary sequence. This development is facilitated by their characteristic *prolepsis*: the principle that they are always selected by their effects on future, unborn, generations. It is not surprising therefore, as we have seen, that when alternative or polymorphic gene complexes come to control them (and to be controlled by them) these complexes should have a hierarchical character and a recognizable sequence of development in time. The earliest and strangest example of such a sequence is still hypothetical. It is that noted by Darwin in *Rhamnus cathartica*. Dioecy, he suggested, had been superimposed on heterostyly in the evolution of the species. It had conflicted with heterostyly and had displaced it. The situation still needs to be experimentally elucidated.

Another example, very thoroughly elucidated, is the series of systems concerned with the control of the sex-ratio, again a problem which exercised Darwin. The simplest of these is inevitably found in dioecious plants since the style as a pollen-filter provides a ready-made device for manipulating all breeding polymorphisms. It is only necessary therefore for the $X-Y$ difference to include a pollen-tube growth-rate differential to ensure a correct adjustment of the sex-ratio.

The campions of the genus *Melandrium* show how this happens. X and Y pollen is produced in equal amounts and sparse pollination gives male and female progeny in equal numbers. But Correns found that the Y pollen tubes grow more slowly than the X so that excess of pollen reduced the frequency of males. The sex-ratio of the population is thus adjusted by a short-term selective feedback. There is also a long term selective feedback for he also found that brothers differed, probably by way of the autosomes, in their relative pollen growth rates. The times of seed germination and flowering in turn are bound to influence the effective sex-ratio and these are directly controlled by the $X-Y$ difference. But the crux of the breeding system, as in all angiosperms, is the pollen-style filter action through which the $X-Y$ difference operates.

In animals more elaborate hierarchies of polymorphism are needed and have been developed. They have to be in part independent of the $X-Y$ segregation device. Thus there is a sex-ratio complex characteristic of a whole group of *Drosophila* species and manifestly superimposed on the $X-Y$ difference (and in classical terms hypostatic to it). The complex

12

causes the death of Y-carrying sperm produced by the sex-ratio male and so reduces the frequency of males in the population. This complex is sheltered by three overlapping inversions in one arm of the X-chromosome. Here the successive inversions have presumably reinforced one another and the end result gives us a prime example of irreversible fixation.

Control of the sex ratio in animals can however be removed a stage further from the internal sex-determining polymorphism. This happens when an infective organism is captured, as we may say, and made to serve as a substitute for an internal genetic polymorphism. Thus a number of species of *Drosophila* have a proportion of strains carrying spirochaetes. These multiply in the tissues of the females and pass into their eggs. They kill, not the Y sperm, but the XY embryos. They thus reduce the proportion of males in the population and they divide the female population into the two alternative matrilinear types, the balance in polymorphism being enforced by a reaction between heredity and infection.

It remains to add that these spirochaetes, in the course of evolution, have lost their capacity for natural infection. They can however be artificially injected into spirochaete-free females. They can even be injected into females of other species which happen to be genetically susceptible: their survival then evidently depends on the genotype of the host.

The spirochaetes, in doing all these things, demonstrate three far-reaching evolutionary principles. For, first, they show how an infecting organism can exploit and enlarge a pre-existing chromosomal polymorphism. Secondly, they show how, through the decay of infection, the ordinary process by which a provirus becomes a virus can be reversed. And thirdly, they show that two different and perhaps competitive systems of sex-ratio control can come into being in different species of fly, the one by heredity, the other by infection, both having developed through processes of selection so as to create a balanced polymorphism.

The incorporation of infectious particles into the genetic system of a fly has the effect of binding the species together. In bacteria this influence, of course, can go much further. An infectious phage can not only modify but transform the breeding habit of a bacterium. This is by no means irrelevant for there now seem to be connecting links between these systems.

In both *Paramecium* and yeast, killer particles or organisms are known whose obvious and immediate effect would seem to be to split each species into two parts, the killers and the sensitives. The situation is complicated, however, hierarchically. There are neutral strains intermediate between killers and sensitives. All strains differ in their internal genetic reaction with their parasites. And the parasites which are no longer naturally infective themselves vary through their own virus infections. We cannot therefore be sure how likely killer infection is to follow the evolutionary patterns of a strictly internal polymorphism in the higher organisms.

9. SUPERNUMERARY CHROMOSOMES

These observations drag us away from Ford's original definition but we have to go even farther afield (or astray) when we think of the role of supernumerary or B chromosomes which have puzzled three generations of experimenters. These chromosomes are too small to form chiasmata and pair regularly at meiosis: they are therefore distributed irregularly in the population. They are widespread in Hemiptera, Coleoptera and Orthoptera and they occur generally in perhaps ten or twenty per cent of diploid species of short-lived flowering plants. Their first characteristic is that individuals with and without these chromosomes exist side by side and breed indiscriminately together. Their second characteristic is that this perverse habit is maintained by a special version of two-track heredity. Losses at meiosis are made good by accumulations and by advantages at other stages. Frequently the gain is differential as between the sexes and favours the male.

The frequency of the extra chromosomes is thus balanced by natural selection. The selection, as with allelic polymorphisms is one favouring diversity, a diversity in the growth rates of germ cells, of embryos or of adults. But the range and variation in frequency is not due to mutation. Nor is it the result of Mendelian segregation at meiosis. It is the result of a defect in the pairing at meiosis which should otherwise lead to this segregation.

Here we have a sharp enough contrast in behaviour but the gap can be bridged. All *Trillium* species are polymorphic for segments of heterochromatin whose variation in size seems to have the same character as the variation in supernumerary chromosomes. For, as with the supernumeraries, the polymorphism exists in parallel in different species and is therefore evidently older than the species themselves.

Another device which reveals the parallelism of the supernumerary to allelic polymorphism is displayed by a favourite botanic garden plant, *Nicandra physaloides*. Here one of the ten chromosomes of the set is an iso-chromosome. The plant, we might say, is thus normally unbalanced and tetrasomic. There are four identical arms which can pair at meiosis. This irregularity simulates that of the supernumeraries and leads to the formation of embryos with nineteen chromosomes. These have a prolonged, sometimes indefinitely prolonged, dormancy. But if they survive they in their turn form a proportion of germ cells with ten chromosomes from which the normal type reappears.

In *Nicandra* we see how a balanced polymorphism can continually adjust and re-adjust itself to the conditions of survival of the seed and of the species. And it depends not on mutation or on Mendelian regularity but on a controlled and exploited irregularity. In many plant species, especially grasses, there are even more exaggerated devices where a lineage of individuals will run up and down the polyploid scale in response to, or selected by, conditions which we do not always understand. But in principle they are all certainly balancing their adaptive polymorphisms.

By considering supernumerary types of chromosome as an aspect of polymorphism we are enabled to see why several generations of experimenters have been bemused by this problem. In all polymorphisms we have the situation in which what is bad for the individual is good for the population. Or, more precisely, what is harmful for most individuals under most conditions is useful to particular and unusual individuals under particular and unusual conditions. It is in this sense of fitness being sacrificed to flexibility that all these polymorphisms are the salvation of the species.

10. VARIATION: COMMITTED AND UNCOMMITTED

It now appears that an unlimited range of genetic materials and devices is used in developing polymorphic variations within species. In the course of evolution these variations have been combined in mechanical and physiological sequences and hierarchies. We know that only a small part of the vast flow of natural variation in the range between the single nucleotide and the whole chromosome set is embodied in evolutionary change. Of this small part, we may ask in conclusion, how much of it comes to be committed, and for what reasons, to the eventual functions of binding or splitting species? Also are these functions opposed and fixed in their opposition?

How variations become committed either to splitting a species or to holding it together and thus becoming a polymorphism we can judge by the fate of chromosome inversions and interchanges. In general either course is open to them. But where, as a heterozygote, the inversion or interchange does more good than harm it binds the species and becomes a polymorphism. Conversely, where it does more harm than good it splits the species by setting up genetic isolation. The good that it does will be in holding its own genes or chromosomes together and the harm that it does will be in lowering fertility. Here the actions of the genes and the regulation of meiosis will both play their part. And in different species, on account of their differences in gene arrangement and in chromosome behaviour at meiosis, the balance that is struck is quite different.

TABLE I.3. Predominant occurrence of types of structural change within and between species (see Darlington 1965, especially Table 81).

	Within races	Between races or species
Inversions	*Drosophila*	*Lilium*
	Sciara	
	Chironomus	*Tulbaghia*
Interchanges	*Oenothera*	*Datura*
	Isotoma	*Triturus*
	Campanula	
	Paeonia	
	Periplaneta	
	*Otocryptops**	

* Heterozygous sex

Next, we have to ask, have the variations of all kinds arising and surviving in nature only the two possible destinies open to them of polymorphism and splitting. Or, is it possible that a variation captured to serve one purpose can be recaptured to serve the other? The evidence that offers itself is uncertain. It is seen where a difference between two species is inherited as a block as though it had arisen from the super-gene of a polymorphism. This is the case with the morphological division of the species of *Rubus*. It is also perhaps the case with the meiotic control mechanisms in the ancestral species of *Triticum aestivum*.

In general, then, we must suppose that the opposition is complete and the choice is final. But how it is made will depend on the total properties of the species, its genetic system and its environment during a certain period of initiation.

Ford's view of polymorphism has thus lifted the study of variation on to a new level. It is a level where we are allowed and even compelled to take the problem of variation as a whole. We then see it as forced into two divergent channels, those tending to hold a community together and those tending to force it apart. At this point a new and wider definition of polymorphism itself becomes necessary. We have to say that it is that variation which is committed to binding rather than to splitting a species.

We may therefore think of polymorphism, not as a mechanism but rather as a movement, and not in isolation from other movements but always as contrasted with species formation. In this way we shall find that the contrast allows us to make connections which in the past we have failed to make.

The idea of balanced polymorphism has indeed kept us in mind of evolutionary issues that we should otherwise have overlooked. We have been examining the origins of large discontinuities from small ones. We have been using breeding systems, studies of heterostyly, of sex ratio controls, and of sex determination itself to show these origins. In all these arguments we have been attempting to bridge a gulf. It is the gulf between the divergent positions of Darwin, of Bateson and de Vries, and of Weismann and Morgan. I remember how this gulf yawned before Bateson when, in 1925, an American correspondent asked him to say what he meant by variation. He wanted a definition. The great naturalist, the pioneer of genetics, and the prophet of discontinuity, was confounded. He gladly referred the question to the several opinions and corporate verdict of his pupils.

During those intervening years, the gulf in our knowledge, which we can now bridge with our knowledge of polymorphic systems, others have tried to jump with the help of *a priori* models mathematical, physiological and philosophical, models based on largely false analogies between evolution and development, or culture and heredity, models taking a short cut through the tiresome paradoxes which the chromosomes intercalate between individuals and societies as well as between genes and organisms. Among these paradoxes long recognized have been Fisher's evolution of dominance and Sewall Wright's small populations. But the list of those to be recognized under the heading of polymorphism is longer, their character more various, and their effects more pervasive.

The gap in our knowledge could not, of course, have been charted in Bateson's time. And it can now be bridged (however provisionally) only with the help of the great range of instruments and organisms that have been brought to bear on it during the years that have elapsed. Breeding experiments, chromosome studies, systematic embryological and physiological comparisons in plants and animals, men and microbes, have all helped to make the connections. It was only by taking the whole living world and its evolution into account, that we could discover the models we now have for the unity and the diversity of genetic systems.

REFERENCES

BELFIELD A.N. & RILEY R. (1969) The relationship of the genomes of allohexaploid wheat. *Chromosomes Today* 2, 5–11.

CARSON H.L. (1955) The genetic characteristics of marginal populations of *Drosophila. Cold Spring Harb. Symp. quant. Biol.* XX, 276–287.

CARSON H.L. (1962) Fixed heterozygosity in a parthenogenetic species of *Drosophila. Studies in Genetics* II, 55–62. University of Texas, Publ. 6205.

CORRENS C. (1928) Bestimmung, Vererbung und Verteilung des Geschlechtes bei den höheren Pflanzen. *Hb. Vererb. Wiss.* 2, 1–138.

CROWE L.K. (1964) The evolution of outbreeding in plants: 1. The angiosperms. *Heredity* 19, 435–457.

DARWIN C. (1877) *The Different Forms of Flowers on Plants of the Same Species.* John Murray, London.

DARLINGTON C.D. (1949) On an integrated species difference (*Rubus*). *Heredity* 3, 103–106.

DARLINGTON C.D. (1956) Natural populations and the breakdown of classical genetics. *Proc. R. Soc. B.* 145, 350–364.

DARLINGTON C.D. (1958) *Evolution of Genetic Systems,* 2nd edition. Oliver & Boyd, Edinburgh.

DARLINGTON C.D. (1963) *Chromosome Botany,* 2nd edition. Allen & Unwin, London.

DARLINGTON C.D. (1965) *Cytology.* J. & A. Churchill, London.

DARLINGTON C.D. (1968) *The Evolution of Man and Society.* Allen & Unwin, London.

DARLINGTON C.D. & JANAKI-AMMAL E.K. (1945) Adaptive isochromosomes in *Nicandra. Ann. Bot.* 9, 267–281.

DARLINGTON C.D. & SHAW G.W. (1959) Parallel polymorphism in the heterochromatin of *Trillium* species. *Heredity* 13, 89–121.

DARLINGTON C.D. & WYLIE A.P. (1956) *Chromosome atlas of flowering plants*. Allen & Unwin, London.

DOBZHANSKY Th. (1951) *Genetics and the Origin of Species*. (3rd edition). Columbia U.P., New York.

DOWRICK V.P.J. (1956) Heterostyly and homostyly in *Primula obconica*. *Heredity* 10, 219–236.

FISHER R.A. (1930, 1958) *The genetical theory of natural selection*. Oxford U.P. and Dover Pubs. Inc., New York.

FOGWILL M. (1958) Differences in crossing over and chromosome size in the sex cells of *Lilium* and *Fritillaria*. *Chromosoma* 9, 493–504.

FORD E.B. (1963) Evolutionary processes in animals. Appendix to C.D. Darlington, 1963.

FORD E.B. (1965) *Genetic polymorphism*. Faber, London.

FRANCA Z.M. & DA CUNHA A.B. (1968) Crossing-over between heterozygous inversions and its relation with polymorphism in *Drosophila willistoni*. *Revta brasil. Biol.* 28, 495–497.

HALDANE J.B.S. (1922) Sex ratio and unisexual sterility in hybrid animals. *J. Genet.* 12, 101–109.

JAMES S.H. (1970) Complex hybridity in *Isotoma petraea* II. *Heredity* 25, 53–77.

JOHN B. & LEWIS K.R. (1965) The Meiotic System. *Protoplasmatologia* 6, 1–335.

MATHER K. (1950) The genetical architecture of heterostyly in *Primula sinensis*. *Evolution* 4, 340–352.

NODA S. (1968) Achiasmate bivalent formation by parallel pairing in PMCs of *Fritillaria amabilis*. *Bot. Mag. Tokyo* 81, 344–5.

NOVITSKI E. *et al.* (1965) Cytological basis of 'sex ratio' in *Drosophila pseudoobscura*. *Science* 148, 516–517.

NUR UZI. (1969) Harmful B-chromosomes in a Mealy Bug (*Pseudococcus obscurus*). *Chromosoma* 28, 280–297.

PREER J.R. & PREER L.B. (1967) Virus-like bodies in killer *Paramecia*. *Proc. Nat. Acad. Sci.* 58, 1774–1781.

SAKAGUCHI B. & POULSON D.F. (1963) Interspecific transfer of the 'sex-ratio' condition from *Drosophila willistoni* to *D. melanogaster*. *Genetics* 48, 841–861.

SAKAGUCHI B. *et al.* (1965) Interference between 'sex-ratio' agents of *Drosophila willistoni* and *D. nebulosa*. *Science* 147, 160–162.

SHARMAN G.B. & BARBER H.N. (1952) Multiple sex-chromosomes in the marsupial, *Potorous*. *Heredity* 6, 345–355.

SOMERS J.M. & BEVAN E.A. (1968) The genetics of the killer character in yeast (*Saccharomyces cerevisiae*). *Heredity* 23, 475.

VOSA C.G. (1966) *Tulbaghia* hybrids. *Heredity* 21, 675–687.

WEISMANN A. (1887) On the signification of the polar globules. *Nature* 36, 607–609.

WESTERGAARD M. (1940) Cytology and sex determination in polyploid forms of *Melandrium album*. *Dansk. Bot. Ark.* 10 (5).

2 ❀ Plant Evolution in Extreme Environments

A.D.BRADSHAW

THE VALUE OF THE EXTREME

Extreme habitats have a fascination of their own. They are occupied by only a few very characteristic species, and they are easy to recognize ecologically and spatially. In extreme environments a single physical factor is usually dominating and other factors subsidiary. This makes it easy to understand what is going on, and to define the intensity of the factor. Extreme habitats are so distinctive that it is also easy to see where they begin and end. Because of this they have important advantages for the study of evolution.

There are innumerable extreme environments: there will be twice as many extreme environments as there are ecological factors. There are extremes of wetness, dryness, salinity, heat, cold, exposure, toxicity, for which we each have our own preference. Here we are going to deal with three extremes, all different.

In each of these three environments evolution has occurred, so that the plants within them are now better adapted than they were previously to withstand the conditions of the environment. These are not special cases; almost all extreme environments are colonized by species of one sort or another, and whenever the extreme populations are examined, they are very different from populations coming from more normal environments. Often they are so different that the normal populations cannot survive in the environments inhabited by the extreme populations e.g. *Trifolium repens* in upland pasture (Bradshaw 1967) and *Achillea millefolium* in alpine habitats (Clausen, Keck & Hiesey 1948). The evolution in these environments permits species to have a much wider distribution than they would otherwise have. This provides us with an important principle that there is a genetical component to ecological amplitude.

So the processes of evolution in extreme environments have a particular interest to the ecologist. But they have a considerable contribution to make to understanding evolutionary processes in general. From their time of inception all species have had to face alien environments, where alien means extreme as far as the plant is concerned, if not extreme in a physical sense. So evolution in extreme environments is merely part of the normal processes of evolution.

ARTIFICIAL SELECTION EXPERIMENTS

To understand how such evolution can come about we need to appreciate the power of selection. A starting point is the famous long term selection experiment, the Illinois corn experiment, which was started in 1900 and has continued ever since. Artificial selection for low and high oil and protein content has been carried out on an ordinary unselected open pollinated population, the Burr White variety of maize. The population size has been maintained at several thousands of plants and in each generation the twelve highest or lowest plants out of sixty analysed are selected to give rise to the next. The course of selection for oil content is shown in fig. 2.1. The significant feature is the long and continued

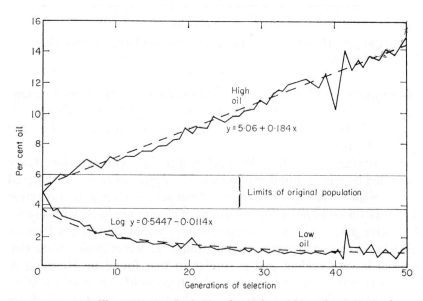

FIGURE 2.1. The outcome of selection for high and low oil content in the Illinois corn experiment (from Woodworth *et al.* 1952).

response to selection which has taken the selected populations to levels of oil completely outside that of the range of the original population (Woodworth, Leng & Jugenheimer 1952). The rate of progress under selection has continued almost undiminished over the whole period.

A similar experiment has been carried out in ryegrass *Lolium perenne* where flowering time has been selected (Cooper 1960) (fig. 2.2). There is the same picture of change although in this case the basic population is

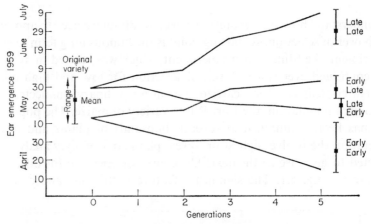

FIGURE 2.2. The outcome of selection for heading date in ryegrass (from Cooper 1960).

only 125 plants and only four are selected for the next generation. One aspect of the experiment is that in the first generation two different sets of parental plants were chosen as starting points. The effect of these has continued throughout.

The characters that have been selected in these two experiments are controlled by a large number of genes. The degree of response achieved concurs with what might be expected from genetic theory. It also agrees with work on *Drosophila*. Such patterns of change are likely with a series of additive genes, scattered through an initial population (Lee & Parsons 1968). In the restricted ryegrass populations the effect of choice of starting material is very evident and is similar to that found by Hosgood & Parsons (1967). There is little sign of restriction of release of variation due to linkage: this can be explained by chromosome numbers which are effectively at least twice that of *Drosophila*.

The Illinois corn experiment with its relatively large population is the model most relevant to what may happen in a natural population. Plant populations are usually large in size. And it must also be remembered that the size of a natural population cannot be measured solely in terms of the number of observed adjacent individuals: it must take into account individuals farther away supplying new genetic material by gene flow. In plants, gene flow due to pollen movement by wind or insects is leptokurtic, so that there is a small amount of gene flow over relatively large distances (Bateman 1947) sufficient to be important in supplying new variability (see p. 41).

Another way we can understand how a population changes under the influence of selection is to look at situations in which various sorts of artificial populations have been put into natural conditions, and the change of their constitution followed over a number of generations. One of the most interesting models was that set up by Charles (1961). In this three different genotypes of ryegrass *L. perenne* (in fact three different cultivated varieties) were mixed and sown in a field and subsequently subjected to a number of different treatments. After sowing the sward

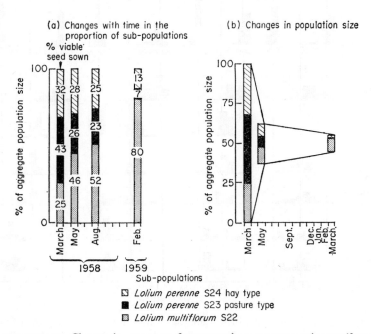

FIGURE 2.3. Change in genotype frequency in a ryegrass mixture (from Charles 1961).

was sampled at regular intervals and the absolute and relative amounts of the three different genotypes determined. The rate of change (fig. 2.3a) is remarkable. In the course of a single year during which no sexual replacement has occurred frequencies of the components have altered by a factor of three, implying very high coefficients of selection. The degree of change depends upon the treatment imposed. The reason for the changes becomes clear when the absolute numbers of the components in the mixture are examined (fig. 2.3b). The original high numbers of seedlings fell markedly through the period of observation, allowing startling changes to occur in the relative frequencies of the components without any of them having increased in absolute number. This may seem to be a very artificial situation, but it is not. Plants are overgenerous in their seed production and the changes that occurred in this experiment are just the changes that could occur among the offspring of plants inhabiting a particular environment.

In 1952 samples of a hybrid population of rice, formed from crossing two varieties Noren 20 and Zuiho which have very different dates of

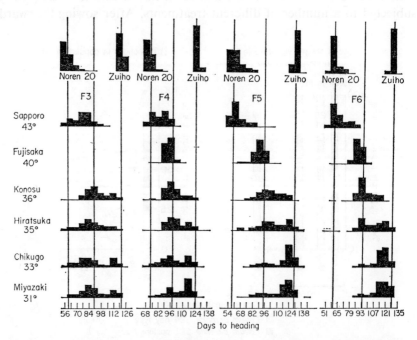

FIGURE 2.4. Histograms of heading dates in rice populations grown for successive generations in different parts of Japan (from Allard & Hansche 1964).

maturity, were set out in a range of environments extending from the extreme north to the extreme south of Japan (Akemine & Kikuchi, in Allard & Hansche 1964). Each year the hybrid populations were harvested and re-sown in their particular habitats. At the same time a certain amount of each material was taken to a central situation (Hiratsuka) and grown for observation. Figure 2.4 shows the changes in flowering time that were observed in comparison with the behaviour of the original parents used as controls. Over the course of four generations the populations in the extreme environments have come to resemble very closely the characteristics of the original parental varieties. The populations in intermediate habitats have taken on their own distinctive characters. Only one population, Hiratsuka, appears to remain in a very variable state after four generations.

In 1937 a paper was published on the changes in the composition of the species of a pasture which had been sown three years previously in Maryland and divided into two halves, one of which was grazed and one of which was cut for hay (Kemp 1937). Very extensive changes in the characteristics of blue grass (*Poa pratensis*), orchard grass (*Dactylis glomerata*) and clover (*Trifolium repens*) were recorded. There was a much higher frequency of prostrate individuals of all three species in the grazed part than in the mown part. However, despite all the interest in the formation of ecological races or ecotypes in plant species, almost nobody appears to have paid any attention to the results or realized their significance. We have had to wait until Stebbins discussed them in 1950. It is perhaps very easy to dismiss the results as an accident or faulty observation. Yet taken now with the support of all the evidence for the power of selection that more recent experiments have given us, they appear very prophetic of the way in which natural populations can change. It is fascinating that the grass plots that have been differently fertilized in the Park Grass experiment at Rothamsted have recently been shown to contain radically different populations of *Anthoxanthum odoratum* differing not only in size and growth rate, but in growth habit, disease resistance and flowering time (Snaydon 1970).

At the same time as Kemp reported what happened in a pasture a paper appeared describing work carried out in Sweden on the effects of propagating varieties of herbage plants in regions away from where they were originally bred (Sylven 1937). One example was Bottnia meadow fescue which was bred for north Sweden. The effects on its fitness in north and south Sweden of propagating it for one generation in south

Sweden is given in table 2.1; more radical changes in the characteristic of a population cannot easily be imagined. Now a great deal more evidence of this sort has accrued (e.g. Beard & Hollowell 1952) and plant breeders realize that shift in characters of varieties due to natural selection is a major problem in seed production.

TABLE 2.1. Changes in yielding ability of Bottnia meadow fescue (bred for grass production in north Sweden) after one generation of seed production in north and south Sweden (Sylven 1937).

Present Site	S. Sweden		N. Sweden	
Seed previously grown in	N. Sweden	S. Sweden	N. Sweden	S. Sweden
Yield (Tonnes/Ha)	44·4	109·1	58·9	55·9
per cent change from original	100	245·7	100	94·9

When Fisher wrote *The Genetical Theory of Natural Selection* in 1930, he envisaged selective advantages of up to about 1 per cent. This attitude to the power of selection prevailed for a long time and is probably why Kemp's and Sylven's results were disregarded. Another reason is that the study of plant evolution has been dominated by a taxonomic approach: the main concern has been to document distinctive populations, or ecotypes. But now attitudes to selection have changed, and Ford (1964) in particular has given us overwhelming examples of much higher selective advantages in animals, so that we realize that we must look at animal populations with a very different perspective. In plants too we now have plenty of examples showing how far Ford's arguments must be applied to plant populations. The phase when we can think of plant populations as more or less static entities is over. We are in a good position to find out how evolution proceeds in natural plant populations in natural situations.

EVOLUTION ON SEA CLIFFS

Sea cliffs are an attractive and distinctive part of the range of environments of a maritime island like Britain. In the northern cyclonic belt they can be an environment almost overwhelmingly extreme to plant life. Winds of 100 km per hour hurling themselves full force against rocky bluffs provide an environment demanding very distinctive adaptations by any plant species which attempt to survive there. Plants on cliffs are characteristically extremely prostrate, such as the Buckshorn Plantain (*Plantago coronopus*); if they grow upwards at all, they do so in compact tufts so that separate shoots provide each other with protection, such as the Sea Thrift (*Armeria maritima*). Species that occur both on cliffs and inland always have genetically very distinctive populations on cliffs characterized by extreme dwarfness and compact growth. This was first realized by Turesson (1922). He was so struck by the parallel evolution in many contrasting species that he recognized a distinctive ecotype, *campestris* (Turesson 1925). More recently very distinct cliff populations have been found in *Achillea millefolium* (Clausen, Keck & Hiesey 1948), *P. maritima* (Gregor 1946) *Agrostis stolonifera* (Aston & Bradshaw 1966), and many other species.

The cliff environment is not always exposed. There are long periods particularly throughout the summer months when wind speeds are very moderate. So cliffs and the more protected habitats close-by form a complex of environments varying both in space and time. These are conditions under which there could well be selection of phenotypic plasticity, where the species would produce a dwarf growth form only when exposed, at other times producing a more normal growth form. However, contrary to popular belief, this is not found. Dwarf plants on cliffs are nearly always genetically and permanently dwarf. This can be explained because the environment is so unpredictable. It is impossible for a species growing on a cliff to adjust phenotypically to the startling environmental changes that it has to face. To change from a tall to a dwarf growth form on the onset of high winds would involve shedding part of the larger growth form before growing a new dwarf one, which would involve a wasteful loss of biomass. It would also mean that at the onset of the severe conditions the plant would be caught in an inappropriate state, and would be irrevocably damaged before changing to the new better adapted phenotype. But the evolutionary significance of phenotypic plasticity is a subject in its own right and has been discussed elsewhere (Bradshaw 1965).

What is the genetic basis of cliff adaptations? An examination of the range of variability in and between cliff populations shows that it is continuous. This is particularly clear in *Achillea millefolium* (Clausen, Keck & Hiesey 1948), (fig. 2.5). From this we must presume that the characteristics are determined by many genes, but how many it is difficult to say because there is insufficient evidence. In *Achillea* and other species some genetical work has been done which suggests that many genes are involved in all cases (Clausen & Hiesey 1958). In *Cytisus scoparius* several genes must be involved (Gill & Walker 1971 and personal communication). Distinctive cliff populations appear wherever there are appropriate habitats. These separate populations could have all been derived from an original single population, but because the coast lines involved are so long and many of the species have their seeds distributed

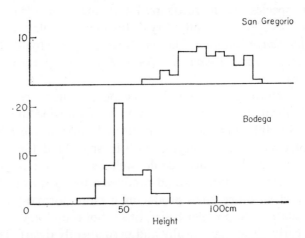

FIGURE 2.5. Histograms of height in two coastal populations of *Achillea millefolium* (from Clausen *et al.* 1948).

only locally by wind etc. it is likely that some of the populations have evolved separately and have different genotypes despite similar phenotypes. There is evidence of this in *C. scoparius*.

How well adapted are these cliff populations to their environments? To determine this it is necessary to subject the plants to the sort of environmental conditions that they experience on their cliffs, either by transplant experiments or in other ways that simulate cliff environments. When *Agrostis stolonifera* populations were subject to wind exposure the cliff population showed an approximately 25 per cent selective advantage over inland populations (Aston & Bradshaw 1966). Similar differences were found for *A. tenuis* (Bradshaw 1960) and *C. scoparius* (Gill & Walker 1971). In all these cases the plants were not subject to extreme cliff conditions for practical reasons, so these differences in selective advantage are underestimates, perhaps gross underestimates, of the real values. The selective forces which have been at work to pick out the genotypes which make these extreme populations are therefore very powerful indeed.

The cliff environment is very heterogeneous. Protected and exposed habitats can be in close proximity. What happens in these sorts of situations? Can the exposed habitats each have their highly adapted dwarf populations and the protected habitats similarly have their own adapted more normal populations? If the distances involved are large enough, clearly this should be so, but if two habitats are close together it is possible that gene flow by pollen between them would prevent the occurrence of the extreme population. Turesson (1922) found that sand dune and pasture populations of *Hieracium umbellatum* were separated by only a few hundred metres; in *P. maritima* (Gregor 1930) the distances separating the distinctive populations are smaller. The pattern of differentiation found in *A. stolonifera* (Aston & Bradshaw 1966) shows that the distances separating distinctive populations can be very small indeed (fig. 2.6). In one transect where plants were sampled over a sharp cliff edge, typical cliff and pasture, populations were found to be separated by only 10 metres. The two environments concerned are utterly distinct: tatter flags showed that the exposed habitat receives the full force of all winds, and that the protected habitat adjoining in the lee of gorse bushes is totally protected. The differences between the selective pressures operating in the two habitats are therefore very extreme and could be expected to maintain the distinctiveness of the adjacent populations. In another part of the same cliffs a set of populations were collected on a transect where the environment changed more gradually. Here there is a correspondingly gradual change in the characteristics of the populations. The precise relationship between patterns of differentiation and the small scale patterns of the environment is very clear.

If these populations are close together then we should expect at least some gene flow between them. Is there any evidence for this, and is there consequently any evidence that selection is actively maintaining the distinctive characteristics of the populations? Some evidence of what is going on in *A. stolonifera* was obtained by sampling the populations not only by taking tillers, but also by taking seed produced *in situ*. This material was then grown on and put into trial when in an adult state. The results (fig. 2.7) reveal a complicated situation. There is every indication in the stream population that a lot of the seed that is produced is the result of crossing with dwarfer, presumably cliff, plants: the mean of the stream seed sample is far less than that of the stream tiller sample. In the case of the cliff population the situation is not so readily understandable, for the cliff seed population is even more dwarf than the cliff tiller population. The cliff tiller population is rather variable and it is possible that the only seeds that were available came from short plants. Whatever is the proper explanation for the cliff population there is certainly evidence of a dynamic situation and of selection acting in the stream population to maintain the characteristics of the population in both a directional and a stabilizing manner. We shall see in the second example that this is also happening in an extreme habitat.

The degree of localization of individuals adapted to extreme conditions can be remarkable. When material of *A. stolonifera* was being collected in the area of gradual environmental change, plants were found

FIGURE 2.6. The topography and the mean stolon lengths of the populations of *Agrostis stolonifera* at Abraham's Bosom (from Aston & Bradshaw 1966).

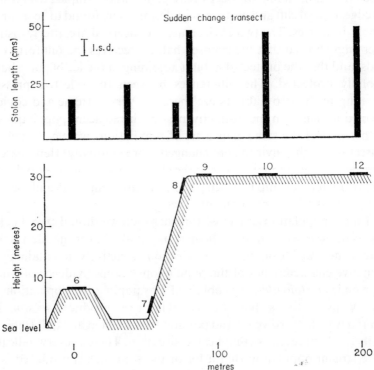

growing in the bed of a stream that ran through the area. These were phenotypically very distinctive indeed. When put into trial with the other populations, although they lost some of their distinctiveness, they were still found to be quite different from the populations adjoining (fig. 2.6). Their distinctiveness is very remarkable since they were collected only one metre distance from the neighbouring collections. They can hardly form a breeding population: but their presence can be explained because *A. stolonifera* is a strongly vegetative perennial plant. These cliff stream populations must have been formed by the vegetative propagation of a few selected genotypes. The possibility that one or a few genotypes can colonize a large area has been very elegantly demonstrated by Harberd (1961). Vegetative propagation is therefore a way in which selection can favour a few highly distinctive genotypes and the species can thereby exploit a highly localized environment. Vegetative propagation also allows the spread of extreme individuals whose characteristics are determined by non-additive gene action, which therefore cannot give offspring like themselves. Populations of *Holcus mollis* have been found

FIGURE 2.6—*continued*

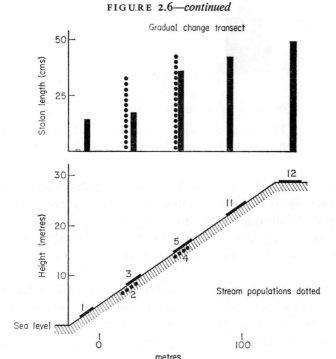

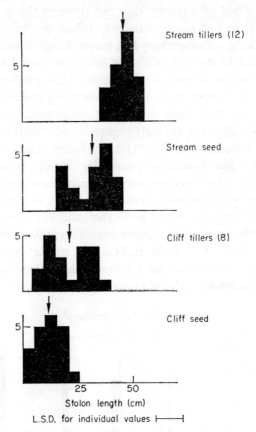

FIGURE 2.7. Histograms of stolon length of two populations of *Agrostis stolonifera* sampled as seed and as tillers (from Aston & Bradshaw 1966).

which are composed entirely of hybrid sterile individuals. These have remarkable vigour and may colonize the whole of an area of woodland, but they certainly do not in any way give offspring like themselves (Jones 1958). Vegetative propagation is a remarkable evolutionary strategy which can be of considerable importance in extreme environments.

It is reasonable to presume that cliff populations have evolved perhaps a long time ago from ordinary inland populations. We cannot see this occurring, but it is interesting to ask how easily it could have occurred. An analysis of the distribution of variability in a range of populations of *Agrostis stolonifera* shows that each population has a considerable range of variability. But these ranges do not encompass one another completely (fig. 2.8).

In particular the cliff population contains no individuals whose size is that of individuals in the protected inland populations. The samples are small, but further experience suggests that this is a true representation of

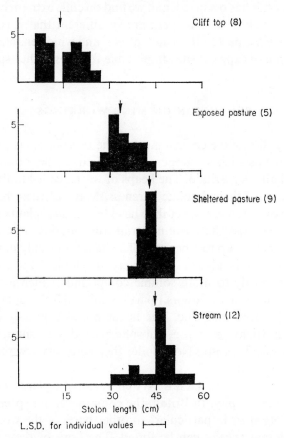

FIGURE 2.8. Histograms of stolon length of four populations of *Agrostis stolonifera* (from Aston & Bradshaw 1966).

the real state of affairs. In this case then, the cliff top population must have evolved by recombination permitting the formation of individuals not normally present in the original parental population, in a manner analogous to that occurring in the Illinois corn experiment. It would be interesting to know just how difficult was the evolution of a cliff population from a normal inland population. Even if we consider only one character, that of size, it is unlikely to have been achieved in the space of one or two generations. But cliff populations differ from inland populations in many other respects, for instance in the shape and texture of their leaves, and in the compactness of their panicles. The process of evolution of the cliff population must have been a complex one. But complex or not, it has occurred, and we find on cliffs extremely distinctive populations which in *A. stolonifera* are so different that it is difficult to credit that they are part of the same species, and which enable the species to colonize rocky exposed sites impossible for most other species.

EVOLUTION ON MINE WORKINGS

Another very distinctive environment is the derelict areas produced by mining for metals such as copper, lead and zinc. These mine workings are scattered all over Britain. The heaps of waste, called tailings, assault the eye because of their desolate bareness. Many of them have been in existence for over 100 years, yet they have few, if any, plants growing on them. This is because the waste material still contains quite appreciable quantities of metals, up to 1 per cent. The metal is usually in the form of a sulphide which is almost completely insoluble. But the sulphide weathers very easily to more soluble compounds which are extremely toxic in concentrations as low as 1 part per million for copper or 10 parts per million for lead and zinc. The heaps are also grossly deficient in major plant nutrients, nitrogen, phosphorus and potassium, and have a very poor physical texture. No wonder therefore very few plants can be found growing on them.

But it is interesting that a few very characteristic plants do occur and may grow quite happily. In Britain the commonest are species of grass, *Festuca* and *Agrostis* in particular. In other parts of the world a whole range of different species can be found. The flora of these metal contaminated sites is so distinctive that in the past prospectors have used the plants to help their exploration for new ore bodies (see Antonovics, Bradshaw and Turner 1971).

All the species that grow on metal contaminated soils also grow on normal soils where they are mixed with a lot of other species. How is it possible for a few species to grow in this very distinctive extreme environment when others are not able to do so? There are two alternative possibilities: firstly, that these few species possess throughout their range

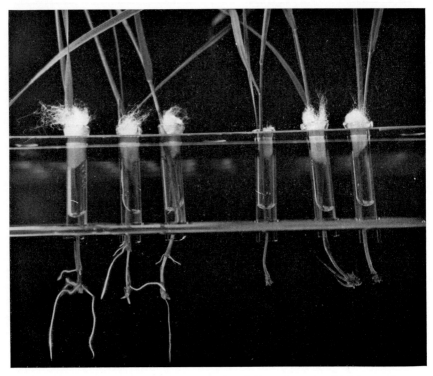

PLATE 2.1. Root growth of copper mine (left) and pasture (right) populations of *Agrostis tenuis* in 0·5 ppm copper solution.

the ability to tolerate the metal toxicity, or secondly, that they possess the ability to tolerate metal toxicity only where they are growing on toxic soils. The second possibility implies that the species can evolve tolerance under the influence of natural selection. In 1934 Prat showed that the second alternative was true for *Melandrium sylvestre* growing on copper mine workings in Central Europe. Since then all the cases that have been investigated have shown the same thing.

An elegant technique for testing the tolerance of mine populations has been developed. It was found that the major effect of metals is on root growth and that tolerant mine populations possess the ability to continue rooting in much higher concentrations of metal than normal populations (Bradshaw 1952). This difference can readily be demonstrated in solution (see plate 2.1), and tolerance expressed as the ratio of root growth in toxic solution to growth in normal solution (Wilkins 1957). There are many different mines containing different metals either singly or in combinations. Tolerance is specific for individual metals so that a

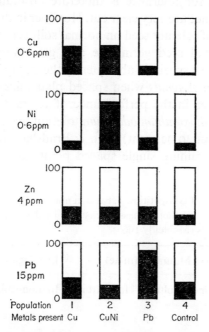

FIGURE 2.9. The indices of tolerance to different metals of four populations of *Agrostis tenuis* from areas contaminated by different metals (from Jowett 1958) (1, Parys Mountain, Anglesey; 2, Black Forest, Germany; 3, Goginan, Cardiganshire; 4, Capel Bangor, Cardiganshire).

population that grows on a copper soil is tolerant to copper only while a population that grows on a soil containing copper and nickel is tolerant to both copper and nickel (Jowett 1958, Gregory & Bradshaw 1965) (fig. 2.9).

What is the genetic basis of this character ? All gradations of tolerance can be found. Some of the variation in tolerance could be environmental in origin, but critical examination of a range of individuals shows little sign of discontinuity of variation. This is particularly so in *A. tenuis* (McNeilly & Bradshaw 1968). In *F. ovina* some discontinuity is found (Wilkins 1960). More work is necessary but at the moment it appears that several genes are involved in whatever species metal tolerance is found.

The selective factors evoking metal tolerance are very easily demonstrable. If normal seed of *A. tenuis* is sown on metal contaminated soil it germinates, produces a coleoptile but no roots. After a few weeks the seedlings die. Tolerant material however germinates and establishes normally. Material of intermediate tolerance behaves in an intermediate fashion. Selection for tolerance is therefore absolute and the factors maintaining tolerance are very clear cut. However in the reverse situation the performance of tolerant seed on normal soils is not the reverse. Both tolerant and normal seed germinate and grow very successfully on ordinary soils in most species. Subsequently there may be some difference in performance. In *A. tenuis* when spaced plants are examined there is very little difference in the performance of tolerant and non-tolerant material but in *Anthoxanthum odoratum* tolerant material is at a distinct disadvantage (about 40 per cent) (Jain & Bradshaw 1966). But fitnesses determined by examining single spaced plants are of limited use. In

TABLE 2.2 Effect of competition between a mine and a pasture population of *Agrostis tenuis* on their fitness, measured as dry wt. yield per plant, grown on two soils (McNeilly 1968).

Populations	Parys Mountain (mine)		Llandegfan (pasture)	
	In monoculture	In mixture	In monoculture	In mixture
Yield				
On acid soil	1·44	0·59	1·42	1·37
On basic soil	0·46	0·24	0·46	0·54
	l.s.d. 5% prob. 0.28			

natural non-toxic situations there will always be very severe competition from other species and it is therefore necessary to determine fitness under the full impact of competition. When this is done quite different values of fitness may be obtained (McNeilly 1968) (Table 2.2).

The operation of these selective factors in nature can readily be seen in mine populations when the tolerance of adult material is compared with the tolerance of seed material produced by those adults when on the mine and when in isolation. Isolation seed shows the same mean as adult material, but a greater variance. Seed produced on the mine shows a distinctly lower mean which must be due to gene flow from neighbouring non-tolerant populations. Since the adult population represents a stable situation, selection must be acting between seed and adult in a directional and stabilizing manner to maintain the characteristics of the adult population (McNeilly 1966) (fig. 2.10). The power of selection is even more clearly seen when looked at in a whole series of populations being subject to gene flow (see fig. 2.12).

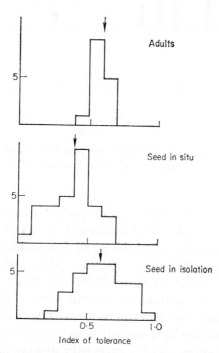

FIGURE 2.10. Histograms of copper tolerance of a mine population sampled as adults, and as seed collected *in situ* and in isolation (from McNeilly 1966).

Mine habitats are usually extremely restricted. Even large mine workings are seldom more than 500 metres across, many mine workings spread for only 100 metres and some may consist only of a dump of material a few metres across. They will always be surrounded by normal habitats very often containing extensive amounts of the species which has evolved tolerant populations on the mine material. Despite this, tolerant populations can be found on extremely small areas of mine waste. If transects are made across the boundary between contaminated and normal areas the tolerance of the populations sampled is found to change precisely at the

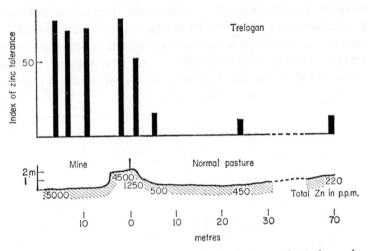

FIGURE 2.11. The zinc tolerance of populations of *Anthoxanthum odoratum* at the boundary of Trelogan mine (from Putwain, in Jain & Bradshaw 1966).

boundary. There is very little sign of gradation on either side of the boundary. This is very clear in *A. odoratum* (fig. 2.11). In many ways such a steep cline, if it can be called this at all, is surprising, because we are dealing here with a plant which is outbreeding and wind pollinated. Gene flow by pollen movement would certainly be expected to blur the distinction of the populations on the two sides of the boundary. The patterns of differentiation that can be expected with different degrees of gene flow and selection have recently been examined (Jain & Bradshaw 1966). Very sharp clines are to be expected with normal amounts of gene flow and selective advantages on each side of the boundary of about 50 per cent. But the sharpness of the transition shown by *A. odoratum* is remarkable. It is possible the gene flow is not as high as expected because of perenniality, but this does not seem to be so since the average life of *Anthoxanthum* individuals is only about two years (Antonovics, in Harper 1967).

What then can be happening to allow such localized differentiation of populations? There will be selection for any mechanism which reduces gene flow across the boundary. One major way that this can be achieved is by differences in flowering time. In an extreme environment such as a mine habitat there is likely to be selection for earliness of flowering so that seed production is completed before severe summer conditions occur. There is plenty of evidence of evolution of such differences for ecological reasons, differences which will have the incidental effect of reducing gene flow. But there is also evidence of enhancement of flowering time differences at the boundary of mine populations which can best be interpreted as arising as a result of selection specifically for isolation (McNeilly & Antonovics 1968). Another way in which gene flow can be reduced is by self-fertilization. It is very interesting that in *Anthoxanthum* and *Agrostis* much higher levels of self-fertility are found in mine populations than in normal populations (Antonovics 1968). In metal contaminated sites elsewhere in the world there are suggestions of a variety of ways in which reduced gene flow has been achieved (Antonovics, Bradshaw & Turner 1971). Although the colonization of an extreme habitat does not demand the isolation of its emergent population from the original populations surrounding it, there will be selection for isolation mechanisms, and it looks as though we can find this isolation in progress of evolution in metal contaminated habitats.

If gene flow is likely to be occurring between mine and normal populations we should be able to pick it up quite easily, particularly since the populations are so distinctive and we have an easy way of measuring the character even if its genetics is not clear yet. A small mine population surrounded by normal populations is an ideal situation in which to see what is going on. The mine at Drws-y-coed in Caernarvonshire is of this sort, and it has the added advantage that it is situated in a U-shaped valley in which the wind is highly polarized and normally blows from the west. The gene flow has been investigated by examining the difference between the tolerance of adult populations and the seed they produce *in situ* (McNeilly 1968). Two transects were made, one in a predominantly upwind and the other in a predominantly downwind direction. In the upwind direction the adults show the sort of very sharp transition in tolerance at the boundary that we have already seen for *Anthoxanthum*, but in the down wind direction the transition is by no means so sharp and populations some distance away from the mine in completely normal pasture have a quite considerable level of tolerance (fig. 2.12). The difference between the two transects is surprising. It certainly cannot be explained by differences in the sharpness of the environmental boundary:

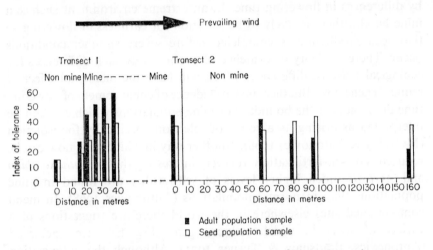

FIGURE 2.12. Copper tolerance of populations of *Agrostis tenuis* across Drws-y-coed mine sampled as adults and as seed (from McNeilly 1968).

the transition from mine to normal conditions occurs over the same very short distance in both transects. The explanation appears when we look at the tolerance of the seed populations. In the upwind transect the seed populations on the mine are less tolerant than the adult populations that gave rise to them. In the down wind transect the seed populations in the pasture are much more tolerant than the adult populations that gave rise to them. This fits in with the suggestion that the gene flow is highly polarized by the wind direction. In the up wind transect where there is flow of non-tolerant genes into tolerant populations this gene flow is kept in check by the very severe selection occurring on the mine. But in the down wind transect the flow of tolerant genes into the non-tolerant populations is not kept in check since the selection pressures here are not so severe, as we have already seen. The result is that genes move out of the mine populations more readily than they move in.

Selection retains the upper hand and, despite the gene flow, the distinctiveness of the mine population is maintained. The situation is therefore very similar to the one that we have already seen in *Agrostis stolonifera*. An extreme habitat, by the very fact that it generates high selection pressures, can maintain a distinctive population even if there is gene flow into it. The reverse gene flow from the mine populations into surrounding normal populations is not very great and is kept in check by selection. But it does occur and it must be remembered that gene flow in a plant

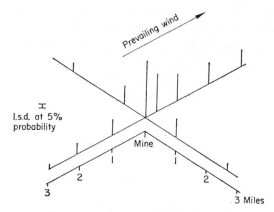

FIGURE 2.13. Frequency of copper tolerant individuals in seed samples of populations of *Agrostis tenuis* around Parys Mountain copper mine (measured as number of survivors on copper soil slightly diluted with ordinary soil) (from Khan 1969).

such as *Agrostis* is markedly leptokurtic: a small amount of pollen is carried a large distance. Low frequencies of tolerant individuals in *Agrostis* populations can be determined by sowing seed samples on appropriately toxic soils: tolerant individuals survive, non-tolerant ones do not. When this was done for a series of seed samples taken at distances of several kilometres round the large mine at Parys Mountain in Anglesey, tolerant individuals were found in frequencies above average in populations a long distance away from the mine, particularly down wind (Khan 1969) (fig. 2.13). The down wind end of the transect was limited by the sea, but the results are sufficient to show that gene flow at an extremely low level can carry genes a very long way away from the original habitat. Genes can therefore migrate long distances and be available for selection in new extreme habitats when they arise.

In *A. stolonifera* we saw that it was possible for a highly localized population to arise consisting of only a few individuals with very distinctive characteristics. If this can happen in a cliff situation where there is selection for morphological characters, it can surely happen in metal contaminated areas where the selection is probably even more severe. There is evidence of this from the remarkable observation of Snaydon (Bradshaw, McNeilly & Gregory 1965) that the plants of *Festuca ovina* and *A. canina* immediately underneath a zinc coated iron fence were significantly more zinc tolerant than those about 30 cm away. Here, as for the *A. stolonifera* growing in the stream, we can hardly consider that there is an independent population, but high selection pressures, together with vegetative propagation, has enabled this extremely localized habitat to be colonized by a few adapted individuals.

What are the origins of metal tolerance? Mine workings are often not very old, so that it is tempting to believe that the mine populations are

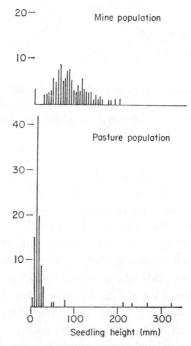

FIGURE 2.14. Height of seedlings of a copper mine and a pasture population of *Agrostis tenuis* (grown for three months on copper soil slightly diluted with ordinary soil) (from Khan 1969).

themselves not very old. But many mine workings were preceded by ore bodies exposed at the surface giving toxic soils which could well have been in existence since the glacial period. This is very clear in other parts of the world where the ore bodies have not been disturbed (Nicolls *et al.* 1965). So natural mine populations could have a very long history. But if non-tolerant seed is sown on to metal contaminated soil, although nearly all the seedlings die, one or two per thousand survive and grow very well and a few more grow rather more weakly (Khan 1969). A histogram of the size of the plants in such an experiment is given in fig. 2.14. The few large individuals are the only survivors, the rest are dead. If survivors of varying heights are grown on and tested for tolerance it is found that there is an extremely good correlation between size and tolerance. The few large individuals have tolerances similar to that of plants from mine populations. Metal tolerant individuals in *Agrostis* are therefore to be found in normal non-tolerant populations, although at very low frequency. With very severe selection a tolerant population can therefore be produced from a non-tolerant one in the course of only one or two generations. The time scale for the evolution of individuals to colonize this very extreme environment is therefore very short.

EVOLUTION ON SHIPS' BOTTOMS

Two preceding examples have a very great deal in common. The third example may or may not have a great deal in common with them. The evidence so far is insufficient to say, but the situation is so interesting that it needs to be looked at if only briefly.

The bottom of a ship is a very special and rather curious environment for various reasons, but particularly because it is always treated with an antifouling paint which is designed to keep living organisms from settling on the surface. The antifouling paint does this by the slow release of toxic compounds, usually copper but more recently other metals. As a result of antifouling compounds, fouling by animals such as barnacles is now a thing of the past. But fouling still sometimes occurs due to various species of algae particularly *Ectocarpus*. It is rather easy to believe that it is inevitable that something will grow on the bottom of a ship; but it is a very extreme environment, and it is profitable to ask what it is that permits *Ectocarpus* to grow there. Recently Russell & Morris (1970) have shown that samples of *E. siliculosus* taken from two ships which had been

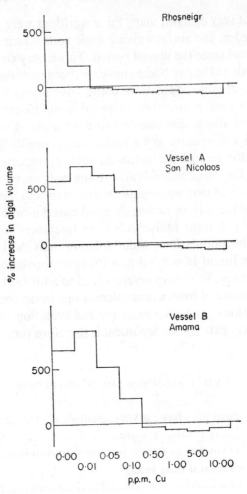

FIGURE 2.15. Increase in volume in different copper concentrations of *Ectocarpus* samples from an unpolluted rocky shore and from two ships' bottoms (from Morris & Russell 1970).

properly treated with antifouling paint were tolerant of 10 times the concentration of copper which a population from a normal rocky shore could stand (fig. 2.15).

At the moment we have only these few facts but preliminary work (Hodgson 1969) suggests that other algal populations, of *Enteromorpha* sp., can be found which are tolerant of mercury. There is also some suggestion that plants of *Ectocarpus* from a ship's bottom reach reproductive maturity in half the time taken by populations from normal habitats (Skelton 1969) which would be expected of a colonist (Lewontin 1965). There is therefore strong evidence of distinctive populations of algae evolving in adaptation to the extreme environment of a ship's bottom.

But what is the evolutionary history of these populations? Are they created out of normal populations by a screening process which picks up a low frequency of tolerant individuals already existing in normal populations, or have tolerant populations evolved slowly over a period in one or two centres which are now foci of contamination of ships' bottoms? This is not known, but the problem is being investigated. It certainly shows that evolution can occur in unlikely places.

SELECTION AND THE GENETIC POTENTIAL
FOR CHANGE

Extreme environments have a few very simple principles to teach us, the most important of which is that because they are extreme they can generate very powerful selection pressures. These can cause spectacular changes in the constitution of a population. They can cause the changes to occur with extreme speed. Gene flow from neighbouring populations may be important, but it can be readily overcome by the selection pressures operating. In other words, selection is all-powerful.

But evolution depends not only on selection, but also on the availability of the appropriate variability. In all three situations that we have considered, evolution has occurred. It follows logically that the variability has been available on which selection could act. But what of the cases where the appropriate variability has not been available? We do not usually study them because there is nothing to study. This is unfortunate, because once we accept the power of selection, then our interest must turn to the availability of variation and we should see the extent to which evolution is dependent not on selection but on the processes releasing new variation into a population.

This is not a new topic. It is merely one that in our enthusiasm to understand the power of selection we have forgotten about. Plant breeders have certainly not forgotten about it. They have to go to elaborate ends in order to transfer desirable genes from one population (often a distantly related species) to another population. Good examples are the red delphinium (Legro 1965) and Compair wheat (Reilly, Chapman & Johnson 1968). How far does the same situation occur in wild populations? The best evidence comes from the distribution of Warfarin resistance in rats. Warfarin is used widely throughout the world, but the places where Warfarin resistance is known are extremely

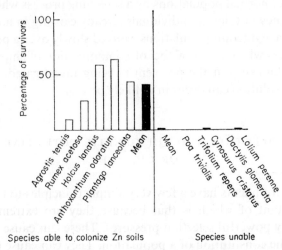

FIGURE 2.16. Frequency of zinc tolerant individuals in seed samples of normal populations of various species which are able, or are unable, to colonize zinc contaminated soils (from Khan 1969).

few and isolated. It seems pretty clear that the occurrence of Warfarin resistance is nothing to do with localization of selection but to localization of the occurrence of the appropriate gene. In this respect the situation seems different to the occurrence of DDT resistance in insects where there seem to be no good cases of populations that have not evolved resistance when selected.

In plants we have little useful information. But it was suggested a long time ago that the failure of some species to spread into the higher altitudes of the Alps, when they did so in the Urals, was due to a loss of genetic variability during the migration of the species from the Urals into Europe (Turesson 1931). There are a number of species which do not colonize mine habitats although they seem generally well adapted to do so. If normal populations of these species and those that do colonize mine habitats are screened for metal tolerance, there is an indication that it is only the latter species which possess tolerant individuals at low frequency (Khan 1969) (fig. 2.16). Although the evidence is incomplete it does suggest that examples can be found which the availability of variability determines the potential for change.

JUSTIFICATION

Finally we can ask whether the potential for change shown by populations in extreme environments is at all appropriate to populations in more normal situations. There is little difference in essence between normal situations and extreme environments from an evolutionary point of view. It is clear that normal environments are less extreme from a physical point of view. But it is debatable whether they are less extreme from the point of view of selection. In normal environments extreme selection pressures caused by extremes of physical factors will be replaced by strong selection pressures caused by biotic factors from the competition arising from a large number of species all attempting to grow in the same environment at the same time. The fact that in normal environments species still show very precise and often localized distributions and can be excluded from very small areas suggests that selection pressures are potentially just as strong in normal environments as in extreme ones. All the work that has been done on population differentiation in plant species in normal environments suggests just the same sort of rapidity and precision of evolution as in extreme environments. There is for instance the localized differentiation found in *Galium pumilum* (Ehrendorfer 1953) and *Potentilla erecta* (Watson 1969), and the remarkable localized differentiation that has developed rapidly in *Anthoxanthum odoratum* in the fertilizer plots established at the end of the last century of the Park Grass experiment at Rothamsted (Snaydon 1970).

Extreme environments are usually rather odd and often very nasty. As a result they have usually been disregarded. But their simplicity and distinctiveness mean they can provide very valuable models for the study of evolution.

REFERENCES

ALLARD R.W. & HANSCHE P.E. (1964) Some parameters of population variability and their implications in plant breeding. *Adv. Agron.* **16**, 281–325.

ANTONOVICS J. (1968) Evolution in closely adjacent plant populations. V. Evolution of self fertility. *Heredity* **23**, 219–238.

ANTONOVICS J., BRADSHAW A.D. & TURNER R.G. (1971) Heavy metal tolerance in plants. *Adv. Ecol. Res.* **7** (in press).

ASTON J.L. & BRADSHAW A.D. (1966) Evolution in closely adjacent plant populations. II. *Agrostis stolonifera* in maritime habitats. *Heredity* **21**, 649–664.

BATEMAN A.J. (1947) Contamination in seed crops. III. Relation with isolation distance. *Heredity* **1**, 303–336.

BEARD D.F. & HOLLOWELL E.A. (1952) The effect on performance when seed of forage crop varieties is grown under different environmental conditions. *Proc. 6th Int. Grassl. Cong.* 860–866.

BRADSHAW A.D. (1952) Populations of *Agrostis tenuis* resistant to lead and zinc poisoning. *Nature, Lond.* **169**, 1098.

BRADSHAW A.D. (1960) Population differentiation in *Agrostis tenuis* Sibth. III. Populations in varied environments. *New Phytol.* **59**, 92–103.

BRADSHAW A.D. (1965) Evolutionary significance of phenotypic plasticity in plants. *Adv. Genet.* **31**, 115–155.

BRADSHAW A.D. (1967) An ecologist's viewpoint. In *'Ecological Aspects of the Mineral Nutrition of Plants.'* I.H. Rorison (ed.), *Brit. Ecol. Soc. Symp.* **9**, 415–427.

BRADSHAW A.D., MCNEILLY T.S. & GREGORY R.P.G. (1965) Industrialization, evolution and the development of heavy metal tolerance in plants. In *'Ecology and the Industrial Society'*. *Brit. Ecol. Soc. Symp.* **5**, 327–343.

CHARLES A.H. (1961) Differential survival of cultivars of *Lolium*, *Dactylis* and *Phleum*. *J. Br. Grassld Soc.* **16**, 69–75.

CLAUSEN J. & HIESEY W.M. (1958) Experimental studies on the nature of species. IV. Genetic structure of ecological races. *Carnegie Inst. Washington Publ.* **615**, 312 pp.

CLAUSEN J., KECK D.D. & HIESEY W.M. (1948) Experimental studies on the nature of species. III. Environmental responses of climatic races of *Achillea*. *Carnegie Inst. Washington Publ.* **581**, 129 pp.

COOPER J.P. (1960) Selection for production characters in ryegrass. *Proc. 8th Int. Grassld. Cong.* 41–44.

EHRENDORFER F. (1953) Okologisch-geographische Mikro-Differenzierung einer Population von *Galium pumilum* Murr. s.str. *Osterreich. Bot. Z.* **100**, 616–638.

FORD E.B. (1964) *Ecological Genetics*. Methuen, London, 335 pp.

GILL J.J.B. & WALKER S. (1971) Studies on *Cytisus scoparius* (L) Wimmer with particular reference to the prostarate forms. *Watsonia*, in press.

GREGOR J.W. (1930) Experiments on the genetics of wild populations, *Plantago maritima*. *J. Genet.* **22**, 15–25.

GREGOR J.W. (1946) Ecotypic differentiation. *New Phytol.* **45**, 254–270.

GREGORY R.P.G. & BRADSHAW A.D. (1965) Heavy metal tolerance in populations of *Agrostis tenuis* Sibth. and other grasses. *New Phytol.* **64**, 131–143.

HARBERD D.J. (1961) Observations on population structure and longevity of *Festuca rubra* L. *New Phytol.* **60**, 184–206.

HARPER J. L. (1967) A Darwinian approach to plant ecology. *J. Ecol.* **55**, 247–270.

HODGSON M.B. (1969) Mercury resistance in ship-borne *Enteromorpha*. Hons. thesis, Dept. Botany, Univ. Liverpool.

HOSGOOD S.M.W. & PARSONS P.A. (1967) The exploitation of genetic heterogeneity among founders of laboratory populations of *Drosophila* prior to directional selection. *Experientia* **23**, 1–5.

JAIN S.K. & BRADSHAW A.D. (1966) Evolutionary divergence in adjacent plant populations. I. The evidence and its theoretical analysis. *Heredity* **21**, 407–441.

JONES K.J. (1958) Cytotaxonomic studies in *Holcus*. I. The chromosome complex in *Holcus mollis* L. *New Phytol.* **57**, 191–210.

JOWETT D. (1958) Population of *Agrostis* spp. tolerant to heavy metals. *Nature, Lond.* **182**, 816–817.

KEMP W.B. (1937) Natural selection within plant species as exemplified in a permanent pasture. *J. Hered.* **28**, 329–333.

KHAN M.S.I. (1969) The process of evolution of heavy metal tolerance in *Agrostis tenuis* and other grasses. M.Sc. Thesis, University of Wales.

LEE B.T.O. & PARSONS P.A. (1968) Selection, prediction and response. *Biol. Rev.* **43**, 139–174.

LEGRO R.A.H. (1965) Delphinium breeding. In 'Genetics Today,' *11th Int. Cong. Genet.* **2**, li–lv, 1963. Pergamon Press, Oxford.

LEWONTIN R.C. (1965) Selection for colonizing ability. In '*The Genetics of Colonizing Species*', H.G. Baker & G.L. Stebbins (eds.), pp. 79–94. Academic Press, New York.

MCNEILLY T. (1966) The evolution of copper tolerance in *Agrostis*. Ph.D. thesis, Univ. Wales.

MCNEILLY T. (1968) Evolution in closely adjacent plant populations. III. *Agrostis tenuis* on a small copper mine. *Heredity* **23**, 99–108.

MCNEILLY T. & ANTONOVICS J. (1968) Evolution in closely adjacent plant populations. IV. Barriers to gene flow. *Heredity* **23**, 205–218.

MCNEILLY T. & BRADSHAW A.D. (1968) Evolutionary processes in populations of copper tolerant *Agrostis tenuis* Sibth. *Evolution, Lancaster, Pa.* **22**, 108–118.

NICOLLS O.W., PROVAN D.M.J., COLE M.M. & TOOMS J.S. (1965) Geobotany and geochemistry in mineral exploration in the Dugald River Area, Cloncurry District, Australia. *Trans. Instn. Min. Metall.* **74**, 695–799.

PRAT S. (1934) Die Erblichkeit der Resistenz gegen Kupfer. *Ber. dt. bot. Ges.* **52**, 65–67.

REILLY R., CHAPMAN V. & JOHNSON R. (1968) Introduction of yellow rust resistance of *Aegilops comosa* in wheat by genetically induced homeologous recombination. *Nature, Lond.* **217**, 383–384.

RUSSELL G. & MORRIS P. (1970) Copper tolerance in the marine fouling alga *Ectocarpus siliculosus*. *Nature, Lond.* **228**, 288–289.

SKELTON M. (1969) Copper tolerance in ship-borne *Ectocarpus siliculosus*. Hons. thesis, Dept. Botany, Univ. Liverpool.

SNAYDON R.W. (1970) Rapid population differentiation in a mosaic environment. I. The response of *Anthoxanthum odoratum* populations to soil. *Evolution* **24**, 257–69.

STEBBINS J.L. (1950) *Variation and Evolution in Plants*. Columbia Univ. Press, New York, 643 pp.

SYLVEN N. (1937) The influence of climatic conditions on type composition. *Imp. Bureau Plant Genetics, Herbage Bull.* **21**, 8 pp.

TURESSON G. (1922) The genotypical response of the plant species to the habitat. *Hereditas* **3**, 211–350.

TURESSON G. (1925) Plant species in relation to habitat and climate. *Hereditas* **6**, 147–234.

TURESSON G. (1931) The geographical distribution of the alpine ecotype of some Eurasiatic plants. *Hereditas* **19**, 329–346.

WATSON P.J. (1969) Evolution in closely adjacent plant populations. VI. An entomophilous species, *Potentilla erecta*, in two contrasting habitats. *Heredity* **24**, 407–422.

WILKINS D.A. (1957) A technique for the measurement of lead tolerance in plants. *Nature, Lond.* **180**, 37–38.

WILKINS D.A. (1960) The measurement and genetic analysis of lead tolerance in *Festuca ovina*. *Rep. Scott. Pl. Breed. Stn* **1960**, 95–98.

WOODWORTH C.M., LENG E.R. & JUGENHEIMER R.W. (1952) Fifty generations of selection for protein and oil in corn. *Agron. J.* **44**, 60–65.

3 ❀ Polymorphism in a Polynesian Land Snail
Partula suturalis vexillum

BRYAN CLARKE AND JAMES MURRAY

INTRODUCTION

Several species of land snails belonging to the genus *Partula* occur on the island of Moorea in French Polynesia. They were studied in great detail by Crampton (1932), who considered their polymorphic varieties of shell colour and pattern to represent 'indifferent characters' unaffected by natural selection. To account for the variations found in the genus, Crampton invoked the pressures of recurrent mutation and migration. More recently they have been ascribed to the consequences of random genetic drift (Huxley 1942, Carlquist 1965).

Partula offers several advantages for the study of micro-evolutionary change. First, it is highly variable, both within and between populations. Second, its mobility is low, so that genetic differences are found between populations only short distances apart. Third, it is ovoviviparous, and easy to rear in the laboratory, so that the variation can be investigated experimentally.

Since 1962 we have been studying the population genetics of the Moorean species of *Partula*. We now present the results of a series of random samples taken from populations of *P. suturalis* Pfeiffer (subspecies *vexillum* Crampton) in the northwestern part of the island, and discuss the factors that may be responsible for the observed patterns of variation.

Because of his distinguished contributions to the study of genetic polymorphism, we are very happy to dedicate this paper to Professor E.B.Ford, F.R.S. on his seventieth birthday.

MATERIALS AND METHODS

Moorea is a volcanic island, in shape approximately an equilateral triangle, with sides about 15 km long. It lies about 16 km to the northwest of Tahiti in the Society Islands (at $17°32'$S and $149°50'$W). Tohivea, its central mountain, attains a height of 1207 m. Details of the Moorean topography can be found in earlier publications (Crampton 1932, Clarke & Murray 1969), which also describe the distributions on the island of the various races and species of *Partula*.

Partula suturalis Pfeiffer is widespread in the higher and more humid parts of Moorea, although sometimes it is found as low as 40 m above sea level. *P.s.vexillum* is the most extensive of its subspecies, occurring everywhere except the north central area, which is the home of *P. dendroica* Crampton, and the eastern and southeastern valleys, which are occupied by *P. s. strigosa* (see Clarke & Murray 1969).

The present study is restricted to the northwestern region of Moorea because elsewhere the genetic state of *P.s.vexillum* is complicated by hybridization with other forms, and by variation in the coil of the shell (Clarke & Murray 1969). In the northwest there is no evidence of hybridization and the coil is uniformly sinistral. From the ecological point of view also the situation is simpler than elsewhere because only two other species of *Partula* are found in the region. These are *P. taeniata* Morch and *P.solitaria* Crampton. The former is found with *P.suturalis vexillum* in every sample except one (No. 81). The latter is found only in a single sample (S1).

In 1962 and 1967 we collected 103 samples of living *P.s.vexillum* from populations in northwestern Moorea. The sampling areas were kept to a minimum (usually about 15 m in diameter or less) because we have found that in a few localities populations 10–15 m apart may show significant differences in the proportions of their morphs. The areas were searched carefully to avoid selecting preferentially those snails that were obvious to the eye, although none of the morphs is notably cryptic. Because *P.s.vexillum* is arboreal in habit, often living as high as 15 m above the ground, our samples are likely to represent less than 25 per cent of the population, and we cannot absolutely exclude the possibility that differences between the morphs in their tendency to climb could give rise to bias in sampling. A number of collections were stratified according to height, and showed no evidence of such differences. They did, however, suggest that the young *vexillum* tend to live lower on the trees than the adults. These factors have caused our samples to be generally small and to be predominantly composed of young snails.

We established the position of each sampling locality by triangulation with a prismatic compass, and recorded details of its altitude, aspect, slope and background. We also recorded at each locality the presence or absence of 6 other species of snails and 45 species of plants.

The shells were scored for age (presence or absence of adult peristome), morph, and shell-length (adults only, measured according to the criteria given by Murray & Clarke 1968).

The polymorphism of *P. suturalis* has been described in detail elsewhere (Crampton 1932, Murray & Clarke 1966). In the region of this study *P.s. vexillum* shows five clear-cut morphs: *strigata*, in which the shell is unbanded and pale brown in colour, but the whorls are crossed by numerous darker brown striations; *frenata*, in which each pale brown whorl is encircled by two dark brown bands (in some there is also a band near the umbilicus and/or near the suture); *bisecta*, in which the dark brown (nearly black) whorl is encircled by a single central pale brown band; *cestata*, in which the pale brown whorl is encircled by a single central dark brown band; and *atra*, in which the shell is entirely dark brown (again nearly black), These variations are known to be inherited (Murray & Clarke 1966), and it is unlikely that the environment has any direct effect upon the coloration of the shell. The exact mode of their inheritance, however, is not yet known, although it seems that more than one locus is involved (Murray & Clarke 1966). Further genetic studies are in progress.

RESULTS

The composition of our samples is given in table 3.1. The first five columns record the numbers of each morph, and the sixth gives the sample size. The seventh, eighth and ninth columns show the altitude of the collecting locality (in metres), the area sampled (in square metres) and the mean adult shell-length (from apex to base, in millimetres). The seventh column records the presence of the seven major *Partula*-bearing plants. The plant species are recorded by numbers as follows: (1) *Inocarpus edulis* Forst. (the Tahitian chestnut), (2) *Hibiscus tiliaceus* L. (the purau tree), (3) *Coffea arabica* L. (the coffee bush), (4) *Canthium* (*Plectronia*) *barbatum* (Forst.) Seem. (a small tree), (5) *Pandanus tectorius* Solander (the screw pine), (6) *Freycinetia demissa* (the climbing pandanus) (7) *Angiopteris* sp. (a large fern).

The positions of the sampling localities are recorded on fig. 3.1. The figure gives all the localities at which *Partula* were found, including those

TABLE 3.1. Samples of *Partula suturalis vexillum* from N.W. Moorea

Sample No.	strigata	frenata	bisecta	cestata	atra	Total (adults in brackets)	Alt. in metres ± 20 m.	Sampling area in m²	Mean shell length in mm (adult only)	Plant species and remarks (see text)
4	8	25	0	0	0	33 (4)	260	—	20·2	7
7	4	7	0	0	0	11 (2)	220	—	19·3	I, 2, 6, 7
8	5	13	0	0	0	18 (3)	270	2500	21·5	2, 4, 6
9	15	12	0	0	I	28 (9)	220	79	20·6	7
10	3	3	0	0	0	6 (0)	250	—	—	I, 3, 7
11	6	2	0	0	0	8 (0)	240	—	—	I, 7
12	3	10	0	0	0	13 (6)	380	—	20·2	
13	5	15	0	0	0	20 (5)	540	—	19·4	6
14	0	5	0	0	I	6 (1)	320	225	21·1	I, 2, 3, 7 *atra* pale at suture
15	5	2	0	0	0	7 (1)	140	—	20·1	I, 2, 3, 7
20	13	10	0	0	0	23 (8)	240	700	20·5	2, 3, 7
21	31	41	0	0	0	72 (50)	340	—	20·2	2, 7, two *strig.* are yellow
23	10	20	0	0	0	30 (7)	140	707	20·0	Bamboo grove
24	4	14	0	0	0	18 (2)	200	707	20·4	2, 3, 7
27	0	9	0	0	0	9 (3)	100	314	20·4	2, 6, 7
28	2	5	0	0	0	7 (1)	200	—	19·7	I, 2, 3, 4, 7
29	0	6	0	0	0	6 (2)	160	900	21·5	I, 2, 4, 7
30	0	10	0	0	0	10 (3)	140	314	21·0	I, 2, 3, 7
31	I	2	0	0	0	3 (0)	140	157	—	I, 2, 3, 4
33	0	4	0	0	0	4 (1)	240	314	—	I, 2, 6, 7
35	I	I	0	0	0	2 (1)	210	177	21·6	I, 2, 3, 4, 7
36	0	10	0	0	0	10 (4)	150	314	20·8	2, 7
37	0	2	0	0	0	2 (0)	90	400	—	I
39	0	11	0	0	0	11 (3)	60	—	21·3	2, 7
40	3	19	0	0	0	22 (7)	80	700	19·7	Bamboo grove
41	11	29	I	0	0	41 (13)	80	900	21·2	I
42	8	11	0	0	0	19 (10)	100	200	21·1	2, 3, 4, 6
43	12	18	0	0	0	30 (4)	140	300	20·5	I, 2, 4
44	0	4	0	0	0	4 (0)	150	177	—	2, 4
45	0	54	0	0	0	54 (34)	170	79	21·1	I, 2
46	0	9	0	0	0	9 (4)	260	20	20·6	2
47	I	11	0	0	0	12 (4)	170	600	21·2	2, 4, 7
48	0	19	0	0	0	19 (5)	80	1500	21·4	I, 2, 3, 7

TABLE 3.1—*continued*

Sample No.	*strigata*	*frenata*	*bisecta*	*cestata*	*atra*	Total (adults in brackets)	Alt. in metres ± 20 m.	Sampling area in m²	Mean shell length in mm (adult only)	Plant species and remarks (see text)
49	1	2	0	0	0	3 (1)	40	—	20·7	1, 2, 3
52	0	4	0	0	0	4 (0)	220	200	—	5
53	0	11	0	0	0	11 (2)	200	177	18·6	2, 7
54	0	11	0	0	0	11 (1)	170	314	20·3	1, 3
55	0	3	0	0	0	3 (2)	130	177	21·7	2, 3, 7
56	0	7	0	0	0	7 (6)	70	300	20·6	1, 2, 3
57	0	2	0	0	0	2 (1)	120	—	21·5	2, 3, 6
58	0	22	0	0	0	22 (13)	160	1000	20·7	1, 4, 7
59	0	13	0	0	0	13 (6)	210	100	20·5	3, 4
60	0	2	0	0	0	2 (2)	150	50	21·3	4
63	0	3	0	0	0	3 (0)	160	20	—	1
64	0	9	0	1	0	10 (0)	250	20	—	1, 5
65	0	11	0	0	0	11 (6)	300	200	20·5	1, 2
66	0	11	0	0	0	11 (1)	280	20	21·0	4, 5
67	0	18	0	0	0	18 (2)	240	79	20·3	1, 2, 7
68	0	5	0	0	0	5 (1)	200	43	20·3	1, 2, 7
69	0	2	0	0	0	2 (0)	120	79	—	3, 4, 7
70	0	6	0	0	0	6 (1)	180	300	22·0	1, 2, 3
71	0	4	0	0	0	4 (1)	200	20	19·7	6, 7
72	2	22	0	1	0	25 (5)	260	177	20·1	5, 7
73	2	20	0	2	0	24 (8)	140	79	20·1	2, 7
74	3	14	0	0	0	17 (14)	220	314	20·5	2, 3, 4
75	4	56	0	1	0	61 (43)	240	20	20·8	2, 4
76	3	8	0	0	0	11 (6)	220	79	20·1	2, 4, 5
77	42	16	0	8	0	66 (21)	260	20	19·6	4
78	23	64	0	5	0	92 (47)	240	79	20·0	2, 4, two *strigs.* have dark bases
79	8	70	0	4	0	82 (46)	210	—	20·0	2, 4, two *strigs.* have dark bases
80	1	36	0	1	0	38 (13)	180	225	20·4	2, 7
81	0	18	0	0	0	18 (7)	100	600	20·0	2, 3, 4
82	0	18	0	0	0	18 (1)	200	300	18·3	2, 3
83	0	20	3	0	0	23 (7)	80	300	20·5	1, 2, 3, 7
84	0	12	1	0	0	13 (6)	90	225	20·2	1, 3, 4, 7
85	0	11	0	0	0	11 (5)	160	314	20·4	1, 2, 6, 7
86	0	21	6	1	0	28 (14)	120	707	19·6	2, 5, 7

TABLE 3.1—*continued*

Sample No.	*strigata*	*frenata*	*bisecta*	*cestata*	*atra*	Total (adults in brackets)	Alt. in metres ± 20 m.	Sampling area in m²	Mean shell length in mm (adult only)	Plant species and remarks (see text)
87	0	29	11	1	0	41 (22)	210	400	20·4	1, 2, 3, *cestata* tends to *atra*
88	0	10	6	0	0	16 (7)	300	150	19·4	2, 3, 6, 7, two *bisecta* tend to *frenata*
89	0	11	2	0	0	13 (3)	340	225	20·0	4 ,5, 7
90	0	35	1	0	0	36 (13)	360	225	19·9	2, 6
91	0	23	4	0	0	27 (12)	340	120	20·2	6
92	0	20	0	0	0	20 (6)	200	50	19·3	Bamboo Grove
93	0	12	0	0	0	12 (5)	220	200	19·7	1, 2, 7
94	1	11	0	1	0	13 (4)	240	20	19·1	4, 6, 7
95	7	29	0	0	1	37 (7)	260	20	20·1	4, 6, 7, *atra* pale at suture
96	2	36	0	0	1	39 (19)	320	20	19·6	6, 7, *atra* has pale suture
169	0	3	0	0	0	3 (0)	60	20	—	2, 6, 7
170	0	64	0	1	0	65 (15)	140	79	20·2	1, 2, 3, 7, *cestata* has thin band
171	5	81	0	0	0	86 (32)	180	79	19·2	2, 6, 7
247	12	26	0	0	0	40 (27)	320	100	19·5	2, 5, 6
281	0	1	2	0	1	4 (2)	80	314	20·7	2, 3
282	0	5	2	0	0	7 (0)	80	314	—	1, 2, 3
283	0	2	2	0	0	4 (1)	90	314	20·5	1, 2, 3
284	0	12	0	0	0	12 (2)	150	79	20·0	2, 6
285	0	4	3	0	1	8 (5)	170	314	21·5	1, 3
401	1	21	2	0	0	24 (8)	180	100	20·4	2, 3, 5
402	4	39	7	0	1	51 (19)	290	100	20·2	2, 5
403	0	5	2	0	0	7 (1)	160	100	20·3	1, 2, 3
404	0	5	1	0	0	6 (2)	240	100	19·1	3, 5, 6
476	3	6	0	0	7	16 (9)	180	100	18·9	1, 2, 6, 7
477	14	17	2	4	14	51 (50)	220	225	18·7	2, 6, *cestata* have dark sutures (=*zonata* Crampton)
478	3	36	0	0	1	40 (36)	300	100	19·0	2, 6, 7
479	3	16	0	0	0	19 (15)	90	150	20·1	2, 6, 7

TABLE 3.1—*continued*

Sample No.	*strigata*	*frenata*	*bisecta*	*cestata*	*atra*	Total (adults in brackets)	Alt. in metres ± 20 m.	Sampling area in m²	Mean shell length in mm (adult only)	Plant species and remarks (see text)
480	11	28	0	0	0	39 (24)	180	100	19·5	1, 2, 6, 7
489	4	2	0	0	0	6 (3)	70	100	20·3	1, 2, 3, 7
490	9	21	0	0	0	30 (11)	140	100	20·3	1, 6, 7
491	5	10	0	0	0	15 (0)	160	100	—	1, 2, 3
492	7	6	0	0	0	13 (3)	250	225	20·9	2, 3, 7
493	16	10	1	0	0	27 (5)	230	225	20·6	2,3
494	5	5	0	1	0	11 (3)	230	225	20·9	1, 2, 3
S1	1	20	1	1	0	23 (16)	280	150	—	5, 6, 7, *cestata* has wide band
S2	8	13	0	0	1	22 (15)	360	79	—	2, 6, 7, *atra* has pale suture
S3	1	5	0	0	0	6 (5)	300	100	—	2, 5, 6
TOTAL	362	1644	60	33	32	2141 (893)				

without *P. suturalis* (i.e. where *P. taeniata* occurred alone). It also records the names of the major valleys.

Figure 3.2 gives the proportions of *strigata* in the samples. It shows a concentration in the western part of Urufara Valley and the eastern part of Faatoai. There are also high frequencies of the morph in southern Tehaoa and in the valleys south of Pafatu. It appears to be absent in Aareo, eastern Urufara, western Faatoai, and in the far northwest.

Figure 3.3 shows the proportions of *bisecta*, which is clearly concentrated in southeastern Urufara, although there are sporadic occurrences elsewhere.

Figure 3.4 shows a concentration of *cestata* at low frequencies in Tehaoa, again with sporadic occurrences elsewhere.

The distribution of *atra* is not shown, but there is a slight concentration in the extreme southwest, with sporadic occurrences in the south and southeast.

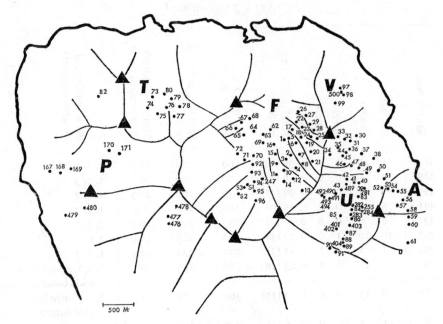

FIGURE 3.1. A sketch map of the northwestern part of Moorea, showing the places at which samples of *Partula* were taken (numbered spots). The branching lines represent mountain ridges, and the triangles represent peaks. The names of valleys are shown by their initial letters, as follows: A = Aareo, F = Faatoai, P = Pafatu, T = Tehaoa, U = Urufara, V = Vaitapi. The sites numbered 1–171 were sampled in 1962, those numbered 280 and above were sampled in 1967. The map covers an area of about 25 sq km.

We have found no statistically significant associations between any of these morphs and any of the physical or biotic factors that we have recorded (altitude, aspect, topography, background, the occurrence of other species of snails, the occurrence of individual species or communities of plants, the density of *P. taeniata* or *P. suturalis*, the mean shell-length of *P. taeniata* or *P. suturalis*, morph-frequencies in *P. taeniata*). Some of these factors are, however, significantly associated among themselves. The notable relations between altitude, density of *P. suturalis*, density of *P. taeniata*, mean shell-length of *P. suturalis* and mean shell-length of *P. taeniata* are beyond the scope of this paper, and will be discussed elsewhere (Clarke & Murray, in prep.).

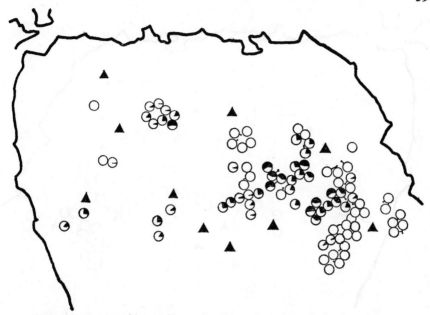

FIGURE 3.2. *Partula suturalis vexillum* in northwest Moorea. The distribution of the *strigata* morph. Three degrees of arc represent one per cent.

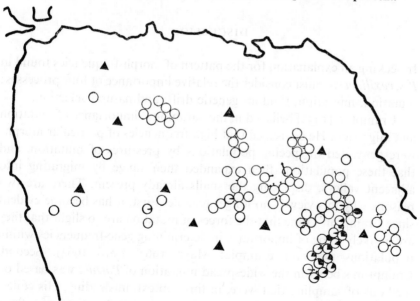

FIGURE 3.3. *Partula suturalis vexillum* in northwest Moorea. The distribution of the *bisecta* morph. Three degrees of arc represent one per cent.

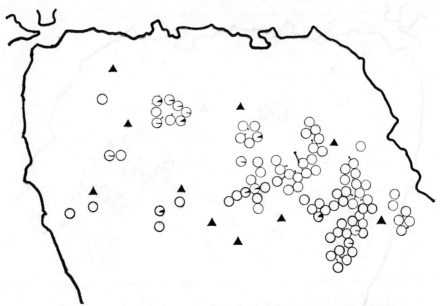

FIGURE 3.4. *Partula suturalis vexillum* in northwest Moorea. The distribution of the *cestata* morph. Three degrees of arc represent one per cent.

DISCUSSION

In seeking an explanation for the pattern of morph-frequencies found in *P. s. vexillum* we must consider the relative importance of four processes; mutation, migration, random genetic drift, and natural selection.

Crampton (1932) believed in the paramount importance of mutation and migration. He envisaged that high frequencies of particular morphs were generated in specific populations by pressures of mutation, and that these populations then expanded their range by migrating into adjacent valleys, displacing the snails already present. There are two reasons why this view is no longer tenable. First, it has become evident since Crampton's time that the forces of mutation are so slight that they are unlikely to be of importance in determining gene-frequencies within populations (see, for example: Mayr 1963, Ford 1964). Second, Crampton's belief in the widespread migration of *Partula* was based on methods of sampling that were, in this context, misleading. His collections were taken from whole valleys or from large subsections of valleys, and he was unaware that significant changes of morph-frequencies could be found within much smaller areas (as is shown by figs. 3.2, 3.3 and 3.4). He visited Moorea several times between 1907 and 1924, and sampled some of the valleys at intervals throughout the period. Because successive samples differed not only in the proportions of morphs within a species but also in the proportions of species he was led to postulate extensive migrations. The true explanation of these differences is almost certainly that the pattern of his sampling within a heterogeneous area differed on each occasion. When we collected on a finer scale, first in 1962 and then in 1967, we found no changes of the type reported by Crampton.

Although Crampton's samples represent large and heterogeneous areas it is nevertheless possible to make a general comparison between his samples and our own. This comparison suggests that the morph-frequencies have maintained a roughly constant distribution during the period from 1907 to 1967. Crampton's data (1932, tables 10 and 11), like ours, indicate concentrations of *strigata* in western Urufara, eastern Faatoai and the valleys south of Pafatu. They similarly show concentrations of *bisecta* in southern Urufara and of *atra* in the extreme southwest. He did not collect in Tehaoa.

The patterns of gene-frequencies in *P. suturalis vexillum* resemble the patterns found in *Cepaea nemoralis* and designated 'area effects' by Cain & Currey (1963). As in *Cepaea*, we find the preponderance of one or a few morphs over areas very much larger than the panmictic population, apparently regardless of habitat or background. Between such areas the morph frequencies may change violently within distances of 200 metres or less, often in apparently uniform environments. Following Cain and Currey, we might argue that random genetic drift cannot be the present cause of the patterns because the areas are too large, and that undetected selective factors are of paramount importance. Following Goodhart (1963) and Wright (1965), however, we might suggest that the first groups of snails arriving in the region were small populations differing by chance from each other, and that the initial differences between these populations persisted because of the evolution of balanced gene-complexes. When the groups expanded and came together, inter-penetration would then be prevented by the relative inviability of hybrids between the balanced systems.

The difficulty with the first argument is that we would expect selective agents capable of bringing about the large observed changes of frequency to be associated with at least some of the environmental variables that we have recorded. The 'high *strigata*' areas, for example, extend over a wide range of habitats. The eastern area stretches from one valley to the next over a high mountain ridge, encompassing a change of more than 450 m in altitude and related changes in temperature, rainfall, sunshine, flora and fauna. Yet its limits do not appear to correspond to any environmental discontinuity.

The possibility that we have overlooked some cryptic environmental factor can never, of course, be excluded. There is an alternative possibility that associations with known factors do exist, but that our samples were not sufficiently large or extensive to detect them. If so, the associations cannot be very strong. In this context it is worth examining any apparent associations that border on, but do not reach significance. Curiously, there is only one; between the frequency of *strigata* and the occurrence of climbing pandanus, *Freycinetia demissa*. If the samples are considered to be statistically independent, an assumption that is not strictly valid because of the geographical grouping, the probability of this association lies between 0·1 and 0·05 (by the median test). Although it is not easy to see why striated shells should be favoured on climbing pandanus, it is perhaps worthy of note that other species and races of *Partula* associated with the plant elsewhere on Moorea (*P. mooreana* Hartman, *P. olympia* Crampton and *P. tohiveana* Crampton) all show relatively high frequencies of *strigata*, or of visually similar morphs.

This association must remain only an interesting possibility until further work has been done. In a large series of comparisons we must expect that by chance one or two will approach significance. There is, as yet, no clear evidence of direct selection by the external environment.

Applied to *Partula* the 'founder' hypothesis of Goodhart (1963) and Wright (1965) also presents difficulties. Not only does *P. suturalis vexillum* show a series of area effects for different genes, but so does every other race and species on the island. We are forced to postulate literally thousands of isolates, arranged in an apparently arbitrary geographical pattern (Clarke & Murray 1969). This is not impossible, but the situation becomes more credible if we assume that a balanced gene-complex can evolve without the prior necessity of isolation. An algebraic analysis of morph-ratio clines has suggested that such a process is theoretically possible (Clarke 1966), and a study of *P. taeniata* has provided evidence that the area effects do indeed represent coadapted gene-complexes (Clarke 1968). This interpretation also helps us to understand the curious patterns of speciation within the genus (Clarke & Murray 1969).

At present, therefore, the evidence available suggests that the selective values of the colour morphs in *P. s. vexillum*, as in other species of *Partula*, are determined more by the gene-complex in which they are found than by the direct selective pressures of the external environment. Our findings are consistent with the predictions of Ford (1940, 1942) that natural selection will prove to be the dominating influence in the majority of genetic polymorphisms.

SUMMARY

One hundred and three samples were taken from populations of the land snail *P. s. vexillum* in the northwestern region of the island of Moorea, French Polynesia. An attempt was made to correlate the frequencies of shell-colour morphs with a large number of environmental factors. It was not successful. The geographical pattern of morph-frequencies resembles the patterns described by Cain & Currey (1963) as 'area effects'. It is suggested that the selective values of the shell-colour morphs are determined more by the internal genetic environment than by the direct effects of external physical or biotic factors.

ACKNOWLEDGMENTS

We are very grateful to Mrs Ann Clarke and Mrs Elizabeth Murray for their help at all stages of the work. Dr James Bishop, Dr Michael Carter and Mr Howard Smith kindly helped us in the field. Mrs Elizabeth Kater, Miss Lorna Stewart and Miss Yvonne Pryde have given invaluable technical assistance. We are greatly indebted to the Royal Society, the National Science Foundation, the Science Research Council, the Carnegie Trust for the Universities of Scotland and the Percy Sladen Trust for financial support.

REFERENCES

CAIN A.J. & CURREY J.D. (1963) Area effects in *Cepaea*. *Phil. Trans. R. Soc. (B)*, **246**, 1–81.
CARLQUIST S. (1965) *Island life*. Natural History Press, New York.
CLARKE B. (1966) The evolution of morph-ratio clines. *Am. Nat.* **100**, 389–402.

CLARKE B. (1968) Balanced polymorphism and regional differentiation in land snails. In E.T.Drake (ed.), *Evolution and Environment*, pp. 351–368. Yale University Press.

CLARKE B. & MURRAY J. (1969) Ecological genetics and speciation in land snails of the genus *Partula*. *Biol. J. Linn. Soc.* **1**, 31–42.

CRAMPTON H.E. (1932) Studies on the variation, distribution, and evolution of the genus *Partula*. The species inhabiting Moorea. *Publs Carnegie Instn* **410**, 1–335.

FORD E.B. (1940) Polymorphism and taxonomy. In J.Huxley (ed.), *The New Systematics*, pp. 493–513. Oxford University Press.

FORD E.B. (1942) *Genetics for Medical Students*, Methuen, London.

FORD E.B. (1964) *Ecological Genetics*. Methuen, London.

GOODHART C.B. (1963) 'Area effects' and non-adaptive variation between populations of *Cepaea* (Mollusca). *Heredity* **18**, 459–465.

HUXLEY J.S. (1942) *Evolution, The Modern Synthesis*. Allen and Unwin, London.

MAYR E. (1963) *Animal Species and Evolution*. Harvard University Press.

MURRAY J. & CLARKE B. (1966) The inheritance of polymorphic shell characters in *Partula* (Gastropoda). *Genetics* **54**, 1261–1277.

MURRAY J. & CLARKE B. (1968) Inheritance of shell size in *Partula*. *Heredity* **23**, 189–198.

WRIGHT S. (1965) Factor interaction and linkage in evolution. *Proc. R. Soc. (B)* **162**, 80–104.

4 ✳ Colour and Banding Morphs in Subfossil Samples of the Snail *Cepaea*

A. J. CAIN

INTRODUCTION

Since Ford's careful definition (1940) of polymorphism, the interest to population geneticists and evolutionists of this remarkable and widespread phenomenon needs no emphasizing. Its interest is enhanced when studies of its stability or otherwise can be made over long periods of time. The late Cyril Diver realized the importance of subfossil snail shells for this purpose as long ago as 1929. Currey & Cain (1968), using *Cepaea nemoralis* (L.) and *C. hortensis* (Müller), two species with an extensive shell-character polymorphism, were able to show a considerable alteration in southern England in the proportions of the bandless form in *C. nemoralis* from the hypsithermal (*c.* 4500 BC) to the present day, consistent with climatic selection; *C. hortensis*, however, showed no consistent signs of change. They showed also in *C. nemoralis* that the type of geographical distribution of morph frequencies which they called area effects (Cain & Currey 1963a) was probably more widespread in pre-Iron Age times than now, and some area effects may have lessened in intensity. They were able to give considerable evidence concerning these changes for a district including the western Marlborough Downs and the lowlying country adjacent, which is famous for its antiquities and in which several important excavations have taken place.

A new excavation in the middle of this district has produced exceptionally well-preserved material, in which it is possible to score with confidence both banding morphs (as before), and shell colour, as brown or not-brown, in *C. nemoralis*. Lip colour has also been examined, and using this material as standard, two previously reported samples have been re-scored. Additional material of *C. hortensis* from the same excavation adds to our knowledge of this enigmatic species.

65

It is a pleasure to offer this paper to a population geneticist whose abilities are equally outstanding in archaeology.

MATERIAL

The conditions under which subfossil material can be accepted for the purposes of population genetics have already been discussed by Currey & Cain (1968). The specimens reported in the present paper are from Bishop's Cannings Long Barrow and West Overton, Round Barrow, previously scored for banding only by Currey & Cain, and new material from tufa at Caerwys and Ddol, Flintshire. The principal new material comes from a salvage excavation by Dr J.G.Evans of an early Neolithic long barrow at South Street, Avebury Trusloe, Wiltshire. The barrow was made by piling chalk, dug from a deep enclosing ditch, on to the existing surface, and the resulting buried turf line could be easily distinguished. The ditch was cut deep into solid chalk, and gradually filled in with chalk washing down from the mound. There would be little temptation for rabbits, moles, or other disturbers to colonize it, and in fact, the filling was well stratified.

Much of the material from Avebury Trusloe is of complete shells, often in excellent condition, since they were buried in nearly pure chalk. Unfortunately, a large sample from the turf line had to be returned to the excavators to be destroyed in the process of radio-carbon dating, and this was not scored in detail for shell colour, the other samples first making it clear that such scoring was practicable. The available samples are as follows.

A   'Shells from buried soil beneath barrow. Primary Neolithic.' (Returned.)

B ⎫ Two small samples, also from the buried soil. 'Neolithic *circa*
C ⎭ 3000 BC.'

D   'Shells from ditch. Neolithic to Late Neolithic, *circa* 3000–2200 BC.'

E   Late Neolithic.

F   'Shells from the ditch. Mainly Beaker and Bronze/Iron but some Neolithic and some Roman.'

G   Bronze Age.

Samples E and G were collected by myself and Professor J.D.Currey during a visit to the excavation; the others are all by Dr J.G.Evans, and the quotations are his labels. The compositions of the samples are given in tables 4.1, 4.4, 4.5 and 4.6.

The Caerwys material is from a large area of white tufa, extensively quarried for lime, and in such good preservation that several shells actually show yellow pigment at the apex. The samples are given in table 4.7, as is a small one from Ddol, another tufa deposit about a mile from Caerwys.

SCORING FOR COLOUR OF SHELL AND LIP

In the Avebury Trusloe samples a clearcut difference was found in the *C. nemoralis* material (of various degrees of goodness of preservation) between shells with a definite light brown pigmentation, occasionally slightly bluish on the upper whorls, and those with none at all, which were more or less white. Only 3 shells out of 144 uncoloured and 63 coloured seemed doubtful, and their scores were confirmed on rescoring several weeks later. The same shade of brown can be seen in old modern specimens of the dark brown morph. The colour of this morph is produced by the combination of a yellowish periostracum overlying a dull lead-blue shell. Modern shells that have laid out in the open soon lose the periostracum and appear dirty blue; but on further exposure the colour changes (probably by oxidation) to a light brown of exactly the shade, though greater intensity, of that seen in the coloured subfossil shells. No other colour morph shows anything like this hue, and it seems reasonable, therefore, to regard the coloured subfossil shells as belonging to the dark brown morph. It is not possible to say with confidence what the white shells were; they could have been any shade of yellow or pink.

No corresponding colour has been seen in *C. hortensis*. This is to be expected, as brown *hortensis* is far less pigmented than dark brown *nemoralis*, and is often scorable in fresh material only with care.

In modern *C. nemoralis* it often happens that the tip of a shell heterozygous for shell-colour alleles shows the colour appropriate to the recessive allele of the particular combination (Cain, King & Sheppard 1960). In several of the subfossil shells, the same appearance is seen, the tip and more or less of the adjacent whorls being white (i.e. non-brown) and it has been possible to score for tip colour in this material.

In nearly all modern material of fully adult shells of *C. nemoralis* the lip and callus are darkly pigmented, but in various peripheral regions a white lip (normal in *C. hortensis*) is found. White lip (albolabiate, *al*) is recessive to normal lip; pale lip, with decreased pigmentation, appears to be heterozygous (Cain, Sheppard & King 1968) but is often scored with difficulty, since only shells with fully thickened lips can be used. A white lip is also seen in hyalozonates, and a pale one in orange-bandeds, but in these cases the normal band pigmentation is absent or very dilute respectively. In the subfossil material of *C. nemoralis*, almost all lips are either clearly pigmented, or clearly not as are all those of *C. hortensis* and *Arianta arbustorum*. Only in a few shells of *nemoralis* was the lip pigment so faded as to be present only in the callus and nearby; but in consequence, two specimens, fragments carrying only a part of the arch of the lip and apparently but not convincingly unpigmented, have been scored as indeterminable for lip colour. Almost all the subfossil material is unbanded, and formally cannot be assigned to either white lip or hyalozonate. Hyalozonate at the present day is even more restricted and sporadic than white lip, and Currey & Cain (1968) saw none in all their subfossil material. It is absent from the Marlborough Downs and Avebury district. Consequently, it is much more likely that the subfossil white-lipped material is of abolabiate (white lip).

RESULTS

I. Avebury Trusloe

(A) SPECIES COMPOSITION

The species composition of the subfossil samples from Avebury Trusloe, and of modern ones from the vicinity is given in tables 4.1 and 4.2. The high frequency of *C. nemoralis* in the subfossil samples and the vast preponderance of *C. hortensis* in the modern ones agrees with a change already shown by Currey & Cain (1968, p. 492) for the district; *nemoralis* was formerly more widespread at lower altitudes and has retreated towards the upper downs. It is now sporadic and difficult to find in the lower areas, which are primarily *hortensis* country.

TABLE 4.1. Species composition of subfossil snail samples, South Street Long Barrow, Avebury Trusloe (National Grid Reference 090692).

Sample	C. nemoralis	C. hortensis	Cepaea sp.	A. arbustorum
A. Main Neolithic sample	94	22	4	5
B. Buried soil (1)	9	—	—	—
C. Buried soil (2)	7	—	—	—
D. Neolithic to late Neolithic	54	19	1	32
E. Late Neolithic	7	4	—	1
F. Mainly Beaker & Bronze/Iron	140	64	7	46
G. Bronze	2	5	—	4

All the shells shown as *Cepaea* sp. are unbanded small juveniles or body-pieces. As can be seen, their addition to either species produces little effect.

The composition of samples A, B and C (Primary Neolithic) already suggests an open grassy environment, distinctly dry (very few *Arianta*) and with some bushes here and there (since some *hortensis* are present, and mixed colonies, as Professor B.C.Clarke first observed, are usually associated with shrubs or trees). The alternative would be dense wood-land, which, from the humus line shown by the excavation, is very unlikely. The later large samples (D and F) suggest that the ditch became overgrown with long herbage which tended to stay wet (more *Arianta*), and with more bushes and trees, or alternatively *hortensis* moved in even on grassland, where it occurs at lower altitudes at the present.

TABLE 4.2. Species composition of modern snail samples from the Avebury Trusloe area.

Sample	C. nemoralis	C. hortensis	A. arbustorum
(1) SU 091 691	1	25	1
(2) SU 092 693	—	101	1
(3) SU 087 692	—	11	1
(4) SU 091 690	2	22	1
(5) SU 090 693	3	31	1

(B) MORPHS OF *C. nemoralis*

(a) *Shell colour*

Figure 4.1 shows the present distribution of frequency of brown to not-brown in the district. Most of the samples are those reported by

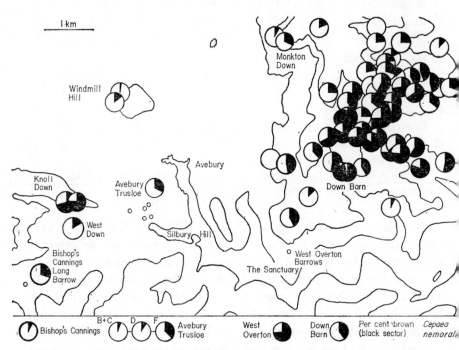

FIGURE 4.1. Frequency of brown (black sector) in modern and subfossil samples of *C. nemoralis* in the Avebury district, Wiltshire. Contours at 100-ft intervals. Subfossil samples at the foot of the diagram. Localities of the modern samples in the vicinity of Avebury Trusloe, and subfossil ones from Bishop's Cannings Long Barrow and West Overton Barrows shown by small circles. All the modern shells from the vicinity of Avebury Trusloe are shown as a single sample. Locality of Down Barn subfossil sample shown in fig. 4.2 for clarity.

Cain & Currey (1963a) and Currey & Cain (1968, rescored for colour). A few additional ones (or additional information) are given in table 4.3. Brown is abundant on the southern part of the higher downs, and common towards the valleys eastward, but rapidly decreases in frequency to north, west and south. It varies considerably in the few places in which it

Locality	C. nemoralis									C. hortensis					
	Yellow			Pink			Brown			Yellow			Others		
	0	3	5	0	3	5	0	3	5	0	5	5fus	0	5	5fus
Copse S. of Herepath West Down SU 076 688	28	1	—	4	1	—	32	—	—	—	—	—	—	—	—
N. of West Overton, Barrow SU 119 691	7	6	6	2	—	—	4	—	—	18	71	5	—	—	—
Roadside nr. Bishop's Cannings Long Barrow SU 066 678	12	5	—	3	5	—	17	2	—	2	2	—	—	—	—
Rough herbage patch, nr. Bishop's Cannings Long Barrow SU 068 676	—	—	2	3	1	3	1	1	2	7	21	2	—	—	—
Devil's Den 2	4	18	2	58	20	1	90	21	2	—	1	1	—	—	—
Windmill Hill SU 084 713	1	—	—	5	—	2	1	—	—	19	11	7	—	2	4
Windmill Hill SU 084 715	2	—	1	16	6	15	1	—	—	11	29	6	—	—	—
Roadside nr. Avebury SU 089 698	—	—	—	4	—	—	2	—	—	19	10	7	—	—	—
Combined sample Avebury Trusloe nemoralis	(nemoralis previously reported)									—	—	—	—	—	—
West Kennet Long Barrow	—	—	—	—	—	—	—	—	—	37	17	8	—	—	—
Clatford Down 8 SU 145 719	—	—	—	—	—	—	—	—	—	15	3	—	—	—	—
SU 144 720	—	—	—	—	—	—	—	—	—	—	16	—	—	—	—
Middledown Wood SU 146 727	—	—	—	—	—	—	—	—	—	5	20	7	—	—	—
Avebury	—	—	—	—	—	—	—	—	—	11	23	17	—	2	2

TABLE 4.4 Morph scores of subfossil *C. nemoralis*, South Street Long Barrow, all unbanded, ooooo (with single exception as indicated in notes).

| | Uncoloured shells | | Coloured shells | | Tips of | | | |
| | | | | | Coloured shells | | Uncoloured shells | |
Sample and state	Pigmented lip	White lip	Pigmented lip	White lip	Coloured	Uncoloured	Coloured	Uncoloured
Sample A (unscored for shell colour and apices)								
Adults ± whole	66	14						
Large pieces with ± mouth	14	—						
Sample B								
Adults ± whole	5		—		—		—	5
Mouthpieces	1	1*	—		—		—	
Large body-pieces		2						
Sample C								
Adults ± whole	5	1	1		1		—	6
Sample D								
Adults ± whole	47	1	5		2	3	—	48
Large juv., strong callus	1		—		—	—	—	1
Sample E								
Adults ± whole	1		1		—		—	1
Mouthpieces	2		1		—	1	—	
Large body-pieces	2		—		—		—	2

Sample F†						
Adults ± whole	73	54	—	24	25	69
Mouthpieces	3	1?	1	—	—	
Large body-pieces	4		—	—		
Large juv.	2	2				
Sample G						
Adults ± whole	1	—				
Mouthpieces	1	—				

* In sample B, several of the 5 adults have very faded lips, and as the mouthpiece lacks most of the pillar and callus (on which colour is best retained) it is only doubtfully unpigmented.

† Sample F includes one adult uncoloured five-banded, 12345, with a far better polish than any of the other shells of *nemoralis*, which may perhaps be recent and is therefore not shown above.

has been found on lower ground, and from some of these only very small samples have been obtained. The few and mostly very old shells of *nemoralis* found in the localities close to the excavation have been lumped as a single sample (of six shells!) in fig. 4.1.

The subfossil material from Avebury Trusloe shows a considerable increase in proportion of browns from the earlier to the later material. The main primary Neolithic sample (A), as explained above, was not scored for colour but contained very few coloured shells, certainly not more than half a dozen in 94 shells, if that many. This is supported by the two small samples (B and C) with in all 15 uncoloured and 1 coloured, and sample D (48 uncoloured, 5 coloured) does not differ significantly from them (exact $P = 1$). Sample F (Beaker and Bronze/Iron), however, with 81 uncoloured to 57 coloured is clearly different. B, C and D can be combined, and against F give an exact P of 0.0000005. Browns, therefore, have increased greatly in frequency from Neolithic to Beaker/Bronze/Iron Age times at this site, and are still locally common in the district.

(b) *Colour of shell tip*

It was shown by Cain, King & Sheppard (1960) from bred material that heterozygotes for shell colour sometimes show the recessive colour at the tip of the shell. In samples from the wild, including three from the southern part of the Marlborough Downs, they showed that the penetrance in dark brown shells of pink and yellow (pale and deep pinks and deep yellows in these samples) was about equal. Slightly more than half the brown/pink and brown/yellow heterozygotes showed the appropriate colour at the tip, the rest being all brown or with only an indeterminate pallidness at the tip.

In the present subfossil material, a number of coloured shells have white at the tip or spreading down over the first few whorls, exactly as does heterozygous colour in modern material from the district. The numbers of shells with coloured or uncoloured tip are given in table 4.4. In all uncoloured shells the tip was white. A form in which the tip is brown but not the rest of the shell, known in some pale browns, is absent from the region now, and appears to have been then. The numbers available for the Neolithic are very small but suggest that both white and brown tips were common. Those from sample F show that they were in about equal proportions (28 white to 25 coloured in coloured shells) in Beaker/Bronze/Iron Age times. In the Neolithic samples all the coloured shells should be heterozygous for white; with 3 coloured and 4 white tips, penetrance would be 0·57 very roughly. In sample F, using the Hardy–Weinberg formula (as did Cain, King & Sheppard) since selective discrepancies are likely to be small compared with sampling errors, the occurrence of 69 uncoloured to 53 coloured shells with apices preserved suggests frequencies of p and q of very nearly 0·25 and 0·75 respectively—by maximum likelihood (Smith 1956), 0·248 and 0·752. In fact, of the putative heterozygotes, 28 show white tips and 17 do not, a penetrance of 0·62. In the largest Marlborough sample reported by Cain, King & Sheppard, the same calculation, using maximum likelihood estimates for p, q and r, gives the penetrance of pink in browns as 0·59 and of yellow in browns as 0·61; in the next largest they are 0·67 and 0·77 respectively. In view of the sample sizes, these may all not be far from 0·5, which might suggest a maternal effect, but only breeding experiments can settle this point.

(c) *White lip*

This morph occurs not infrequently in Western Ireland and at high altitudes in the Pyrenees (e.g. Arnold 1968, 1969) but is rare or absent elsewhere in western Europe. Cain, Sheppard & King (1968) and Cook (1966) have shown it to be recessive to normal lip pigmentation and closely linked to banding and shell colour. It is allelic to hyalozonate (Cook 1967, Cain, Sheppard & King 1968). A form with a paler lip than the normal appears to be the heterozygote but is scorable only on shells with fully thickened and in no way abnormal lips.

In the Avebury Trusloe subfossil samples, white lip occurs (table 4.4) in 14 out of 80 scorable shells in sample A, and 1 in 12 of B and C combined; these three samples can be combined. In D and E it occurs in 1 out of 59, which possibly differs from the primary Neolithic samples (exact P is 0.0713) but not from the later ones (F and G combined, none in 133). As the normal pigmentation in the subfossil forms is weaker than in life, no attempt to recognize pale lip can be made.

White lip has not been seen in any of the numerous samples recorded by Cain & Currey (1963a) from the Marlborough Downs. It is absent from the lowland samples except near Bishop's Cannings Long Barrow where it reaches a high frequency in a small sample (4 in 13 shells). It occurs on Milk Hill and Tan Hill, just south of the district on the northern edge of the Vale of Pewsey. It has been found (Cain & Currey 1963b) with

4

certainty from only one locality out of 70 on Salisbury Plain. Pale lip has occurred very infrequently on the Marlborough Downs (three shells in widely separate localities in the area of higher downs shown in fig. 4.1). It occurs with white lip in the Bishop's Cannings sample. Remarkably, four out of the six shells, from Avebury Trusloe, including two fairly fresh ones, show it. It also occurs in the larger Windmill Hill sample but this contains a number of old faded shells, and although its frequency appears to be high, this may be misleading. It is certainly there, in two virtually fresh shells. It would seem, therefore, that white lip was present at Avebury Trusloe in the primary Neolithic, and then became much rarer, perhaps absent in Beaker/Bronze/Iron Age times; at present it is very limited and sporadic, but does occur in the district.

II. Bishop's Cannings and West Overton

Using the Avebury Trusloe material for comparison, it has been possible to rescore these samples for colour (table 4.5). The Bishop's Cannings Long Barrow is Neolithic, the West Overton round barrows were dated by the excavator, Dr Isobel Smith, at about 1650 BC.

TABLE 4.5. Subfossil material from Bishop's Cannings Long Barrow and West Overton Barrow.

	Uncoloured shells		Coloured shells	
I. *C. nemoralis*	Pigmented lip	White lip	Pigmented lip	White lip
Bishop's Cannings Long Barrow (all unbanded)	26	1*	3	—
W. Overton, Barrow	4	—	11	—
II. *C. hortensis*	Unbanded	5-banded	5-banded with fusion	
Bishop's Cannings Long Barrow (omitted from Currey & Cain 1968)	7	12	7	
W. Overton	—	2	—	

*A fragment with a small portion only of lip, perhaps *C. hortensis*.

Rescoring for colour of shell gives 27 uncoloured and 3 coloured for the Neolithic sample, 4 uncoloured and 11 coloured for the round barrow sample. The Neolithic one does not differ significantly from samples B, C and D combined, also Neolithic, from Avebury Trusloe (exact P, 1·0), nor from its modern sample (exact P, 0·17). The West Overton Barrow sample is slightly different from the Avebury Trusloe sample F (exact P, 0·027), and less different from the nearest modern sample (exact P, 0·072) but differs sharply from the Avebury Trusloe Neolithic samples (exact P, 0·0000007). These subfossil samples (especially West Overton) are small; nevertheless, it is satisfactory to find that they agree with those of the Avebury Trusloe samples that one would expect on grounds of date. Correspondingly, a small Romano-British sample (Down Barn) already reported by Currey & Cain (1968) does not differ from modern ones in its vicinity.

Few shell tips are preserved in these samples. Bishop's Cannings has 16 uncoloured shells, with uncoloured tips and a single coloured one with white tip. West Overton has only coloured shells with tips, 3 coloured and 2 white. Such as they are, they agree very well with the Avebury Trusloe material. No white lips occur in either sample with certainty, although there is a single white-lipped fragment of doubtful species, which might be *nemoralis*, from Bishop's Cannings. Whether it is or not, the sample is too small to differ significantly in lip colour from any of the Avebury Trusloe sample-groups, and this is even more true of West Overton. The Bishop's Cannings modern sample has both white and pale lip; the nearest to West Overton has a single pale lip in 37 scorable shells. The West Overton subfossil sample does not differ significantly from its modern one, but no conclusion can be drawn from this, as the sample size is low. The Bishop's Cannings sample does just differ from its modern one (exact P, 0·045).

III. Cepaea Hortensis

Only three of the Avebury Trusloe samples, A, D and F, have useful numbers of *C. hortensis*, but they are well spaced in time (table 4.6) and the species is common in the vicinity at the present day (table 4.2). The modern distribution of unbanded against banded is given in fig. 4.2, and of fusion in banded shells in fig. 4.3.

(a) *Unbanded*

There is considerable local variation in the frequency of unbandeds at present, with perhaps some indication of area effects. For example, the

TABLE 4.6. Subfossil samples of *C. hortensis*, Long Barrow, Avebury Trusloe.

Sample	Unbanded	5-banded	5-banded with fusions
A	13	6	3
D	6	9	4
E	—	4	—
F	9	30	16
G	—	2	3

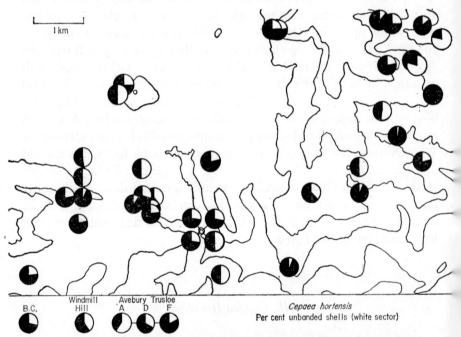

FIGURE 4.2. Frequency of unbanded shells (white sector) in modern and subfossil samples of *C. hortensis* in the Avebury district. Samples recorded by Clarke (1960) for Silbury Hill arbitrarily grouped around that locality, and connected to it by lines. Contours at 100-ft intervals. Locality of subfossil sample from Windmill Hill shown by small circle. Knoll Down samples displaced for clarity into a vertical column.

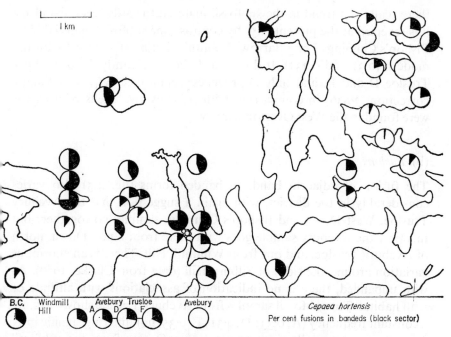

FIGURE 4.3. Frequency of shells with fusions of bands (black sector) in the class of banded shells, in modern and subfossil samples of *C. hortensis* in the Avebury district. Conventions as in fig. 4.2.

Bishop's Cannings and West Down samples agree quite well with two of those from Knoll Down, the Avebury Trusloe samples, the Avebury sample, and three out of four from Silbury Hill; whereas two of the Knoll Down samples, that from just north of Avebury Trusloe, and one from Windmill Hill seem to go together. There seems little pattern, however, in the downland samples. There is no indication of any association of unbanded with altitude nor with habitat. The subfossil material from Avebury Trusloe shows a decrease in frequency of unbanding from earlier to later, A, 0·59; D, 0·32; F, 0·16. The difference between A and F is significant (exact P, 0·00048), and the heterogeneity χ^2 gives P less than 0·001. The five modern collections from the vicinity have frequencies of 0·16, 0·37, 0·09, 0·23 and 0.29 respectively, not significantly different except between (1) and (2) (exact P, 0·036), and not heterogeneous overall ($P > 0·05$, omitting the small sample (3)). The overall modern frequency is 0·31, most like that from sample D (Neolithic to late Neolithic), significantly different from that for A, just not for that for F. If, therefore, there really is a trend in the subfossil material towards less unbandeds, it is reversed at the present day. By contrast, the subfossil material from Bishop's Cannings Long Barrow (Neolithic) is almost identical with the modern (0·31 and 0·237). It does not differ significantly from Avebury Trusloe D and F, and only approaches significance ($\chi^2_{(1)}$ with Yates's correction, 3·836) in comparison with A. Only two shells of *hortensis* were found in the West Overton barrow.

(b) *Fusions*

The fusion of adjacent bands in banded forms (fig. 4.3) also varies considerably in the district, with again a suggestion that the Avebury Trusloe, West Down and Bishop's Cannings samples go together; but this time the Avebury sample goes with those from Knoll Down, north of Avebury Trusloe, and one from Windmill Hill. There is an enormous variation around the sides of Silbury Hill (data from Clarke, 1960). As with unbanded, there is no indication of associations with altitude or with habitat. The subfossil samples from Avebury Trusloe show virtually a constant frequency (A, 0·33; D, 0·31; F, 0·35) not differing significantly from the modern collections (exact P, 0·076), nor from the subfossil sample from Bishop's Cannings; which itself is just not different from its modern one (exact P, 0·055). The Windmill Hill Neolithic sample recorded by Currey & Cain (1968) is intermediate between the two modern ones, as it was for unbanded also. Nearly all the modern samples from Avebury Trusloe and Bishop's Cannings show a decrease in frequency of fusions as compared with the subfossil ones (5, decrease; 1, effectively the same; 1, increase).

DISCUSSION

Since the end of the Late-Glacial period, about 8000 BC, there appears to have been a fairly steady improvement of the mean temperature up to the hypsithermal or 'climatic optimum' of the late Boreal and early Atlantic periods, from about 5000–4000 BC, as shown by pollen analysis (Godwin 1956; Johnson & Smith 1966). In the early Atlantic period (pollen zone VIIa) the climate became wetter. Drier and progressively cooler conditions held in the sub-Boreal (VIIb) which includes the Neolithic and Bronze Ages in Britain, and about 4–500 BC (early Iron Age) there was a marked deterioration to the present damp and cool Subatlantic period. The distributions of banding morphs in subfossil samples and over western Europe at present were shown by Currey & Cain (1968) to be consistent with each other and with experimental work, and to suggest the prevalence of unbandeds in hotter times and regions, five-bandeds in cooler and wetter ones. The additional subfossil material reported here supports their conclusions. Arnold (1968, 1969) finds a definite association of unbanded with warmer, drier conditions in the Pyrenees.

The consistent increase in browns in *C. nemoralis* from primary Neolithic to Iron Age times (including the present day) which is shown by the samples from Avebury Trusloe, Bishop's Cannings Long Barrow, and West Overton is supported by material from elsewhere in Britain. Of the samples ascribable to the hypsithermal recorded by Currey and

TABLE 4.7. Subfossil material, tufa areas, Flintshire.

| I. *C. nemoralis* | Uncoloured shells | | | Coloured shells |
	Unbanded	Midbanded	Midbanded heavy spread	Unbanded
Caerwys, white tufa matrix	60*	—	—	—
In tufa, near surface	17†	—	—	—
Reworked tufa, surface and soil above	11‡	8	9	3
Semi-modern	2	3	8	1

II. *C. hortensis*	Unbanded	5-banded	5-banded with fusion
White tufa matrix	5	4	1
Near surface	10	5	4
Reworked tufa, tufa surface and soil above	1	2	—
Semi-modern	—	4	1

* 5 white lipped, 54 pigmented lip, 1 lip unscorable.
† 2 white lipped.
‡ 1 may be white lipped.
Later collections have produced a single shell in tufa at Caerwys which may be coloured.

Cain, three—in one case datable to 4490 ± 150 BC and associated with a Mesolithic culture—are from white tufa and in good condition for scoring colour. These are from Blashenwell near Corfe, Dorset; Brook, Kent; and Wateringbury, Kent. To these can now be added tufa material from Caerwys and Ddol, Flintshire (tables 4.7 and 4.8). In all the truly Mesolithic subfossil specimens from these sites available to me for scoring colour, there is only one possible brown, from Caerwys. In

TABLE 4.8. Modern *Cepaea* samples, tufa areas, Flintshire.

	C. nemoralis									C. hortensis					
	Yellow			Pink			Brown			Yellow			Others		
Locality	0	3	5	0	3	5	0	3	5	0	5	5fus	0	5	5fus
Caerwys near road	—	—	—	—	—	—	—	—	—	24	60	30	—	8	2
Far side of quarry	—	24	—	—	16	—	1	5	—	10	33	33	1	15	12
Ddol (1)	—	—	—	—	—	—	—	—	—	2	6	9	—	—	—
(2)	—	—	—	—	—	—	—	—	—	1	6	4	—	—	—

A large number of banded *nemoralis* (Caerwys) are heavily spread-banded.

modern collections dark browns occur at three out of the four sites (and may be present at the fourth since the modern sample is small). At Caerwys it shows an increase from the white tufa samples to the reworked material near the top, the surface and soil, and the modern sample (frequencies <0·013, 0·097, 0·071, 0·130). This strongly supports the impression given by the Neolithic and later samples in good condition for scoring colour, that dark brown was rare when the climate was markedly warmer than at the present day. As with banding, the modern distribution supports this hypothesis. Brown is rare or absent over nearly all France, and was not seen by Arnold during extensive collecting in the Pyrenees (Arnold 1968). It is sporadic but locally common in England, Wales and Ireland, and occurs also in the Netherlands. Its distribution is, in short, cool Atlantic (northwestern), neither very cool as in most of Scotland nor warm as in South French coastal districts. Cain & Currey (1963a) suggested that dark brown was associated with cooler summers, from their studies of morph distributions on the Marlborough Downs. Carter (1968) supported this suggestion from work on the Berkshire Downs, and Cain (1968) found an association of dark browns with topographical features on sand dunes which would also agree with it.

In the Oxford district, in which visual selection by predators accounts for most of the variation between sites (Cain & Sheppard 1950, 1954), dark brown shows far less frequency variation between habitat types than does pink, and is sporadic in its occurrence; this suggests that selective factors other than visual selection are important even when the other colour morphs are responding vigorously (Cain, in preparation).

White lip in *C. nemoralis* occurs in about 16 per cent of the shells in the main primary Neolithic sample, 2 per cent in the later Neolithic one, and not at all in 133 shells from the Beaker/Bronze/Iron Age sample at Avebury Trusloe. This is a highly significant decrease. Yet there is considerable evidence that the gene is now present at some frequency in the vicinity. The West Overton sample is too small to be useful here. The Bishop's Cannings ones, although small, indicate a significant increase in frequency between the Neolithic sample and the present day. It would seem that white lip was present in the vicinity in primary Neolithic times, decreased or vanished later, and is now sporadically present in the lowland area. The modern Bishop's Cannings sample suggests a gene frequency of 0·23, while that in the fossils can at best have been less than 0·07. That in Avebury Trusloe sample A, however, was about 0·40. Counting pale lips as evidence, as well as white lips, the small modern samples suggest gene frequencies of 0·33 and 0·31 for Avebury Trusloe and Bishop's Cannings respectively. White lip occurs in the tufa at Caerwys at about 10 per cent (gene frequency about 0·30) but is absent from or rare (not more than $2\frac{1}{2}$ per cent) in samples from the surface, where subsequent reworking by water has washed in later material, and from the soil above, and is not present in fresh modern shells; it still occurs a few miles away on Halkyn Mountain, the highest part of this carboniferous limestone district. The Wateringbury sample is small, and in that from Brook there is some fading of a few lips, so white lip has not been scored.

White lip has been studied by Arnold in the Pyrenees (1968, 1969). He found it associated in three regions with riverside as against warmer and drier hillside localities and its distribution generally in the Pyrenees, England and Ireland 'appears to follow regions where rainfall is relatively high, or where temperatures are seldom elevated'. Cameron (1969a) found white lip to increase in frequency with altitude on Slieve Carran (northwestern Clare, Eire), but to be absent from the much lower Slieve Elva. Clarke, Diver & Murray (1968) record it from sand dunes at Bundoran (Sligo), Cook & Peake (1962) from Mullaghmore (Sligo), and Burke (in Cain, Sheppard & King 1968) from Streedagh nearby. Cook & Peake (1960) and Cook (1966) record it in some abundance from collections in the Dartry Mountains (Sligo, Eire) at heights of 800–1000 ft. All these western Irish samples are at much lower altitudes than those with white lip in the Pyrenees, as would be expected if there is an association with climate. Cook (1966) records it from Fountains Fell, near Malham Tarn, Yorkshire, and Arnold (1968) from Cross Fell, Cumberland. I have seen it on Halkyn Mountain (Flintshire, Wales) at 900 ft, and it occurs at Minera, Flintshire (collection by Dr R.W.Arnold in Cain, Sheppard & King 1968). I have seen it also from South Cornwall, and Arnold (1966) records it from North Cornwall. By contrast, white lip is absent from the Oxford district and the Marlborough Downs, and nearly so from the Larkhill area of Salisbury Plain. Carter (1968) does not mention it in his extensive survey of the Berkshire Downs, and generally in southern England it is rare or absent. One locality for it is the Pentridge, Dorset. The small district from Windmill Hill to near Bishop's Cannings in which it occurs sporadically at the present day does not seem to be marked out by any obvious climatic features; but as Cain & Currey (1963a) pointed out for the distribution of *C. nemoralis* and *C. hortensis* on the Marlborough Downs, local climate can differ greatly in the geographical distribution of temperatures and humidity from climate as determined by standard meterological stations and can be heavily dependent on topography. A local combination of climatic factors may allow a morph to persist locally in an otherwise unfavourable region. Certainly, the general impression from the surveys mentioned above is that white lip occurs mainly in western or high-altitude regions of Britain, i.e. in those with considerable rainfall. The localities given above additional to those mentioned by Arnold (1968) support his association of white lip with greater rainfall or lowered temperature.

The occurrence of white lip in subfossil material from the Mesolithic and primary Neolithic but not in later samples in two regions (Wiltshire, Flintshire) where it does occur at the present day, either in the same vicinity or only four miles away is of special interest. What the present Subatlantic period and the Early Atlantic have in common is wetness, not temperature. Whereas the spread of browns and the decrease in unbandeds correlate well with the continual decrease in mean temperatures from the hypsithermal to the preesent day, the variation in frequency of white lip suggests a much stronger association with dampness than with temperature, which should be testable. Usually in western Europe a cooler climate is also a damper one, and the possible effects of the two components on *Cepaea* morphs cannot be disentangled; but the subfossil material appears to allow us to distinguish.

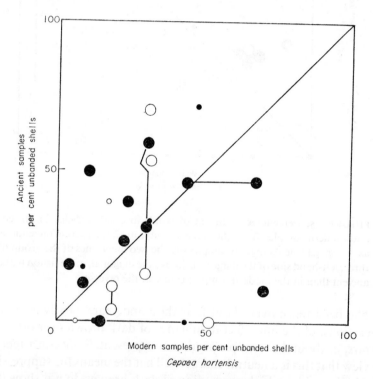

Cepaea hortensis

FIGURE 4.4. Percentage of unbanded shells, in ancient and in modern samples from the same vicinity, in *C. hortensis*. Large circles, both samples of more than 10 shells. Small circles, either sample of 10 or less shells. Black circles, ancient sample from Mesolithic or Neolithic. White circles, ancient sample from Bronze or Iron Age. Lines join circles representing samples from the same locality (two subfossil samples from Maiden Castle, lower left corner; three from Avebury Trusloe; one from Abingdon equally comparable with two modern ones, middle of diagram). Above the diagonal, more unbandeds in the ancient than in the modern sample; below it, the reverse.

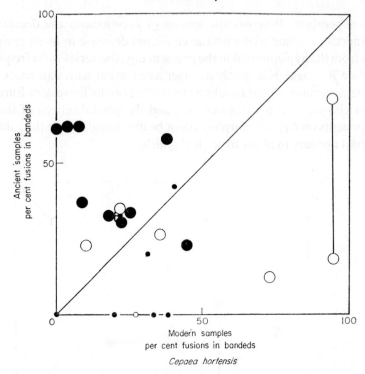

FIGURE 4.5. Percentage of fusion of bands in banded shells, in ancient and modern samples from the same vicinity, in *C. hortensis*. Conventions as in fig. 4.4 (the Abingdon samples are the two inner ones of the group of three, left-hand side of the diagram). Above the diagonal, more fusion in the ancient than in the modern sample; below it, the reverse.

The persistence over about four thousand years, nearly a thousand generations, of the imperfect dominance of dark brown over other shell colours, in about the same proportion as at present, is not consistent with the view that this is a neutral character. That the means for suppressing it exist is clear since a high proportion of dark browns do not show it, yet it persists. Until the genetic basis is known, further discussion is hardly possible.

As usual, the very closely related *C. hortensis* is behaving very differently from *C. nemoralis*, and this is the only definite phenomenon shown by the subfossil material of this species. There is only an indication at Avebury Trusloe of a decrease in unbandeds from primary Neolithic to Iron Age times, and a possibility of increase at the present day, not supported by the samples from Bishop's Cannings.

If all the subfossil and modern samples of 10 or more shells of *hortensis* recorded by Currey & Cain (1968) and in this paper are plotted for percentage unbanded and percentage fusion in old and modern samples, and the old are divided into Mesolithic and Neolithic, and Bronze and Iron Age samples, as in figs. 4.4 and 4.5, there is some tendency for the oldest (Mesolithic and Neolithic) to have more unbandeds and more fusions than the modern ones, while the Bronze/Iron Age ones show a wide scatter. The actual numbers are:

	Unbandeds		Fusions	
	more than modern	less than modern	more than modern	less than modern
Meso & Neolithic	8	3	8	1
Bronze, Iron	2	4	2	4

These trends are not significant, but afford the only indication so far of any temporal trend in this enigmatic species, except under definite visual selection (Clarke 1960). Fig. 4.5 also suggests that there are much fewer fusions in modern shells at sites from which Meso- and Neolithic material was collected.

Currey & Cain (1968) were unable to find any definite trends in subfossil *C. hortensis*. They indicated the existence of area effects involving unbanded in *C. hortensis* on Salisbury Plain (Cain & Currey 1963b) but these were less well marked than in *C. nemoralis*. Cain, in Cain, Cameron & Parkin (1969) found some indication of the association of unbanded yellows with a large hollow in the dunes near Durness. Cameron (1969b) found no close association of unbandeds as such (irrespective of shell colour) with local topography, except that they were significantly less common in hedgerow samples in the bottoms of dry valleys—in his river valleys there was no such effect. There were considerable differences in relation to both topography and habitat of unbandeds in different shell colours. We do not know the colours of subfossil *hortensis*. Carter (1968) found far more pinks and browns in a district on the south coast of England than farther inland. It may well be that Neolithic populations had very different proportions of shell colour, and that a direct comparison of unbandeds in subfossil populations with modern ones is therefore invalidated. The strong interaction between all three shell colours and unbanded that Cameron has found appears to be another point of difference between *C.hortensis* and *C.nemoralis*, in which only brown, and dark brown at that, shows a marked interaction with banding.

The change in species composition in the Avebury district, with more *hortensis* at the present day in the lowland areas, was noticed by Currey & Cain (1968, p. 492) and is confirmed by the material from Avebury Trusloe. The three common large helicids in subfossil material are *C.nemoralis*, *C.hortensis*, and the usually monomorphic *Arianta arbustorum* which greatly prefers damp localities (e.g. Cameron 1969a, b; Cameron in Cain, Cameron & Parkin 1969). For those samples for which *Arianta* as well as *Cepaea* were collected, fig. 4.6 shows the variation in species composition from the ancient samples (all but one, Mesolithic or Neolithic) to the present day, including Avebury Trusloe. The marked diminution of *C. nemoralis* is obvious. (In that pair which seems to have

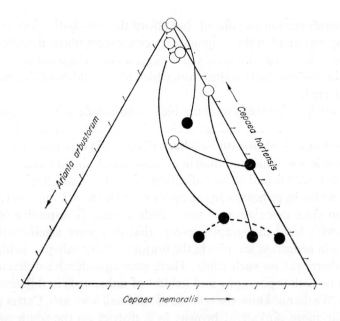

FIGURE 4.6. Species composition of ancient (black circle) and modern (white circle) samples, including *A. arbustorum*. Dotted lines join the three subfossil samples from Avebury Trusloe, and the four modern ones from the same vicinity.

changed least (Bishop's Cannings) we found no live material at the site. A little higher a small population of *C. nemoralis* was found, and a little lower, at the roadside, were only *C. hortensis* and *Arianta*. These two live samples were combined for the purpose of the diagram, to minimize the change. In fact, the very few live shells found on about the same contour were all *C. hortensis*.) There is no obvious change in the proportions of *A. arbustorum*, which need not be considered further. Consequently, all the sample-pairs in which both species occur at least in one sample can be used to investigate the change in relative abundance of *C. nemoralis* and *C. hortensis*. The data are given in Currey & Cain (1968) and in the present paper. Figure 4.7 shows the results; in the Mesolithic and Neo-

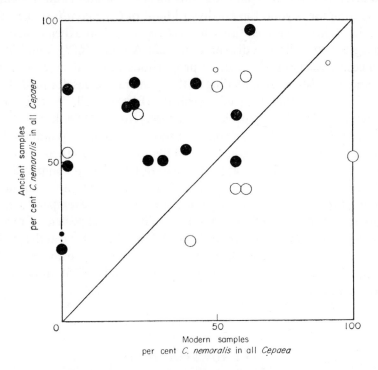

FIGURE 4.7. Proportions of *C. nemoralis* in ancient and modern samples of *Cepaea* from the same vicinity, at least one of which contains both species. Large circles, neither sample less than 20 shells. Small circles, one or both samples less than 20 shells. Black circles, ancient sample from Mesolithic or Neolithic. White circles, ancient sample from Bronze or Iron Age. Above the diagonal, more *nemoralis* in the ancient than in the modern sample; below it, the reverse.

lithic samples, 13 had more *nemoralis* formerly than at present, and only one the reverse (and not to any great extent). Of the Beaker/Bronze/Iron Age samples, the figures are respectively 5 and 5, all samples being considered. The exact P for this difference is very slightly greater than 0·05. If only the larger samples are considered (neither in a pair less than twenty shells) it is 0·045. There is therefore some indication that *C. nemoralis* has contracted its range between the Mesolithic/Neolithic and the present day, but no evidence that it changed between the Beaker/Bronze/Iron Age and now. The Avebury district shows (as does the Thames Valley and the Caerwys area) that *C. hortensis* is abundant in valley bottoms and at lower altitudes at present. Cain & Currey (1963a) have given reasons to believe that this local distribution and its general distribution far beyond the northward limit of *C. nemoralis* suggest that it is at home in cooler conditions than is *C. nemoralis*, and their conclusion is in agreement with the distributions that Arnold (1968) found in the Pyrenees. The change in relative proportions reported here is also in agreement, *C. hortensis* spreading as the temperature has declined. There is no indication that the tufa localities (Blashenwell, Wateringbury, Caerwys, Ddol) differ in species proportions from the Neolithic sites on chalk (Avebury Trusloe, Avebury, Bishop's Cannings, Windmill Hill, Sanctuary) or calcareous gravel and alluvium (Abingdon), although the immediate vicinity of tufa areas must have been damp compared with chalk grassland in relation to ground-water. The most important factor affecting the species proportions, therefore, seems to be temperature, not dampness in respect of which one might have expected a relative abundance of *hortensis* in both the Early Atlantic Period and at the present day, and less in Beaker/Bronze Age times (Avebury Trusloe F, Easton Down, City Farm, West Overton Barrow), which is not at all borne out by the samples.

ACKNOWLEDGMENTS

I am deeply indebted to Dr J.G.Evans for the subfossil material from Avebury Trusloe, to Professor J.D.Currey for permission to use it and many of his collections for this paper, to Mr G.Oxford for further collections, and to Professor P.M.Sheppard F.R.S. for his advice, criticism and help in computation.

REFERENCES

ARNOLD R.W. (1966) Factors affecting gene-frequencies in British and continental populations of *Cepaea*. D. Phil. thesis, Oxford.

ARNOLD R.W. (1968) Studies on *Cepaea* VII. Climatic selection in *Cepaea nemoralis* (L.) in the Pyrenees. *Phil. Trans. R. Soc. Lond., B* 253, 549–593.

ARNOLD R.W. (1969) The effects of selection by climate on the land snail *Cepaea nemoralis* (L.). *Evolution* 23, 370–378.

CAIN A.J. (1968) Studies on *Cepaea* V. Sand-dune populations of *Cepaea nemoralis* (L.). *Phil. Trans. R. Soc. Lond., B* 253, 499–517.

CAIN A.J., CAMERON R.A.D. & PARKIN D.T. (1969) Ecology and variation of some helicid snails in northern Scotland. *Proc. malac. Soc. Lond.* 38, 269–299.

CAIN A.J. & CURREY J.D. (1963a) Area effects in *Cepaea*. *Phil. Trans. R. Soc. Lond., B* 246, 1–81.

CAIN A.J. & CURREY J.D. (1963b) Area effects in *Cepaea* on the Larkhill Artillery Ranges, Salisbury Plain. *J. Linn. Soc. Lond. (Zool)* 45, 1–15.

CAIN A.J., KING J.M.B. & SHEPPARD P.M. (1960) New data on the genetics of polymorphism in the snail *Cepaea nemoralis* L. *Genetics* 45, 393–411.

CAIN A.J. & SHEPPARD P.M. (1950) Selection in the polymorphic land snail *Cepaea nemoralis*. *Heredity* 4, 275–294.

CAIN A.J. & SHEPPARD P.M. (1954) Natural selection in *Cepaea*. *Genetics* 39, 89–116.

CAIN A.J., SHEPPARD P.M. & KING J.M.B. (1968) Studies on *Cepaea* I. The genetics of some morphs and varieties of *Cepaea nemoralis* (L.). *Phil. Trans. R. Soc. Lond., B* 253, 383–396.

CAMERON R.A.D. (1969a) The distribution and variation of *Cepaea nemoralis* L. near Slievecarran, County Clare and County Galway, Eire. *Proc. malac. Soc. Lond.* 38, 439–450.

CAMERON R.A.D. (1969b) The distribution and variation of three species of land snail near Rickmansworth, Hertfordshire. *Zool. J. Linn. Soc. Lond.* 48, 83–111.

CARTER M.A. (1968) Studies on *Cepaea* II. Area effects and visual selection in *Cepaea nemoralis* (L.) and *Cepaea hortensis*. *Phil. Trans. R. Soc. Lond., B* 253, 397–446.

CLARKE B.C. (1960) Divergent effects of natural selection on two closely-related polymorphic snails. *Heredity* 14, 423–443.

CLARKE B.C., DIVER C. & MURRAY J. (1968) Studies on *Cepaea* VI. The spatial and temporal distribution of phenotypes in a colony of *Cepaea nemoralis* (L.). *Phil. Trans. R. Soc., B* 253, 519–548.

COOK L.M. (1966) Notes on two colonies of *Cepaea nemoralis* (L.) polymorphic for white lip. *J. Conch.* 26, 125–130.

COOK L.M. (1967) The genetics of *Cepaea nemoralis*. *Heredity* 22, 397–410.

COOK L.M. & PEAKE J.F. (1960) A study of some populations of *Cepaea nemoralis* L. from the Dartry Mountains, Co. Sligo, Ireland. *Proc. malac. Soc. Lond.* 34, 1–11.

COOK L.M. & PEAKE J.F. (1962) Populations of *Cepaea nemoralis* L. from sand-dunes on the Mullaghmore Peninsula, Co. Sligo, Ireland, with a comparison with those from Annacoona, Dartry Mts., Co. Sligo. *Proc. malac. Soc. Lond.* 35, 7–13.

CURREY J.D. & CAIN A.J. (1968) Studies on *Cepaea* IV. Climate and selection of banding morphs in *Cepaea* from the climatic optimum to the present day. *Phil. Trans. R. Soc. Lond.*, B **253**, 483–498.

DIVER C. (1929) Fossil records of Mendelian mutants. *Nature, Lond.* **124**, 183.

FORD E.B. (1940) Polymorphism and taxonomy. In J.S.Huxley (ed.), '*The New Systematics*', pp. 493–513. Oxford University Press, London.

GODWIN H. (1956) *The history of the British flora. A factual basis for phytogeography.* Cambridge University Press, Cambridge.

JOHNSON C.G. & SMITH L.P. (1966) *The biological significance of climatic change in Britain.* Academic Press, London.

SMITH C.A.B. (1956) Counting methods in genetical statistics. *Ann. hum. Genet.* **21**, 254–276.

5 ❀ Shell Size and Natural Selection in
Cepaea nemoralis

L. M. COOK AND P. O'DONALD

INTRODUCTION

Like many other helicid snails *Cepaea nemoralis* grows to a definitive adult size, at which point a lip is laid down on the edge of the mouth of the shell. The average maximum shell diameter over most of the range of the species is 20 to 22 mm, but in some regions, notably the Pyrenees, Cantabria and parts of Ireland, there are large-shelled races that reach a mean size of 24 to 30 mm. This geographical distribution has a parallel in the related species *Arianta arbustorum*, which is for the most part slightly smaller than the typical *C. nemoralis* but which reaches extremely large dimensions in some mountainous localities (see Cook 1965 where data of C. Oldham are presented). In parts of the range of *C. nemoralis* where dimensions fall within the typical limits there may be striking changes in mean breadth over short distances in uniform habitats (for example, Wolda 1969).

These facts indicate that shell variation may be a useful subject for ecological genetic study. Shell dimensions are in part inherited. Many years ago di Cesnola (1907), using a method devised by W.F.R.Weldon, showed that stabilizing selection acted upon *A. arbustorum* so that individuals of more aberrant shape had a lower expectation of life than ones close to the optimum. In that species, as in *C. nemoralis*, it would be of interest to know what components of the environment act on size and shape. The same components may be responsible both for small local variations and for the substantial increase in breadth of certain races. Alternatively, the large size of the races may be determined by particular epistatic gene complexes and not by direct selection for size. The changes in average breadth over very short distances may indicate changes in otherwise undetected environmental variables that affect fitness through shell size. In a species with visual polymorphism as striking and universal as *C. nemoralis*, it is possible that there is interaction between colour and

93

banding and variability in breadth. The following observations were collected with these general problems in mind.

VARIABILITY IN MORPH SIZE

The possibility that morphs within one colony may differ in size was first investigated while studying the distribution of small-shelled and large-shelled colonies in a region of Ireland where a large-shelled race occurs (Cook & Peake 1960, 1962). The method employed was to measure to the nearest 0·1 mm the maximum shell diameter of each individual in a sample. Every individual in each sample was measured over a short period of time in order to reduce the risk of possible changes in technique. Since there is usually a large fluctuation in mean diameter from one colony to another, comparisons of morph size have to be made within colonies. Each sample may be composed of individuals of a number of different morph types, and so the categories compared may contain no more than a few individuals. Taking only those morphs that were numerous enough for a within-colony comparison, the Irish measurements provided the results summarized in table 5.1. Further details of the colonies will be found in the papers referred to above.

In the Benwhiskin sample, collected like the others in 1958, there are pinks and yellows, unbandeds and bandeds. Analysis of variance shows that there is no difference in size between the ground colour classes $(P > 0·2)$, while unbandeds are significantly larger than bandeds $(P < 0·05)$. In the other six samples comparisons are made between an unbanded class and a banded class. In four of them brown unbanded and yellow banded are compared while in the other two yellow unbanded and yellow banded are compared. In four of the six, unbandeds are larger than bandeds.

For all colonies t has been calculated by dividing the difference in means by the standard error of the difference. The corresponding normal deviate for an infinite number of degrees of freedom has then been found from the t-table, and this is entered in table 5.1 beside each colony. Apart from Benwhiskin only one of the values, that for colony A2, is significant $(P < 0·05)$, but combining the probabilities for all seven colonies by the method of Fisher provides $\chi^2_{(14)} = 23·7$, which is significant at the 5 per cent level and suggests that the variation in morph size within colonies is greater than would be expected by chance.

TABLE 5.1. The breadth of banded and unbanded morphs in some samples of *Cepaea nemoralis* from Ireland.

Colony	Number	Mean breadth (mm)	Variance	t
BENWHISKIN				
Pink unbanded	53	22·0	0·85	
Pink banded	116	21·8	1·25	2·08
Yellow unbanded	29	22·1	1·34	
Yellow banded	90	21·7	1·16	
A2				
Brown unbanded	31	21·3	0·83	2·02
Yellow banded	15	20·8	0·55	
A3				
Brown unbanded	22	21·2	0·89	0·99
Yellow banded	26	20·9	0·87	
A4				
Brown unbanded	17	20·6	0·82	0·90
Yellow banded	33	20·4	0·76	
A5				
Brown unbanded	25	20·2	0·65	0·08
Yellow banded	21	20·3	0·90	
C1				
Yellow unbanded	11	21·0	0·56	1·80
Yellow banded	24	21·5	0·78	
C4				
Yellow unbanded	21	21·4	0·64	0·44
Yellow banded	32	21·3	0·60	

In order to test whether there are any trends in banding and size the values of t have been summed with positive sign when unbandeds are larger than bandeds and negative sign when bandeds are the larger. The sum is then squared and divided by 7 to give a χ^2 value with one degree of freedom which measures the tendency over all samples for unbandeds to be larger than bandeds. The value obtained is $\chi^2_{(1)} = 2·96$, which has a probability of occurring by chance of between 0·1 and 0·05.

To obtain a more precise picture of the possible variation in shell dimensions within a single colony, 200 individuals were measured of each of nine morphs taken from a collection made by Dr R.W. Arnold on the Lambourn Downs, Berkshire. The results are shown in table 5.2. Analysis of variance (table 5.3) indicates that in this case there is no

TABLE 5.2 Breadth of different morphs in a sample from the Lambourn Downs, Berks.

Morph	Number	Mean breadth (mm)	Variance
Brown unbanded	200	20·6	0·94
Brown midbanded	200	20·4	0·74
Brown midspread	200	20·4	0·72
Pink unbanded	200	20·2	0·93
Pink midbanded	200	20·1	0·65
Pink midspread	200	20·2	0·63
Yellow unbanded	200	20·2	0·69
Yellow midbanded	200	20·3	0·70
Yellow midspread	200	20·3	0·71

TABLE 5.3. Variance ratios (F) for comparison of three colour and three banding classes shown in table 5.2.

Component	Degrees of freedom	Variance ratio	P
Between colours	2	19·97	< 0·001
Between band classes	2	2·12	n.s.
Interaction	4	1·76	n.s.
Between groups	8	6·40	< 0·001

difference in size between banding classes but a highly significant difference between colours. There is no interaction between colour and band class. Brown has the largest mean size, while pink has the smallest. A curious feature is the large variance for brown unbanded and pink unbanded but not for the yellow unbanded.

The range of morph size in two further samples was studied. The animals were collected for other purposes, including the investigation of the relation between shell dimension and survival (described below). They come from a sand dune at Point of Ayr on the north Welsh coast, and from a roadside in the Pyrenees. The Pyrenean colony was located at Cledes in the Garonne valley a few kilometres south of the border of France and Spain. The snails there are large-shelled and there is great variation in size. The Welsh individuals are small-shelled with a relatively small variance. Both samples have similar morph frequencies. The

TABLE 5.4. Breadth of morphs in samples from colonies in Wales and the Pyrenees.

Colony and morph	Number	Mean breadth (mm)	Variance
POINT OF AYR			
Pink banded	287	20·5	0·70
Yellow banded	396	20·4	0·75
CLEDES			
Pink banded	475	24·5	1·72
Yellow banded	682	24·5	1·50
Pink unbanded	32	24·6	1·00
Yellow unbanded	62	24·7	1·44

results are shown in table 5.4. Here, there is no significant variation in mean breadth between morphs. In the Pyrenean sample the class with the largest mean is the unbanded, of which, however, few were found.

It appears, then, that variation in shell size may occur between the morphs in one colony. So far as these results are concerned, however, there is no general tendency for a particular category to be larger than the others. In the Irish samples unbandeds were larger than bandeds, but the variation in the Lambourn Downs colony occurs between colours. There is no difference in size between colour classes in either the Point of Ayr or the Cledes sample, but combination of the probabilities obtained for each of the three sets of data shows that the overall variation between morphs within colonies is significant. A possible source of error arises if size and morph frequency both change over a short distance in a colony. A spurious correlation between the two variables would then appear if a sample was taken from an area sufficiently extensive to include patches of different morph and size composition. There is no reason to believe this to be true of any of the present examples, but the danger makes it difficult to establish whether or not the morphs of a particular locality vary in size.

VARIATION IN SURVIVAL WITH SIZE

Three samples of snails from the Lambourn Downs were kept in dry cardboard boxes in a cool room over winter 1966–67. *C. nemoralis* normally hibernates from November to March or April, burrowing a centimetre or two below the surface of the ground. During this period an epiphragm is laid down over the mouth of the shell and the animal does not feed. In the experimental conditions a similar epiphragm was produced to that found in nature, and no food was available; so that, although there was a higher ambient temperature, conditions were otherwise somewhat similar to those experienced in the wild.

These samples were the subject of an undergraduate class experiment, which took place in February 1967. Each animal was scored as live or dead depending on its condition as seen through a small hole made in the shell. The shell size of the individuals in the two categories was recorded. It appeared that for two of the three samples the mean breadth of survivors was greater than that of non-survivors. All the shells were remeasured by one person (L.M.C.). The values obtained on the second scoring are shown as Lambourn I, II and III in table 5.5. They depend

TABLE 5.5 The breadth of survivors and non-survivors in samples of over-wintering *Cepaea nemoralis*.

Colony	Number	Mean breadth (mm)	Variance	t
LAMBOURN I				
Live	356	20·57	0·75	+3·56
Dead	623	20·37	0·65	
LAMBOURN II				
Live	55	22·51	0·89	+1·46
Dead	223	22·30	0·89	
LAMBOURN III				
Live	154	21·70	1·05	−0·44
Dead	752	21·74	1·03	
POINT OF AYR				
Live	528	20·48	0·70*	+0·37
Dead	155	20·45	0·83*	
CLEDES				
Live	702	24·73	1·35†	+5·86
Dead	549	24·31	1·74†	

* This difference in the variances corresponds to a probability of 0·089 which is thus not significant at a probability of 0·05.

† This difference is significant at a probability of slightly less than 0·001.

on the assessment of condition of the animal made during the class investigation, but since the possible basis for a difference in variability with size was not mentioned, there is no reason to suspect bias in favour of one end of the range or the other. In one set of data the difference in diameter between survivors and non-survivors is significant ($P < 0.001$), while in the other two, which go in opposite directions, it is not significant.

In an attempt to confirm the suggestion of differential survival, the Point of Ayr and Cledes samples were both subjected to conditions similar to the Lambourn samples, being kept in boxes over winter in an unheated room. The Point of Ayr sample was set up on 8 September 1968, that from Cledes on 29 September. Both were scored at intervals by examining the animals through the shell under a strong light. Mortality occurred at a much higher rate in the Pyrenean than in the Welsh sample —after 169 days 44 per cent of the larger Pyrenean animals had died, while only 23 per cent of the Welsh ones had died after 199 days. Comparison of shell breadth of survivors and non-survivors was made after a period long enough for an appreciable fraction to have died but not long after the point at which the animals would normally have become active. The Spanish sample was scored on 16 April and the Welsh on 23 April. The results are included in table 5.5. Survivors are significantly larger than non-survivors in the Cledes sample ($P < 0.001$) and larger but not significantly so in the Welsh. When the normal deviates for each of the five trials are combined by the method used above, a $\chi^2_{(1)}$ of 23·38 is obtained. This is highly significant, indicating that under mild winter conditions there is likely to be selection favouring large size. The very great difference in rate of mortality between the small-shelled Welsh snails and the large-shelled Spanish ones shows that this tendency does not hold from one colony to another from a different provenance, but that there are differences in genetic susceptibility or physiological acclimatization, or both, between them.

GENETIC CONTROL AND ENVIRONMENTAL MODIFICATION

Under artificial conditions the heritability of shell dimensions in *C. nemoralis* is about 60 per cent (Cook 1967, Wolda 1969). A more extensive study of the closely related snail *Arianta arbustorum* gave a figure of 70 per cent (Cook 1965). Some evidence from a natural colony, however, indicates that the environmental effect on size may at times be much greater than is suggested by these estimates. In 1962 a second sample was collected from the Benwhiskin colony first examined in 1958 and referred to in table 5.1. The dimensions for the second sample are as follows

Morph	Number	Mean breadth	Variance
Pink unbanded	11	23·00	2·302
Pink banded	45	23·22	2·318
Yellow unbanded	16	23·59	1·343
Yellow banded	29	23·01	1·212

Measurements on both samples were checked together, so as to afford a direct comparison. The later collection contains animals strikingly larger than those in the early one, although both come from a locality identical to within 1 or 2 m. There are no significant differences in size between morphs, but the variances are greater than they were for the first sample. The reason for the change is unknown, but an effect of this magnitude would involve very great selective mortality if it were the result of differential elimination. It is more likely that environmental conditions have changed in such a way as to allow juveniles in the intervening period to reach a larger size. This explanation is consistent with the increase in variance that has taken place. It is also understandable that there should be differences in morph size in the first sample but not the later one if it is assumed that the first adults developed under poorer conditions. The frequency of shells with the umbilicus incompletely closed dropped between the two dates from 62 to 29 per cent. As Cameron (1969) has pointed out, an open umbilicus may be a side effect of lip development at a stage of growth before the potential adult size has been reached.

There is thus experimental evidence for a relatively large genetic component in the determination of shell size and an observation of great variation in nature over a little more than one generation. The relation of shell size to colour and banding class must be due either to the action of genes linked to the morph genes, or to some kind of secondary effect of the major genes themselves. The apparent environmental effects described here point to the second explanation.

The samples from Point of Ayr and Cledes were scored on several successive dates during the period of observation. Dead individuals were separated from living ones and measured. It is therefore possible to study in more detail the progress of the selection indicated by the results in table 5.5. The detailed profile for Point of Ayr merely supports the evidence presented in table 5.5 showing that no significant changes occurred and it will not be discussed further. On the other hand, as mortality increased, the Cledes samples exhibited systematic changes in the moments of the distribution of survivors (table 5.6). Using these

TABLE 5.6. Cledes sample. Statistics of survivors at particular dates during the selection.

Date survivors measured	Mean	Variance	Third moment	Fourth moment	Mortality between dates
29 Sept	24·537	1·5740	−0·2002	7·9241	0
31 Dec	24·587	1·4737	−0·1163	6·5341	0·09
7 Feb	24·630	1·4176	−0·0020	5·7656	0·14
14 Mar	24·678	1·3635	0·0173	5·1479	0·17
16 Apr	24·725	1·3484	0·0743	4·8803	0·14
21 May	24·794	1·3077	0·0851	4·7193	0·30
14 Jun	24·853	1·2348	0·2260	4·1794	0·48

changes it is possible to measure the intensity of the selection acting on the sample. The methods follow directly from those used by Weldon (1901) and di Cesnola (1907) at the beginning of the Century in studies of selection in land snails, and later elaborated by Haldane (1954).

(I) SOME THEORETICAL MODELS

Selective intensity is usually measured by the difference between the fitness of the optimal phenotype, w_0, and the mean fitness, $\bar{w}$. The formula is $I = (w_0 - \bar{w})/w_0$ (Van Valen 1965). This is the formula usually used to define the genetic load of a population, but it is here applied to variations in fitness of phenotypes rather than of genotypes. O'Donald (1968) showed that the intensity of selection can be calculated from observed changes in the mean and variance of a character if fitness is a quadratic function of the character such that $w = 1 - \alpha - K(\theta - x)^2$ where x is a value of the character and θ its optimum value. The value α is the amount by which the optimal type falls short of 100 per cent survival. Another useful fitness function is $w = (1 - \alpha)\exp\{-K(\theta - x)^2\}$ which Cavalli-Sforza & Bodmer (1970) call the 'nor-optimal' model of selection. They use it to predict the effects of stabilizing selection on human birth weight.

O'Donald (1970a,b) showed that when selection is linear, quadratic or nor-optimal the change of mean fitness can also be calculated. This is a more interesting measure of selection than its intensity. The intensity of selection is the proportionate amount by which fitness is depressed because not all individuals have the optimum phenotype. The quantity $\Delta\bar{w}/\bar{w}$ is the proportionate amount by which selection has raised the mean fitness. It is not, therefore, a purely hypothetical quantity like I, but a direct measure of evolutionary change if measured from one generation to the next, and of phenotypic response when measured within a generation. If fitness is a linear function of x, then w_0 can be anywhere up to infinity and I cannot be determined. The change in fitness may still be established, however, since $\Delta\bar{w}/\bar{w} = (\Delta\bar{x})^2/V$ where V is the variance of the values of x. If the fitness function is quadratic then

$$I = \{(\theta - \bar{x})^2 + V\}/\phi \quad \text{where} \quad \phi = (1 - \alpha)/K.$$

The values of θ and ϕ can be calculated from the changes in the mean and variance of x, $\Delta\bar{x}$ and ΔV (O'Donald 1968). The relative change in the fitness is

$$\Delta\bar{w}/\bar{w} = \frac{\{\mu_4 - V^2 + 4V(\theta - \bar{x})^2 - 4\mu_3(\theta - \bar{x})\}}{\{\phi - (\theta - \bar{x})^2 - V\}^2,}$$

where μ_3 and μ_4 are the third and fourth moments of the character about its mean (O'Donald 1970a). If the fitness is nor-optimal, however, it is then only possible to calculate I and $\Delta\bar{w}/\bar{w}$ if the character is normally distributed (O'Donald 1970b). The formulae are

$$I = 1 - \{\sqrt{(V'/V)}\}\exp\{(\Delta\bar{x})^2/2(V' - V)\},$$

and

$$\Delta\bar{w}/\bar{w} = \{V/\sqrt{[V'(2V - V')]}\}\exp\{(\Delta\bar{x})^2/(2V - V')\} - 1$$

where V' is the variance after selection so that $V' - V = \Delta V$.

(II) APPLICATION TO THE EXPERIMENTAL DATA

These formulae can be applied to the selection that took place in the samples of the snails showing differential survival depending on size. The sample from Point of Ayr showed no significant selection for size.

TABLE 5.7. Cledes sample. Comparison of the values of the optimum, intensity of selection and change of relative fitness given by the linear, quadratic and nor-optimal fitness functions.

Period of selection	Quadratic model			Nor-optimal model			Linear model
	θ	I	$\Delta\bar{w}/\bar{w}$	θ	I	$\Delta\bar{w}/\bar{w}$	$\Delta\bar{w}/\bar{w}$
29 Sept–31 Dec	25·4165	0·03807	0·003132	25·3226	0·04438	0·003535	0·001592
31 Dec–7 Feb	25·8106	0·03356	0·001879	25·7279	0·03564	0·001967	0·001285
7 Feb–14 Mar	25·8515	0·03848	0·002313	25·8811	0·03965	0·002273	0·001603
14 Mar–16 Apr	28·8672	0·07221	0·001672	28·9054	0·07536	0·001661	0·001616
16 Apr–21 May	26·7315	0·06524	0·004077	27·0214	0·07164	0·003920	0·003561
21 May–14 June	25·7559	0·05164	0·004474	25·8572	0·05141	0·004110	0·002687

TABLE 5.8. Cledes sample. Comparison of the values of the optimum, intensity of selection and change of relative fitness for the total amount of selection up to each date.

Selection for the period from 29 Sept up to the date	Quadratic model			Nor-optimal model			Linear model
	θ	I	$\Delta\bar{w}/\bar{w}$	θ	I	$\Delta\bar{w}/\bar{w}$	$\Delta\bar{w}/\bar{w}$
31 Dec	25·4165	0·03807	0·003132	25·3226	0·04438	0·003535	0·001592
7 Feb	25·6598	0·06631	0·008963	25·4781	0·07718	0·010074	0·005563
14 Mar	25·8827	0·09728	0·018164	25·5929	0·11233	0·020403	0·012673
16 Apr	26·4223	0·13593	0·027595	25·8496	0·14430	0·030511	0·022497
21 May	27·1244	0·20327	0·047276	26·0586	0·19519	0·051840	0·042117
14 Jun	27·2201	0·24323	0·071016	26·0065	0·23605	0·079197	0·063743

However, the reduction in the variance from non-survivors to survivors shown in table 5.5 approached statistical significance with $P = 0.089$. If selection is really acting to reduce the variation round the mean in this sample, then O'Donald's (1968) simple formula $I = (V - V')/(3V - V')$ for selection acting only upon the variance indicates that $I = 0.0225$. The corresponding formula for the relative change of mean fitness is $\Delta\bar{w}/\bar{w} = \frac{1}{2}(1 - V'/V)^2 = 0.00106$. The formulae for the nor-optimal model give $I = 0.0233$ and $\Delta\bar{w}/\bar{w} = 0.00106$. This agreement is to be expected since the distribution of size is approximately normal. Fitness increased by only about 0.1 per cent in this sample.

The Cledes sample showed a very significant increase of mean in the survivors and a very significant reduction in variance. The data are shown in table 5.6, where it is seen that changes in all four moments of the distribution are remarkably consistent and regular from the beginning to the end of the period. In tables 5.7 and 5.8, the values of θ, I and $\Delta\bar{w}/\bar{w}$ are shown for the quadratic and nor-optimal models and also values of $\Delta\bar{w}/\bar{w}$ for the linear model. An interesting feature of the tables is the comparison of I and $\Delta\bar{w}/\bar{w}$ as measures of natural selection. The period 14 March–16 April shows in both quadratic and nor-optimal models a great increase of θ from 25.9 to 28.9. This is because the mean increased by 0.047 mm in this period but the variance was reduced by only 0.015. Similar increases in mean during the periods of selection before and after this particular period were accompanied by reductions of variance of 0.054 and 0.041. Thus, no doubt as a result of chance, selection appeared to be mainly on the mean from 14 March to 16 April. For this reason the optimum is calculated to be farther away from the mean. In the formula

$$I = (w_0 - \bar{w})/w_0$$

an arbitrary increase in w_0 must necessarily produce an arbitrary increase in I, which does indeed show a sharp increase in this period.

If selection is mainly on the mean—for example if the fitness function is linear—then w_0 will be increased without limit for it will always be advantageous to increase the character. In this case $I = 1$ necessarily. The intensity of selection is therefore unsatisfactory as a measure of selection; for the actual increase of mean fitness, $\Delta\bar{w}/\bar{w}$, is small for the period 14 March–16 April, although the intensity of selection is high. Since the selection is largely on the mean, this is the one period for which values of $\Delta\bar{w}/\bar{w}$ calculated by linear, quadratic and nor-optimal models all agree.

In general tables 5.7 and 5.8 show that there is good agreement between the quadratic and nor-optimal models in the values they give for I and $\Delta\bar{w}/\bar{w}$. For the overall period of selection the quadratic model shows that fitness increased by 7·1 per cent.

The change of mean fitness was greatest during the later stages of the selection. But the mortality was higher during these stages also. The change of fitness for each individual death is more or less constant, except for the first period during which there were few deaths but quite a considerable change of fitness (of about 0·3 per cent). The rate of change by selection is more or less proportional to the number of selective deaths. This is to be expected if size directly determines the chances of survival. It would not be so if the chances of survival depended on the numbers surviving—if for example there was competition between survivors.

DISCUSSION

Under the experimental conditions there is strong selection in favour of large size in over-wintering animals from some colonies. If the important physical factor involved is the relatively high temperature, then this cause of mortality may be significant in mild Atlantic winter climates but much less so in the colder continental conditions such as those prevailing in the Pyrenees from which the Cledes sample comes.

Wolda (1963, 1967) has studied the relation of a number of ecological factors to the fitness of different morphs of *C. nemoralis*. With respect to size, he showed that there is a positive correlation between body size and both the number of clutches laid and the number of eggs per clutch. A difference in breadth of 1 mm between one snail and another is likely to lead to a difference in output of some 15 per cent (Wolda 1963). A large adult size is associated with rapid growth rate in juveniles (Wolda 1969). Wolda also shows clearly that there may be marked variation in response to the same conditions by animals from different colonies, and that this variation has a genetic basis. Like over-winter survival, the effect of egg number acts to favour the larger individuals. The range of breadth within colonies is usually quite small, but fluctuation in mean from one colony to another may sometimes be great. Presumably, therefore, there is selection for an optimum value in each colony due to a balance of factors tending to increase dimensions with those tending to reduce them. At present, no firm suggestions may be made as to what factors favour small size. A comparable situation has been studied in detail in the mouse, where factors tending to increase body size are more readily demonstrated than those tending to reduce it (Falconer 1955). In both small mammals and helicid snails a possible force limiting maximum dimensions in the wild is the action of predators. The reduction of predation in some mountain and island localities may in part account for the existence in them of races of large individuals.

The central problem in the ecological genetics of *C. nemoralis* concerns the factors affecting the almost universal visual polymorphism. There is much information showing that differences in fitness between morphs occur under particular conditions, but it is still not clear whether the stability of the system arises from the operation of a polymorphism-maintaining factor, which all colonies have in common, or whether a balance is achieved by different combinations of forces in different places. The fact that the species remains polymorphic over an extra-ordinary range of environments strongly suggests the existence of a common factor, but its nature remains obscure. The evidence presented here emphasizes the fact that all aspects of the phenotype must be considered together when assessing relative fitness. Quite small differences in mean breadth between individuals may lead to appreciable differences in survival and output. The mean survival rate of individuals from different colonies under the same conditions may vary greatly, possibly due to differences in genetic constitution. The colour and banding genes appear to affect the adult size, and the relative response of one morph compared to another varies with genetic background or environment or both. The visual morphs can therefore be considered to influence output and survival via the effect on size, quite apart from direct effects on fitness of the differences in visibility or thermal properties of shells of different colours. For this reason, apart from any other, relative fitnesses must vary from one habitat to another.

We are very grateful to Dr R.W.Arnold for discussion and for allowing us to measure some of his collections.

REFERENCES

CAMERON R.A.D. (1969) The distribution and variation of *Cepaea nemoralis* L. near Slievecarran, County Clare and County Galway, Eire. *Proc. malac. Soc. Lond.* **38**, 439–450.
CAVALLI-SFORZA L. & BODMER W.F. (1970) *The Genetics of Human Populations.* Freeman & Co., San Francisco (in press).

DI CESNOLA A.P. (1907) A first study of natural selection in '*Helix arbustorum*' (Helicogena) *Biometrika* **5**, 387–399.

COOK L.M. (1965) Inheritance of shell size in the snail *Arianta arbustorum*. *Evolution* **19**, 86–94.

COOK L.M. (1967) The genetics of *Cepaea nemoralis*. *Heredity* **22**, 397–410.

COOK L.M. & PEAKE J.F. (1960) A study of some populations of *Cepaea nemoralis* L. from the Dartry Mountains, Co. Sligo, Ireland. *Proc. malac. Soc. Lond.* **34**, 1–11.

COOK L.M. & PEAKE J.F. (1962) Populations of *Cepaea nemoralis* L. from sand dunes on the Mullaghmore Peninsula, Co. Sligo, Ireland, with a comparison with those from Annacoona. *Proc. malac. Soc. Lond.* **35**, 7–13.

FALCONER D.S. (1955) Patterns of response in selection experiments with mice. *Cold. Spring Harb. Symp. quant. Biol.* **20**, 178–196.

HALDANE J.B.S. (1954) The measurement of natural selection. *Proc. 9th Int. Congr. Genetics.* 480–487.

O'DONALD P. (1968) Measuring the intensity of natural selection. *Nature* **220**, 197–198.

O'DONALD P. (1970a) Measuring the change of population fitness by natural selection. *Nature Lond.* **227**, 307–308.

O'DONALD P. (1970b) Change of fitness by selection for a quantitative character. *J. Theoret. Population Biol.* **1**, 219–232.

VAN VALEN L. (1965) Selection in natural populations. III. Measurement and estimation. *Evolution* **19**, 514–528.

WELDON W.F.R. (1901) A first study of natural selection in *Clausilia laminata* (Montagu). *Biometrika* **1**, 109–124.

WOLDA H. (1963) Natural populations of the polymorphic landsnail *Cepaea nemoralis* (L.) *Arch. neerl. Zool.* **15**, 381–471.

WOLDA H. (1967) The effect of temperature on reproduction in some morphs of the landsnail *Cepaea nemoralis* (L.). *Evolution* **21**, 117–129.

WOLDA H. (1969) Fine distribution of morph frequencies in the snail *Cepaea nemoralis* near Groningen. *J. Anim. Ecol.* **38**, 305–327.

6 ❋ Evolutionary Oscillations in *Drosophila pseudoobscura*

THEODOSIUS DOBZHANSKY

A RETROSPECTIVE INTRODUCTION

In his masterwork *Ecological Genetics*, Ford (1964) wrote: 'One of the most far-reaching results of recent work on ecological genetics is the discovery that unexpectedly great selective forces are normally operating to maintain or to adjust the adaptations of organisms in natural conditions... That consideration forces us completely to readjust our thoughts on evolution and to recognize that a population may adapt itself very rapidly to changed conditions.' Indeed, the readjustment is a radical one: whereas the pioneers of evolutionism were concerned to prove by indirect evidence that evolution did in fact happen, we are facing the challenge to observe it actually happening, to investigate the causes that bring particular evolutionary events about, and even to reproduce or deliberately alter their course experimentally.

The present writer had the intellectual adventure of going through the readjustment so well characterized by Ford in his own research career. In 1935, I began examining the giant chromosomes in the larval salivary glands in the progenies of wild-collected *Drosophila pseudoobscura*. It became immediately apparent that a large proportion, in some localities a majority, of individuals in nature are heterokaryotypic, i.e., have the two partners of a chromosome pair differing in inversion of one of more gene blocks. Some homokaryotypic individuals also differ in the gene arrangement. Very soon a collection of five or six gene arrangements was accumulated; a study of the chromosomes of flies coming from different parts of western United States and Mexico increased this number to 15 in *D. pseudoobscura* and 7 in the sibling species, *D. persimilis* (Dobzhansky & Sturtevant 1938, Dobzhansky & Epling 1944).

In those days it was regarded as almost self-evident that the inversions in *Drosophila* chromosomes were prime examples of adaptively neutral traits. Inversions were detected as early as 1913 and 1914 by Sturtevant in the chromosomes of various strains of *D. melanogaster* (see Morgan, Bridges & Sturtevant 1925 for references). They were known as C-factors, the presence of which in heterozygous condition caused reduction or suppression of recombination in the particular chromosome which carried such a 'factor'. In 1926, Sturtevant showed that one of the C-factors was an inversion of a block of genes in a chromosome of *D. melanogaster*. He did so by the very laborious method of accumulating mutants in the inverted as well as the supposedly original chromosome, and mapping their orders by recombination studies. Introduction in the early thirties of the technique of examination of the giant chromosomes in the salivary glands made not only the detection but also the precise description of the inversions a relatively simple matter. Most C-factors were soon identified as inversions. A chromosome with an inverted section has evidently the same gene loci, merely arranged in different sequences, as the chromosome which gave rise to it. There seemed to be no reason why the developmental effects of the inverted and uninverted chromosomes should be different.

Population samples from various parts of the distribution areas of *D. pseudoobscura* and *D. persimilis* were found to differ in their chromosomal polymorphisms. They differed most often quantitatively, the same gene arrangements having different frequencies, but sometimes also qualitatively, some gene arrangements being present in certain populations and absent in others. The chromosomal polymorphism thus gives rise to geographic races, connected by frequency gradients, clines, in geographically intermediate localities (Dobzhansky 1937, Dobzhansky & Epling 1944). The situation bears a resemblance to human races that differ in the incidence of various blood group genes. Neither in *Drosophila* nor in man were the causes of the racial differentiation apparent. *D. pseudoobscura* flies with different gene arrangements are morphologically identical, the chromosomal differences being undetectable except by cytological examination. They seem to be also physiologically identical; at least they can be propagated with equal facility in laboratory cultures.

In 1939, there was initiated repeated sampling at approximately monthly intervals during the breeding season, of populations of *D. pseudoobscura* in three ecologically rather different localities on Mount San Jacinto, in California. The purpose of this work was a study of the frequencies of allelism of recessive lethals in the third chromosomes of these populations, in samples collected simultaneously and collected at different times. The samples were submitted also to cytological examination, to determine the frequencies in them of the different gene arrangements in the third chromosomes. Neither in my memory nor in my note books can I find what working hypothesis was the cytological study intended to test. Anyway, the results obtained seemed startling: not only were the populations of the three localities, about 15 miles apart, clearly different in the chromosome frequencies, but in two of the localities the chromosome frequencies were changing significantly from month to month. That these changes could be caused by natural selection seemed hard to believe; although we do not know even now precisely how many generations of *D. pseudoobscura* occur per year in the natural habitats of this fly, the selective forces that were necessary to assume to account for the changes seemed, at that time, too unlikely to operate in these natural habitats. The possibility that the changes may have resulted from random genetic drift appeared more plausible. A coup de grâce to this surmise was administered by the finding that at least some of the changes are regularly cyclic, following the succession of the year's seasons (Dobzhansky 1943, 1947a,b). But even this would not have made the selectional explanation believable to many biologists, had it not been possible to reproduce some of the changes in artificial populations in laboratory experiments (Wright & Dobzhansky 1946).

SEASONAL CHANGES

The most extensive data, derived from the longest continued observations, on seasonal changes in the chromosome frequencies are available for the population of *D. pseudoobscura* of Piñon Flats. This locality lies at about 4000 ft elevation, on the desert-exposed slope of Mount San Jacinto, in California. The population has two maxima, in spring and in autumn, a depression during the hottest summer months, while during the coldest months the flies leave their hibernating places only on exceptionally warm days. Table 6.1 is collated from the data of Dobzhansky 1943, 1947b, Dobzhansky, Anderson, Pavlovsky, Spassky & Wills 1964, Epling, Mitchell & Mattoni 1957, and some hitherto unpublished observations. Percentage frequencies, rounded off to the nearest integer, of the three commonest third-chromosome gene arrangements are shown. The chromosome variants are Standard (ST), Arrowhead (AR), and

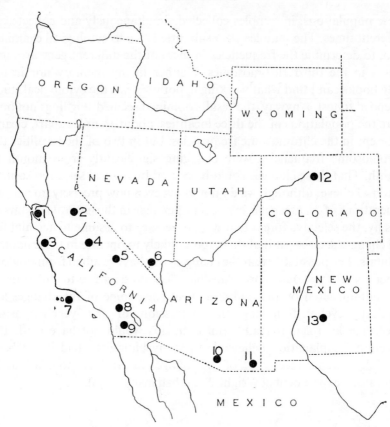

FIGURE 6.1. Map of western United States of America showing the position of some of the localities mentioned in the text. 1—Berkeley; 2—Mather and the Sierran Transect (cf. Table 6·3); 3—Santa Lucia Mts.; 4—Sequoia Park; 5—Panamint Mts. (Wildrose); 6—Charleston Mts.; 7—Santa Cruz Island; 8—San Jacinto Mts. (Piñon, Keen, Andreas); 9—Borrego Valley; 10—Sonoita; 11—Chiricahua Mts.; 12—Rist Canyon and Muggins Gulch; 13—Hondo, Lincoln and Ruidoso.

Chiricahua (CH), described in Dobzhansky & Epling 1944. Within a given year, the frequency of ST chromosomes tends to be high early in the season, to decrease during spring, and to recover during the summer and autumn. The frequencies of AR and CH undergo changes which are the reverse of those of ST. The annual minima of ST, and maxima of AR and CH, are marked in table 6.1 by asterisks. When samples are available for several months of the same year, the asterisks fall usually in June or May. Note should be taken of the fact, discussed in more detail below,

TABLE 6.1. Percentage frequencies of the three commonest gene arrangements in the third chromosomes of *Drosophila pseudoobscura* at the Piñon Flats locality, on Mount San Jacinto, in California. Asterisks (*) mark the minimal annual frequencies of ST, and maximal of AR and CH chromosomes.

Month and year	ST	AR	CH	n	Month and Year	ST	AR	CH	n
Apr 1939	51	29	13	61	Sept 1952	58	19	14	104
May 1939	28*	36*	30	240	Feb 1953	61	19	6	444
Jun 1939	30	35	31*	154	Mar 1953	60	19	5	412
Aug 1939	36	33	26	156	Apr 1953	48	23	14	588
Sept 1939	51	23	23	190	May 1953	42	21	15	342
Oct 1939	55	25	17	284	Jun 1953	28*	37*	18*	60
Mar 1940	45	20	30	386	Nov 1953	70	13	2	180
Apr 1940	35	28	34	176	Feb 1954	61	18	5	164
May 1940	28	27	40	202	Mar 1954	56	19	6	86
Jun 1940	24*	30	42*	170	Apr 1954	56	23	5	304
Sept 1940	35	25	38	104	May 1954	44	27	11	128
Nov 1940	37	33*	26	80	Jun 1954	27*	41*	11	56
Mar 1941	56	11	24	110	Jul 1954	58	20	7	168
Apr 1941	58	20	17	110	Aug 1954	57	18	10	410
May 1941	44	28	24	100	Sept 1954	54	9	30*	106
Jun 1941	33*	32*	32*	192	Jul 1955	51*	15	18	192
Aug 1941	52	21	26	108	Aug 1955	58	22*	12	96
Sept 1941	56	19	16	80	Sept 1955	55	16	20*	146
Nov 1941	45	24	24	100	Oct 1955	57	22	11	250
Apr 1942	51	22	20	102	Apr 1956	47	26	12	248
May 1942	48	17	25	100	May 1956	37	30	16	228
Jun 1942	30*	23*	40*	114	Jun 1956	23*	40*	27*	52
Jul 1942	42	22	31	124	Jul 1956	49	18	20	228
Mar 1946	56	18	18	558	Aug 1956	57	17	15	108
Jun 1946	26*	24*	44*	500	Sept 1956	45	20	18	56
Mar 1952	49	16	19	74	Mar 1963	80	6	3	114
Apr 1952	51	20	21	156	Apr 1963	76	11	3	300
May 1952	30	23	33*	40	May 1963	63*	14*	5*	190
Jun 1952	26*	33*	29	140	Mar 1970	76	18	2	80
					Apr 1970	62*	22*	5*	1000

that the chromosome frequencies also undergo changes from year to year. Nevertheless, the seasonal changes have remained the same during the whole period of observation: annual maxima of AR and CH, and minima of ST, occur in late spring or early summer.

The question that logically presents itself is whether seasonal changes occur in all populations of *D. pseudoobscura*, or at least in those polymorphic for ST, AR, and CH chromosomes. The situation turns out to be quite complex. Andreas Canyon and Keen Camp are two other localities on Mount San Jacinto, about 15 miles distant from Piñon and from each other. Andreas lies at the desert's edge and at an elevation of about 800 ft; few flies can be trapped during the very hot summers, but they are fairly abundant in mid-winter. Observations, continued from 1939 to 1942, disclosed chromosome frequency changes of the same kind as at Piñon (Dobzhansky 1943). The population of Keen Camp behaves differently. This locality lies at about 4300 ft, on the north side of the mountain; it receives relatively abundant snow and rainfall. The flies can be trapped rarely in March and April, but abundantly in mid-summer, until about September. Dobzhansky (1943, 1947b), and Epling, Mitchell & Mattoni (1957) took samples at Keen during the same years as at Piñon, but with altogether different results—no seasonal changes were found.

Mather is about 350 miles north of San Jacinto, at an elevation of approximately 4600 ft, on the western slope of the Sierra Nevada. Observations there extended from 1945 to 1970, but only in 1954 was the collecting period continuous from late May to early October. The results are shown in table 6.2. Here the frequencies of ST chromosomes increase steadily from the early to the late collections, the AR decrease, and those of CH and other chromosomes (TL = Tree Line, PP = Pikes Peak) show no systematic changes. The data from other years, though more fragmentary, are consistent in showing that, in contrast to San Jacinto, only ST

TABLE 6.2. Percentage frequencies of the five commonest gene arrangements in the third chromosome of *Drosophila pseudoobscura* at Mather, California, in 1954.

Date	ST	AR	CH	TL	PP	*n*
30 May–6 June	21	45	11	12	10	532
15 Jun–17 Jun	22	40	18	10	10	136
2 Jul–4 Jul	25	45	10	10	10	196
17 Jul–19 Jul	23	39	13	12	12	296
6 Aug–10 Aug	28	35	12	11	12	346
20 Aug–27 Aug	31	29	13	12	14	338
3 Sept–10 Sept	34	30	13	10	11	416
1 Oct–3 Oct	30	33	9	13	13	46

and AR chromosomes are involved in the seasonal oscillations at Mather. Moreover, the rhythm of the changes is different, ST increasing and AR decreasing during the collecting season; the reversal of the changes takes place apparently during hibernation. Strickberger & Wills (1966) have made monthly samples of the population at a locality near Berkeley, California, from May 1962 to August 1964. There the seasonal oscillations are the reverse of those observed at Piñon and Andreas, on Mount San Jacinto. At Berkeley, the frequencies of ST chromosomes reach maxima in mid-summer (June, July and August) and minima during the late autumn and winter. CH chromosomes show the opposite behaviour—low frequencies in summer and high ones in winter, which is also the reverse of their rhythm in San Jacinto. AR chromosomes do not change significantly. Evidence of seasonal oscillations in chromosome frequencies has also been obtained in some populations on the eastern slope of the Rocky Mountains in Colorado (Crumpacker, private communication).

It is rather more difficult to prove that seasonal oscillations do not occur in a given locality, than to show that they do. The most convincing negative evidence is that for Keen Camp, mentioned above; negative data have also been obtained for some Colorado localities by Crumpacker (private communication). *D. persimilis* is a species closely related to *D. pseudoobscura*, and both species occur sympatrically at Mather and many other localities. *D. persimilis* is also chromosomally polymorphic; seasonal oscillations, though apparently present, are much weaker than in *D. pseudoobscura* (Dobzhansky 1956).

ALTITUDINAL AND SEASONAL VARIATIONS COMPARED

The distribution area of *D. pseudoobscura* has a complex topography and an impressive variety of climates and vegetation. In some places, particularly on mountains, environments to which a population is exposed may differ greatly within distances of the order of 10 miles or even less. It has been shown that some chromosomal polymorphisms are flexible enough to go through genetic reconstructions related to the succession of seasons. One may, therefore, expect them to respond also to ecological variables in their environments. Indeed, this has been observed. The average chromosome frequencies in the three localities on Mount San Jacinto mentioned above are as follows (Dobzhansky 1943):

Locality	Elevation (ft)	ST	AR	CH	n
Andreas	800	58	24	15	3818
Piñon	4000	41	26	28	3443
Keen	4300	31	25	40	5546

One may envisage a parallelism between the elevational and seasonal variations as follows. At Piñon and at Andreas the percentages of ST chromosomes increase, and those of CH decrease during the hottest part of the summer; ST is most frequent in the locality with the warmest (Andreas) and CH in that with the coolest climate (Keen). A flaw in this argument is that the percentages of ST dwindle, and those of CH grow at Piñon and Andreas between March and June, as the weather becomes progressively warmer. A rather better agreement is obtained for the Sierra Nevada populations. The frequencies of the chromosomes in populations on the approximately 60-mile transect of the Sierra Nevada range are shown in table 6.3 (according to Dobzhansky 1948).

TABLE 6.3. Percentage frequencies of the three commonest gene arrangements in the third chromosomes of *Drosophila pseudoobscura* at different elevations in the Sierra Nevada of California.

Locality	Elevation (ft)	ST	AR	CH	n
Jacksonville	850	46	25	16	1146
Lost Claim	3000	41	35	14	760
Mather	4600	32	37	18	1450
Aspen	6200	25	44	16	478
Porcupine	8000	14	45	27	44
Tuolumne	8600	11	55	22	82
Timberline	9900	10	50	20	10

ST chromosomes wane, and AR wax in frequency as one moves up the slope of the Sierra Nevada range. Even though the species is rare at higher elevations (above 7000 ft), and few chromosomes have been sampled, the difference is highly significant. The data in table 6.2 show that, at a given locality, the ST chromosomes become more, and AR less frequent as the summer progresses and the weather grows warmer. Comparison of tables 6.2 and 6.3 leads to the conclusion that the chromosomal composition of a population at a given elevation becomes more like that of populations at lower elevations earlier in the season. One possible interpretation of this may be that ST chromosomes are favoured in warmer, and AR chromosomes in cooler environments.

Strickberger & Wills (see above) have interpreted their observations on the seasonal changes in the population of Berkeley, California, to mean that CH chromosomes are favoured by abundant rainfall and ST by aridity. Such hypotheses are hard to prove or to disprove by observations on populations in nature, especially since the ecology of *D. pseudoobscura* remains quite inadequately known. Variables in the fly's environments which undergo changes with the year's seasons are many; changes in temperature or rainfall may simply coincide in time with quite different modifications in the environment, such as the kind or quantity of food available.

Altitudinal, like seasonal, changes do not occur in all populations. An example of genetic constancy despite drastic environmental changes is provided by the population of the Chiricahua Mountains, Arizona (Dobzhansky 1962). Collections were made at elevations from 3900 ft (Sonoran desert) to 8400 ft (predominantly coniferous forest and fairly abundant rainfall). The two commonest gene arrangements, AR and CH, were uniform in frequency in all samples. We are forced to conclude that some populations of *D.pseudoobscura* are genetically flexible and others are rigid. The former react by genetic changes to seasonal as well as to altitudinal environmental alterations; the latter keep their genetic composition constant despite environmental changes. It may be noted in passing that genetic flexibility or rigidity appears to characterize not only different populations within the species *D.pseduoobscura*, but also different species. Thus, the chromosomal polymorphism in *D. subobscura* is relatively rigid in comparison with that in many populations of *D.pseudoobscura* (Kunze-Mühl, Muller & Sperlich 1958, Krimbas 1965, Anderson, Dobzhansky & Kastritsis 1967).

SELECTION IN EXPERIMENTAL POPULATIONS

Experimental populations of *Drosophila* can be kept for many generations under controlled laboratory conditions in population cages of various sizes and constructions.

In all types of cages containers with fresh food are introduced, and containers with used-up food are removed at desired intervals. A mixture of flies with known proportions of chromosomal or other genetic variants is introduced into the cage, and allowed to breed freely. Samples of the population are taken from time to time, to observe the changes in the proportions that may be taking place. Many experiments on chromosomally polymorphic populations of several species of *Drosophila* have been made by many authors. A review of the resulting literature is not needed here; it will be sufficient to give some examples directly relevant to the understanding of the selectional processes going on in the natural populations discussed above.

As a general rule, experimental populations in which all the competing chromosomal types are derived from a single natural population reach a stable equilibrium, at which the several chromosomes continue to occur with fixed frequencies. The establishment of such stable equilibria is most simply explained if the heterokaryotypes are heterotic, i.e. superior in Darwinian fitness (adaptive value) to the homokaryotypes. The equilibrium point, and the rate at which the equilibrium is attained from the original founders of the experimental population, permit estimation of the fitnesses of the competing karyotypes. The first experiments of this sort, described by Wright & Dobzhansky (1946), dealt with ST and CH chromosomes derived from the Piñon Flats locality on Mount San Jacinto. The estimated fitnesses (W) and the selection coefficients (s) of the karyotypes were as follows:

	ST/ST	ST/CH	CH/CH
W	0·70	1	0·30
s	0·30	0	0·70

In a more recent experiment, involving chromosomes with five different gene arrangements from the Mather locality, the following estimates of W were obtained (Anderson *et al.* 1968):

Homokaryotypes		Heterokaryotypes			
ST/ST	0·35	ST/AR	1·00	AR/PP	0·33
AR/AR	0·58	ST/CH	0·76	AR/TL	0·45
CH/CH	0·26	ST/PP	0·61	CH/PP	0·26
PP/PP	0·45	ST/TL	0·51	CH/TL	0·19
TL/TL	0·35	AR/CH	0·26	PP/TL	0·27

It can be seen that the highest fitness, conventionally taken to be unity, is that of the ST/AR heterokaryotype; other heterokaryotypes containing ST (ST/CH, ST/PP, ST/TL) are also superior in fitness to any of the homokaryotypes. By contrast, several of the homo- and hetero-karyotypes have fitnesses below 0·5, i.e., in the semilethal range. These genetic variants are normal constituents of natural populations, and yet they are subject to impressively strong selective forces.

Another fact of undoubted relevance to the understanding of the situation in nature is that the fitness of the karyotypes is exquisitely sensitive to environmental conditions. The fitness estimates quoted above are obtained in populations kept at a temperature of 25°C, and fed on a nutrient medium seeded with a yeast strain of *Saccharomyces cerevisiae*. Lowering the temperature by only 10°, to 15°C, makes the fitnesses of the karyotypes uniform, within the limits of experimental error. Changes of fitness also result from the introduction of different species of yeasts and modifications of the culture media. The occurrence of cyclic changes in the chromosome frequencies in nature strongly suggests that the fitnesses of the karyotypes undergo alterations in the natural habitats of the flies. Consider again the situation in the Piñon Flats population. In spring the percentages of ST chromosomes decrease, and those of CH increase. In mid-summer the change is reversed. This is best explained if the ST/ST karyotype is less fit than CH/CH in spring, but more fit in summer. This consideration leads to a difficulty. Indeed, all the experiments in the population cages show ST/ST to be either superior in fitness to CH/CH or (at low temperatures) equal in fitness. The puzzle has been resolved by Birch (1955). A common denominator in all population cage experiments is that the larvae, and also eggs and pupae, are subjected to extreme crowding, while the adults are relatively uncrowded. Birch conducted an experiment in which the larval crowding was lessened or eliminated, and observed an increase of the frequency of CH chromosomes at the expense of ST.

DIRECTIONAL CHANGES

DIRECTIONAL CHANGES IN THE SIERRA NEVADA

In addition to the cyclic seasonal changes in the polymorphic systems, changes of a longer duration were also discovered in some populations of *D. pseudoobscura* and *D. persimilis*. The PP gene arrangement in the third chromosome of *D. pseudoobscura* is common, or even predominant, in the populations of the eastern slope of the Rocky Mountains and in

Texas. Until 1946, this gene arrangement was exceedingly rare, though not entirely absent, in California. Among the approximately 20,000 third chromosomes examined, only four PP chromosomes were found in three localities, two north of San Francisco Bay and one in Sierra Nevada (Dobzhansky & Epling 1944). A single PP chromosome was found in 1946 among 336 examined from the Mather locality. In subsequent years the frequency of PP rose spectacularly to a maximum of 11·7 per cent in 1954, and then declined again, though not to the old low value. At the same time the frequencies of other gene arrangements also underwent changes, as can be seen from the summary in table 6.4. In this

TABLE 6.4. Changes in the percentage frequencies of the gene arrangements in the third chromosomes of *Drosophila pseudoobscura* at Mather, California.

Date	ST	AR	CH	PP	TL	SC	OL	n
1945	36	36	17	—	10	0·6	0·3	308
1946	31	37	17	0·3	11	2	2	336
1947	30	39	20	0·7	8	2	0·5	425
1950	20	50	17	3	8	0·7	0·2	812
1951	29	43	11	5	10	1	1	856
1954	27	36	13	12	11	0·5	0·5	1312
1957	45	33	4	10	6	2	—	316
1959	40	36	11	4	7	2	0·3	298
1961	64	14	3	6	11	0·6	—	350
1962	54	27	2	9	7	1	0·2	450
1963 June	54	20	7	7	12	—	—	160
1963 July	55	23	6	7	8	—	0·7	286
1965 April	34	25	11	7	21	0·7	1	400
1965 July	41	25	10	4	9	0·7	—	134
1969 June	38	45	3	2	12	—	0·3	312

table only the seasonal totals are entered for the years 1946–62, since the data for the collections in different months have been published in Dobzhansky 1952, 1956 & 1963.

It is evident from the data in table 6.4 that other changes happened, in addition to the sudden appearance, rise, and decline of PP. A fairly steady decline of CH chromosomes has taken place; ST declined from 1945 to 1950, rose to a very high frequency by 1961, and by 1969 declined again to almost the same frequency as in 1945. The changes in AR counterbalance those in ST; AR rose between 1945 and 1950, declined from 1950 to 1961, and rose again between 1961 and 1969. The low ST frequency in April 1965 is what is expected because of the seasonal oscillations, see table 6.2; the only irregularity in the picture is a sudden rise of TL chromosomes in April of 1965, which was no longer perceptible by July.

TABLE 6.5. Changes in the percentage frequencies of the gene arrangements in the third chromsomes of *Drosophila persimilis* at Mather, California.

Year	WT	KL	ST	Others	n
1945	70	15	9	7	198
1946	77	10	11	2	128
1947	80	7	8	5	434
1950	87	5	5	3	224
1951	72	15	10	2	272
1954	81	13	3	3	2919
1957	73	16	10	1	88
1959	84	6	8	2	402
1961	84	10	4	3	490
1962	83	11	4	2	966
1963	80	11	5	4	100
1965	85	7	5	3	331
1969	80	12	4	4	566

Table 6.5 reports the changes in the gene arrangements in the third chromosomes of *D. persimilis* in the Mather locality. They are less striking than in *D. pseudoobscura*; however, WT (Whitney gene arrangement) had low and KL (Klamath) high frequencies in 1945 and 1951. Ignoring the small sample in 1957 (when the species was rare relative to *D. pseudoobscura*) chromosome frequencies have remained fairly steady since 1954, when the species was very abundant.

DIRECTIONAL CHANGES
ON SAN JACINTO

Genetic changes from year to year have been recorded not only at Mather, but in other localities as well. On San Jacinto, as at Mather, PP chromosomes appeared between 1946 and 1956, rose in frequency in 1953–56 and apparently declined thereafter, but still continue to occur in the populations. The longest continued observations are available for two localities, Piñon Flats and Keen Camp. The early data have been published by Dobzhansky (1947b) and Epling & Lower (1957). The annual totals for the early years, and the data for the more recent collections are reported in tables 6.6 and 6.7. In addition to the changes in PP chromosomes, there has been on San Jacinto, as at Mather, a rather spectacular decline in frequency of CH chromosomes. Around 1940, CH was the commonest gene arrangement at Keen, and second or third commonest at Piñon. In 1966, CH fell to the third place at Keen and the fourth place at Piñon. ST and AR chromosomes also varied significantly but less regularly. At Piñon there has undoubtedly been an increase in ST after 1952 (except for a low value in 1956), and a decrease in AR. This again parallels the events at Mather. The samples at Keen were interrupted in 1956; the single sample taken in June 1966 does show however the highest ever recorded frequency of ST, the lowest of CH, and a moderate one of AR. The TL gene arrangement has fluctuated irregularly.

CHANGES IN OTHER LOCALITIES IN CALIFORNIA

As shown above, parallel genetic changes occurred approximately simultaneously in the populations of Mather and of San Jacinto, some 350 miles apart. This could be an extraordinary coincidence. More likely, the changes were not restricted to the populations of isolated localities but affected those of geographically more extensive areas. A follow-up study was consequently undertaken by Dobzhansky, Anderson, Pavlovsky, Spassky & Wills (1964). They collected in 1957, and again in 1963, population samples from 14 localities in different parts of California, for which comparable samples were available, collected in 1940 or earlier. The results were conclusive—similar genetic changes did occur over the entire state. Specifically, PP chromosomes were found in 1957 and 1963 at every locality in which they were not encountered in 1940. With a single exception (Santa Lucia Mountains) the frequencies of ST chromosomes were higher in 1957 and 1963 than they were in 1940. With three exceptions (Santa Lucia Mountains, Santa Cruz Island, and Sequoia National Park) the percentages of CH chromosomes were lower, in some localities strikingly so, in 1957 and 1963 than in 1940. AR and TL chromosomes showed less consistent behaviour, the former mostly declining and the latter increasing in frequency. A decline has also affected

TABLE 6.6 Changes in the percentage frequencies of the gene arrangements in the third chromosomes of *Drosophila pseudoobscura* at Piñon Flats, California.

Date	ST	AR	CH	PP	TL	n
1939	42	30	24	—	4	1085
1940	36	26	34	—	4	1118
1941	47	23	25	—	5	800
1942	43	21	29	—	7	440
1945	33	31	34	—	2	352
1946	41	21	31	—	7	1058
1952	43	21	25	6	5	514
1953	51	20	12	10	6	2026
1954	55	20	9	11	4	1426
1955	55	18	15	7	5	690
1956	43	26	16	10	5	798
1963 March	80	6	3	4	6	114
1963 April	76	11	3	6	4	300
1963 May	63	14	5	8	8	190
1970 March	76	14	2	2	1	80
1970 April	64	20	5	3	7	1000

TABLE 6.7. Changes in the percentage frequencies of the gene arrangements in the third chromosomes of *Drosophila pseudoobscura* at Keen Camp, California.

Year	ST	AR	CH	PP	TL	n
1939	28	30	38	—	4	1986
1940	31	23	42	—	4	2382
1941	35	24	37	—	4	764
1942	36	16	40	—	7	414
1945	41	22	29	—	8	288
1946	50	15	28	—	7	800
1952	37	24	29	6	5	548
1953	40	26	20	10	4	744
1954	35	28	27	7	3	1538
1955	37	24	32	4	3	730
1956	31	27	33	6	3	1108
1966	48	24	14	3	10	438

SC (Santa Cruz) chromosomes; these chromosomes were fairly frequent in 1940 in the ranges along the coast of the Pacific Ocean and on Santa Cruz Island; in 1963 they were quite rare or had even disappeared entirely.

In recent years, attempts have been made to secure further data on changes in California localities, particularly in those which are climatically, ecologically or geographically divergent from Mather and from San Jacinto. Santa Cruz Island in the Pacific is separated from the mainland by the approximately 20-mile wide Santa Barbara channel. Owing of the kindness of Professor Wyatt Anderson, I was able to examine a sample collected there in March 1970 by Professor Charles Remington, of Yale University. As many as eight gene arrangements occur on the Island. The results obtained are shown in table 6.8.

TABLE 6.8. Changes in the percentage frequencies of the gene arrangements on Santa Cruz Island, California.

Year	ST	AR	CH	PP	TL	SC	EP	OL	n
1936	55	17	—	—	—	29	—	—	42
1940	43	18	7	—	—	32	—	—	72
1963	62	13	10	1	5	7	2	—	400
1970	40	14	16	1	11	13	3	1	205

Although the early samples were rather inadequate, some of the changes indicated by the figures are significant. The SC chromosomes have, like elsewhere on the California coast, decreased in frequency in 1963, but possibly recovered somewhat in 1970. Unlike elsewhere in California, CH chromosomes have not only not gone into eclipse but have become more frequent (the figure given for the years 1936–1937 in table 1 of Dobzhansky et al. 1964 is in error). ST was more frequent in 1963 than in 1970 (paralleling Piñon, see table 6.6 and perhaps Mather, table 6.4). PP chromosomes, not found in 1940, were present but still rare in 1963 and 1970. The EP (Estes Park) gene arrangement is common in the Rocky Mountains but rare in California; its apparent absence in 1936 and 1940 may have been due to the smallness of the samples.

Panamint is a mountain range bordering Death Valley on the west, surrounded on all sides by desert. Samples were taken in Wildrose Canyon, at about 8000-ft elevation. The June 1968 sample was obtained through the courtesy of Professors C.J.Wills and S.Prakash. The chromosome frequencies are shown in table 6.9.

TABLE 6.9. Changes in the percentage frequencies of the gene arrangements on the Panamint Mountains, California.

Year	ST	AR	CH	PP	TL	n
1937	14	67	19	—	—	224
1938	33	39	23	—	2	230
1939	37	46	14	—	3	580
1940	31	44	21	—	3	360
1957	26	59	11	1	3	224
1963	24	41	5	13	17	132
1968	33	45	11	6	4	142

Though living in an environment very different from Mather and San Jacinto, the population of Panamint underwent the same changes as the populations of these latter localities, but with what appears to be a delay of several years. It is certain that PP was either absent or quite rare in the Panamints in 1937–1940. In 1957, when PP reached one of the highest frequencies on record in Mather (table 6.4), and was perhaps already declining on San Jacinto (tables 6.6 and 6.7), only two PP chromosomes were found in a sample of 224. However, in 1963 PP on Panamint had the highest frequency of any locality in California (Dobzhansky *et al.* 1964). The ostensible decline in 1968, though not significant statistically, is in agreement with the observations at Mather and San Jacinto. The highest recorded frequencies of CH in the Panamints occurred in 1938 and 1940, and the lowest in 1963. This agrees as well as could be expected with the observations at Piñon (table 6.6). The oscillations of the frequencies of ST and AR in the Panamints are hard to compare with those elsewhere since the number of years in which samples were taken is too small. The sudden rise of TL in 1963 may or may not be comparable with that at Mather in 1965 (table 6.4).

Borrego State Park lies at the desert's edge south of Mount San Jacinto. The most recent collection there was made in February of 1966 by Professor D.L.Jameson, Mr R.C.Richmond, and the author. The results are shown in table 6.10. Though the comparison is difficult because samples were taken in only three years, the data, as far as they go, parallel those for Piñon and Keen Camp (tables 6.6 and 6.7). ST

TABLE 6.10. Changes in the percentage frequencies of the gene arrangements in the Borrego State Park, California.

Year	ST	AR	CH	PP	TL	SC	n
1938	54	30	14	—	—	2	132
1941	60	36	5	—	—	—	42
1966	70	16	3	2	9	0·5	200

chromosomes were more, and AR and CH less frequent in 1966 than in 1938 and 1941. PP chromosomes were found only in 1966.

CHANGES OUTSIDE CALIFORNIA

Professors W.W.Anderson, M.Druger, and the author have taken, in the summers of 1964 and 1965, samples of the populations of *D. pseudoobscura* at 18 localities in British Columbia, Washington, Oregon, Nevada, Utah, Arizona, Colorado, New Mexico and Texas, for which samples taken earlier, between 1937 and 1941, were available (Dobzhansky, Anderson & Pavlovsky 1966). In the Pacific Coast states (British Columbia, Washington and Oregon) the results paralleled those in California (see above). ST and PP chromosomes were more, and AR and CH are less frequent in 1964 and 1965 than in 1940. However, east of the Sierra Nevada-Cascades mountain chains significant changes were few, and they did not show any general trends like those in the Pacific Coast states. Only on Charleston Mountains (southern Nevada) and at Sonoita (southern Arizona) did the changes resemble those in neighbouring California, immediately to the west. In both localities there were decreases in the frequencies of CH and increases in those of PP chromosomes (Dobzhansky *et al.* 1964, Dobzhansky, Anderson & Pavlovsky 1966).

Professors D.W.Crumpacker, L.Ehrman and the author collected in July 1965 a sample in Rist Canyon, near Fort Collins, Colorado. This locality is only about 15 miles distant from Muggins Gulch, where a sample was taken in August 1941 by the author. The composition of the 1941 and 1965 samples proved to be very different. However, since populations separated by such distances are sometimes genetically quite distinct (e.g., those of the three localities on San Jacinto, see above), it was uncertain whether the difference should be ascribed to temporal or to ecological causes. Accordingly, Professors F.J.Ayala and

D.W.Crumpacker took further samples at Rist and Muggins Gulch in July of 1969. The composition of all four samples is shown in table 6.11.

TABLE 6.11. Changes in the percentage frequencies of the gene arrangements at Muggins Gulch and at Rist Canyon, Colorado.

Locality and year	ST	AR	CH	PP	TL	EP	OL	n
Muggins 1941	5	17	—	58	8	12	—	64
Rist 1965	5	50	—	29	11	5	1	449
Rist 1969	6	32	1	27	16	12	6	180
Muggins 1969	3	34	—	39	16	6	2	262

There is no significant difference between the 1969 samples from Rist and Muggins. Compared with 1941, there has been an increase in the frequency of AR, at the expense of PP chromosomes.

An analogous problem arose with samples collected in the cluster of neighbouring localities in the mountains of New Mexico. These are Hondo (1941, J.T.Patterson, collector), Ruidoso (1964, M.Druger and the author; 1969, L.Ehrman; 1970, M.Wasserman), and Lincoln (1965, W.W.Anderson and the author). The data are summarized in table 6.12. Here the ambiguity of temporal versus ecological variations is still not resolved.

TABLE 6.12. Changes in the percentage frequencies of the gene arrangements at three localities in New Mexico.

Locality and year	ST	AR	CH	PP	TL	EP	OL	n
Hondo 1941	—	56	7	35	1	—	—	142
Ruidoso 1964	2	69	1	26	1	—	—	82
Lincoln 1965	2	42	1	48	2	5	—	200
Ruidoso 1969	0·3	58	3	36	1	0·3	2	376
Ruidoso 1970	0·4	67	3	27	—	0·4	2	244

SELECTIVE PROCESSES AT WORK

The observational and experimental evidence reviewed above clearly shows that many, though possibly not all, natural populations of *D. pseudoobscura* undergo genetic changes rapid enough to be recorded not only within an observer's lifetime but in successive months. It is certain that these changes are brought about by natural selection; random processes may contribute at most minor local fluctuations. The nature of the selective forces, the results of the operation of which are so plainly visible, is still conjectural—despite more than thirty years of study.

The cyclic seasonal and the year-to-year changes present rather different problems. The fact that the same kind of changes are repeated every year, hand in hand with the succession of the seasons, can only be explained if the environments of the flies favour the carriers of chromosomes with different gene arrangements in spring, summer, autumn and winter. Just what aspects of the seasonal environment increase and decrease differentially the Darwinian fitness of the karyotypes remains to be discovered. The work of Birch (1955), referred to above, seems to suggest a plausible explanation for the seasonal genetic changes observed in the populations of Piñon Flats and Andreas Canyon, on Mount San Jacinto. Larval crowding makes the fitness of ST/ST homokaryotypes higher than that of the CH/CH homokaryotypes; absence of larval crowding makes CH/CH superior to ST/ST. During the spring months the populations of *D. pseudoobscura* increase in size, owing to the abundance of moisture and fermenting tree sap. With the advent of hot and dry weather, they dwindle sharply, down to a few survivors in July and August. The changes in the frequencies of the chromosomes are, then, just what one would expect from Birch's results.

Unfortunately, this simple explanation does not fit the situation at Mather or at Berkeley. At Mather (table 6.2), ST chromosomes increase, and AR decrease continuously from May to at least August or September. The population density of *D. pseudoobscura* does not vary anywhere near as much as at Piñon and Andreas; the annual peak is perhaps reached at some time in July or August. At Berkeley (Strickberger & Wills 1966), the annual frequency climax of ST is in summer and CH in winter. Birch's hypothesis may, then, be adjusted if one supposes that at Berkeley the ST chromosomes are favoured when the population expands, and CH when the population contracts.

The year to year changes have proved thus far refractory to explanation. The climate of California is notably variable from year to year, owing chiefly to different amounts of precipitation in winter, greater or lesser parts of which may fall either as rain or as snow in the mountains. The summers are mostly rainless, but some are hotter than others.

Striking variations have been recorded at Mather in the frequency of *D. pseudoobscura* in relation to the closely allied but possible competing species, *D. persimilis*, *D. miranda*, and *D. azteca* (Dobzhansky 1963). For some years it seemed that the rises and falls in the frequencies of ST and AR chromosomes at Mather may be related to the succession of drought and wet years, but continued observations have invalidated this hypothesis. No correlation with any other climatic or biotic factor could be discerned.

It happens that the appearance and increase of PP chromosomes in California coincided in time with the beginning of the large-scale use of DDT and other insecticides in the agricultural parts of that state. Anderson, Oshima, Watanabe, Dobzhansky & Pavlovsky (1968) experimented with laboratory populations polymorphic for ST, AR, CH, TL and PP chromosomes, derived from flies collected at Mather. The adult flies were exposed in every generation to treatment with small amounts of DDT or dieldrin, then returned to a population cage, and allowed to breed. In the experiments carried out at the National Institute of Genetics, Misima, Japan, the amount of the treatment given was adjusted to result in a 20–35 per cent mortality of the exposed flies. In the parallel experiments at the Rockefeller University, New York, the mortality did not exceed 10 per cent. The results were negative in both experiments—no meaningful difference between the treated and control populations was observed.

Professor L. Cory (private communication) nevertheless believes that the insecticide hypothesis cannot be ruled out, and that experiments treating larvae rather than adults with DDT should be made. Cory, Field & Serat (1970) and Cory (unpublished) carried out a most careful study of the concentrations of DDT, and of its derivative in living bodies, DDE, in different parts of the state of California, and collated data for other states in the western U.S.A. The general result is that there is at least a rough correlation between the genetic changes recorded in the populations of *D. pseudoobscura* and the degree of contamination of the environments by insecticide residues. The matter needs further study.

Although the quest for environmental alterations that may have brought about the genetic changes in *D. pseudoobscura* should continue, one may also examine some other possibilities. In general, natural selection can alter the gene pool of a population under the influence of one or both of the following stimuli. First, the Darwinian fitness of some components of the gene pool may be modified by a change in the environment. Genetic variants favoured by the new environment will then increase, and those discriminated against will decrease in frequency. Secondly, the environment may remain constant, but there may appear new genetic variants. Mutation and recombination are the sources of such new variants. It should be made clear that the chromosomal inversion polymorphisms involve not alleles of a single gene but supergenes, complexes of linked gene variants. The evidence on which the foregoing statement rests has been reviewed in Dobzhansky (1970). The cytologically visible changes, inversions, play the role of binding together the components of the supergenes.

The supergenes locked up in the inverted sections in a given natural population are coadapted with other supergenes present in the same population. This is to say, the combinations of the supergenes, taken two at a time, give rise to highly fit, heterotic, heterokaryotypes. The chromosomal polymorphism is maintained by balancing natural selection of at least two kinds. The fitness of the heterokaryotypes is higher than that of the homokaryotypes. This is heterotic balancing selection. Different supergenes, in heterozygous as well as in homozygous condition, may be more or less fit in different subenvironments in the territory which the population inhabits. This is diversifying (or disruptive) balancing selection.

The gene pool of a population of *D. pseudoobscura* contains, then, a system of interdependent coadapted supergenes. The relative frequencies of the components of that system are adjusted by natural selection to fit the environment in which the population lives. New supergenes are formed owing to mutation or recombination. Suppose, for example, that there appears a new PP chromosome, or a new ST chromosome which increases the fitness of the ST/PP heterokaryotype, or of the ST/ST or the PP/PP homokaryotypes, relative to the old PP and ST chromosomes, and to the other components of the gene pool. Natural selection will then favour replacement of the old PP and ST by the new, and also an increase in the incidence of PP and ST chromosomes in the population. Or else, suppose that the new ST chromosome makes a better ST/ST homokaryotype, and less fit CH/ST and SC/ST heterokaryotypes, than the old ST did. This may lower the incidence of CH and SC chromosomes in the population.

It is extremely difficult to obtain conclusive evidence for or against the hypothesis of emergence of new supergenes. Chromosomes extracted from natural populations can be, and have been kept for years in laboratory cultures. The evolution of the supergenes takes place however not only in nature but also in laboratory cultures, as demonstrated among others by Strickberger (1963). Perhaps more convincing are experiments in laboratory population cages, repeated a decade or more apart with samples of the chromosomes extracted from the natural population of the same locality. Possible sources of error are, then, changes in laboratory techniques which may be reflected in different fitness estimates. Pavlovsky & Dobzhansky (1966) compared the outcomes of the experiments in population cages conducted in 1961–65 with those made in 1945–1947. In both series of experiments the chromosomes used were extracted from the population of Mather, California. In the older experiments the estimated fitnesses of the karyotypes ST/ST:ST/AR:AR/AR were approximately 0·8:1:0·6, while in the newer ones the karyotypes did not differ appreciably in fitness. By contrast, the old experiments gave estimates of the fitness of AR/AR and CH/CH homokaryotypes not far from equality; in the newer experiments the estimates were as follows:

AR/AR	AR/CH	CH/CH
0·82	1	0·26

It is tempting to correlate this difference in the outcome of the experiments with the observations on the composition of the natural population of Mather. As shown in table 6.4, the incidence of CH chromosomes was much lower in 1961–65 than it was in 1945–47.

SUMMARY

A natural population of *D.pseudoobscura* can be characterized by the incidence in it of third chromosomes with different gene arrangements. This characterization is, in principle, similar to that of a human population in terms of the relative frequencies of different blood groups and antigens. The composition of *D. pseudoobscura* populations is, however, remarkably inconstant in time. In some populations there occur regular cyclic frequency changes, following the succession of the year's seasons. These seasonal cycles are undoubtedly due to natural selection; the populations respond to seasonal changes in their environments. In addition, some populations, particularly those of California and other Pacific

Coast states, undergo changes from year to year. As far as the presently available information goes, these changes are not regularly cyclic. Their causation is an open problem. One working hypothesis is that the changes are induced by the contamination of the environments of the flies by pesticide residues. Another hypothesis postulates the emergence of new supergenes in the chromosomal inversions; natural selection then reconstructs the gene pool of the population by favouring a new set of relative frequencies of the chromosomes with different gene arrangements. Neither hypothesis can be regarded as proven.

REFERENCES

ANDERSON W.W., DOBZHANSKY TH. & KASTRITSIS C.D. (1967) Selection and inversion polymorphism in experimental populations of *Drosophila pseudoobscura* initiated with the chromosomal constitutions of natural populations. *Evolution* **21**, 664–671.

ANDERSON W.W., OSHIMA C., WATANABE T., DOBZHANSKY TH. & PAVLOVSKY O. (1968) Genetics of natural populations XXXIX. A test of the possible influence of two insecticides on the chromosomal polymorphism in *Drosophila pseudoobscura*. *Genetics* **58**, 423–434.

BIRCH C.L. (1955) Selection in *Drosophila pseudoobscura* in relation to crowding. *Evolution* **9**, 389–399.

CORY L., FIELD P. & SERAT W. (1970) Distribution patterns of DDT residues in the Sierra Nevada Mountains. *Pestic. Monitoring J.* **3**, 204–211.

DOBZHANSKY TH. (1937) Genetics and the origin of species. 1st Ed. Columbia Univ. Press, New York.

DOBZHANSKY TH. (1943) Genetics of natural populations. IX. Temporal changes in the composition of populations of *Drosophila pseudoobscura*. *Genetics* **28**, 162–186.

DOBZHANSKY TH. (1947a) Adaptive changes induced by natural selection in wild populations of *Drosophila*. *Evolution* **1**, 1–16.

DOBZHANSKY TH. (1947b) A directional change in the genetic constitution of a natural population of *Drosophila pseudoobscura*. *Heredity* **1**, 53–64.

DOBZHANSKY TH. (1948) Genetics of natural populations. XVI. Altitudinal and seasonal changes produced by natural selection in certain populations of *Drosophila pseudoobscura* and *Drosophila persimilis*. *Genetics* **33**, 158–176.

DOBZHANSKY TH. (1952) Genetics of natural populations XX. Changes induced by drought in *Drosophila pseudoobscura* and *Drosophila persimilis*. *Evolution* **6**, 234–243.

DOBZHANSKY TH. (1956) Genetics of natural populations. XXV. Genetic changes in populations of *Drosophila pseudoobscura* and *Drosophila persimilis* in some localities in California. *Evolution* **10**, 82–92.

DOBZHANSKY TH. (1962) Rigid vs. flexible chromosomal polymorphisms in *Drosophila. Am. Nat.* **96**, 321–328.

DOBZHANSKY TH. (1963) Genetics of natural populations XXXIII. A progress report on genetic changes in populations of *Drosophila pseudoobscura* and *Drosophila persimilis* in a locality in California. *Evolution* **17**, 333–339.

DOBZHANSKY TH. (1970) *Genetics of the evolutionary process.* Columbia Univ. Press, New York.

DOBZHANSKY TH., ANDERSON W.W. & PAVLOVSKY O. (1966) Genetics of Natural Populations XXXVIII. Continuity and change in populations of *Drosophila pseudoobscura* in the western United States. *Evolution* **20**, 418–427.

DOBZHANSKY TH., ANDERSON W.W., PAVLOVSKY O., SPASSKY B. & WILLS C.J. (1964) Genetics of natural populations XXXV. A progress report on genetic changes in populations of *Drosophila pseudoobscura* in the American Southwest. *Evolution* **18**, 164–176.

DOBZHANSKY, TH. & EPLING, C. (1944) Contributions to the genetics, taxonomy, and ecology of *Drosophila pseudoobscura* and its relatives. Carnegie Inst. Washington, Publ. **554**, 1–183.

DOBZHANSKY TH. & STURTEVANT A.H. (1938) Inversions in the chromosomes of *Drosophila pseudoobscura. Genetics* **23**, 28–64.

EPLING C. & LOWER W.R. (1957) Changes in an inversion system during a hundred generations. *Evolution* **11**, 248–256.

EPLING C., MITCHELL D.F. & MATTONI R.H.T. (1957) The relation of an inversion system to recombination in wild populations. *Evolution* **11**, 225–247.

FORD E.B. (1964) *Ecological genetics.* Methuen, London.

KRIMBAS C.B. (1965) The genetics of *Drosophila subobscura* populations I. Inversion polymorphism in populations of southern Greece. *Evolution* **18**, 541–552.

KUNZE-MÜHL E., MÜLLER E. & SPERLICH D. (1958) Qualitative, quantitative und jahreszeitliche Untersuchungen über den chromosomalen Polymorphismus natürlicher Populationen von *Drosophila subobscura. Zeit. VererbLehre* **93**, 237–248.

MORGAN T.H., BRIDGES C.B. & STURTEVANT A.H. (1925) The genetics of *Drosophila. Bibliogr. Genetica*, **2**, 1–262.

PAVLOVSKY O. & DOBZHANSKY TH. (1966) Genetics of natural populations. XXXVII. The coadapted system of chromosomal variants in a population of *Drosophila pseudoobscura. Genetics* **53**, 843–854.

STRICKBERGER M.W. (1963) Evolution of fitness in experimental populations of *Drosophila pseudoobscura. Genetics* **17**, 40–55.

STRICKBERGER M.W. & WILLS C.J. (1966) Monthly frequency changes of *Drosophila pseudoobscura* third chromosome gene arrangements in a California locality. *Evolution* **20**, 592–602.

STURTEVANT A.H. (1926) A crossover reducer in *Drosophila melanogaster* due to inversion of a section of the third chromosome. *Biol. Zbl.* **46**, 697–702.

WRIGHT S. & DOBZHANSKY TH. (1946) Genetics of natural populations XII. Experimental reproduction of some of the changes caused by natural selection in certain populations of *Drosophila pseudoobscura. Genetics* **31**, 125–150.

7 ✳ Melanism in the Two-spot Ladybird, *Adalia bipunctata*, in Great Britain

E. R. CREED

INTRODUCTION

Geographic variation in the frequency of the melanic varieties of the two-spot ladybird, *Adalia bipunctata*, has been reported or reviewed by several authors, covering much of the insect's range, including continental Europe (Lusis 1961), Britain (Hawkes 1920, 1927; Creed 1966, 1969) and Canada (Smith 1958). Surveys sufficiently detailed to permit conclusions to be drawn about the causes of local, and hence of general, variation have been undertaken by Lusis in Latvia and around Leningrad and by myself in southern Britain and central Scotland. We both found the melanics to be relatively more numerous in industrial areas. Lusis also concluded that melanics were more common in regions with a maritime climate, and I suggested that in Britain some unidentified factor favoured the melanics towards the north and west.

The occurrence of melanic moths in areas affected by atmospheric pollution has received considerable attention, especially in the Peppered Moth, *Biston betularia*, and is rightly regarded as one of the most remarkable examples of evolutionary change actually observed taking place. However, attention has mainly been directed at one aspect of selection, albeit a powerful one; that is selective predation by birds. This in itself is insufficient to account for the balance now reached by this polymorphism, and relatively little attention has been paid to the nature of the opposing selective forces other than the experimental approach by Clarke & Sheppard (1966) and the theoretical deductions of Haldane (1956). A comparison of industrial melanism in moths with the situation in the two-spot ladybird may ultimately help to identify these other selective factors, since there is apparently little selective predation of the ladybird by vertebrate predators.

The idea of 'Industrial Melanism' in the two-spot ladybird is also supported by the findings in central Scotland where an association was found between the frequency of melanics and the local average summer smoke levels, though not with sulphur dioxide, and in Birmingham (Creed 1971) where a decrease in the frequency of melanics has occurred at about the same time as a marked reduction in smoke pollution following legislation; a similar decrease in the frequency of melanic *B. betularia* has been recorded by Cook, Askew & Bishop (1970) in Manchester. The present paper extends the distributional data to cover most of the insect's range in Britain and examines the data for this country with respect to prevailing pollution conditions.

SAMPLES

The insects have been collected both as adults and pupae. The eventual colour patterns cannot be distinguished in the pupae, and they therefore provide a check on the randomness of the adult samples; at no time has a significant difference been found in the frequency of melanics between adults emerging from pupae and those caught in the wild from the same place, thus eliminating selection by the collector as a reason for geographic variation.

Although insects with elytra ranging from almost completely red to completely black may be encountered (Mader, 1926–1937), the great majority can be classified into one of three types, illustrated in fig. 7.1. The distinction between the two melanics, *sexpustulata* and *quadrimaculata*, is not always clear, and in this paper they are grouped together; regional differences in their proportions will be the subject of a further

FIGURE 7.1. The three commonly encountered forms of the Two-Spot Ladybird: the typical red form, *sexpustulata* and *quadrimaculata*.

paper. A fraction of the typicals may have an extra black spot or an extension of the black so that, rarely, up to half the elytron may be black; these are all grouped with the typical patterning.

Most of the localities have been sampled in more than one year. At a number of sites in Birmingham a significant decrease in the frequency of melanics has occurred; the heterogeneity of the successive samples precludes adding them together. Nowhere else has a significant change in frequency been detected, though at Perth, Scotland, the value of P approaches 0·05. Except for the Birmingham sites, all collections from one area have been added; figures given for Birmingham refer to the more recent, lower, frequencies and may be compared with the earlier frequencies in the same areas (Creed 1966).

Most samples come from urban areas since it is difficult to obtain sufficient numbers from the more scattered rural populations. The insect is widespread in southern and eastern England, but elsewhere it is very much more common in towns; in the more mountainous areas it may be almost completely replaced by the ten-spot ladybird, *A. decempunctata*.

GEOGRAPHIC VARIATION

Details of collections are given in table 7.1. This includes sites recorded in my 1966 paper where additional data are available, but excludes those where no further collections have been made. The melanic frequencies are shown pictorially on the map (fig. 7.2). It will be seen that there are three main areas with high frequencies: Birmingham, northern England and the midland valley of Scotland including both Glasgow and Edinburgh. The highest frequencies have been found in these two cities and around Manchester. The insect is uncommon throughout much of southern Scotland, but eight individuals from Kelso and Hawick were all red; these, with the larger samples from Dumfries and Castle Douglas in southwest Scotland suggest that there is a considerable area of low melanic frequency between Carlisle and Newcastle upon Tyne and the more northern, Scottish sites. The frequency falls from 74·5 per cent at Newcastle to 16 per cent 14 miles to the north at Morpeth, roughly half this change taking place in the first two and a half miles to Gosforth. Similarly the melanic frequency falls 67 per cent in the 31 miles from Glasgow to Ayr and 81 per cent in 17 miles from Edinburgh to Haddington.

TABLE 7.1. The numbers of Two-Spot Ladybirds caught, and the frequencies of melanic individuals at localities in Great Britain. The most recent estimates of the mean summer concentrations of smoke and sulphur dioxide (both expressed as μg m^{-3}) are given where available. Only those localities for which data additional to that given in Creed (1966) are included.

Locality	Numbers			Per cent Black	Pollutants	
	Red	Black	Total		Smoke	SO$_2$
Altrincham, Cheshire	36	342	378	90·5		
Appleby, Westmorland	35	15	50	30·0		
Ashbourne, Derbyshire	88	52	140	37·1		
Ayr	80	31	111	27·9		
Bakewell, Derbyshire	21	7	28	25·0		
Banbury, Oxfordshire	138	4	142	2·8		
Barnstaple, Devon	60	0	60	0·0		
Bedford	64	0	64	0·0	22	67
Birmingham, Aston	354	121	475	25·5		
Castle Bromwich	88	17	105	16·2	39	82
Dudley	39	27	66	40·9		
Edgbaston	208	260	468	55·6		
George Road	54	28	82	34·1		
Hall Green	230	104	334	31·1		
Kingshurst	42	4	46	8·7		
Smethwick	104	49	153	32·0	43	97
Sutton Coldfield	51	9	60	15·0	34	96
Ward End	21	7	28	25·0		
West Bromwich	21	12	33	36·4	35	117
Wolverhampton	48	8	56	14·3	30	62
Bridgnorth, Shropshire	97	10	107	9·3		
Bristol	26	0	26	0·0	17	42
Bury, Lancashire	9	68	77	88·3	52	101
Buxton, Derbyshire	24	103	127	81·1		
Cambridge	131	6	137	4·4	28	70
Cannock, Staffordshire	102	15	117	12·8	43	62
Canterbury, Kent	19	0	19	0·0	20	29
Cardiff, Glamorgan	43	0	43	0·0	26	62
Carlisle, Cumberland	67	147	214	68·7	59	48
Carmarthen	66	0	66	0·0		
Castlecary, Stirling	37	89	126	70·6		
Castle Douglas, Kirkcudbright	22	0	22	0·0		
Chester	16	64	80	80·0	65	62
Chesterfield, Derbyshire	23	10	33	30·3	35	88
Cleobury Mortimer, Shropshire	30	3	33	9·1		
Colchester, Essex	14	0	14	0·0		

TABLE 7.1—*continued*

Locality	Numbers			Pollutants		
				Per cent		
	Red	Black	Total	Black	Smoke	SO$_2$
Congleton, Cheshire	35	150	185	81·1		
Coventry, Warwickshire	71	1	72	1·4	29	76
Darley Dale, Derbyshire	47	17	64	26·6		
Darlington, Durham	12	66	78	84·6	64	71
Derby	42	7	49	14·3	45	88
Doncaster, Yorkshire	92	2	94	2·1	50	73
Duffield, Derbyshire	47	24	71	33·8		
Dumbarton	38	115	153	75·2		
Dumfries	60	6	66	9·1		
Dundee, Angus	106	5	111	4·5	36	63
Dunfermline, Fife	61	78	139	56·1		
Edinburgh	35	306	341	89·7	50	52
Exeter, Devon	117	1	118	0·8	24	18
Falkirk, Stirling	24	34	58	58·6	33	81
Forfar, Angus	10	0	10	0·0		
Glasgow, Baillieston	1	38	39	97·4		
Giffnock	3	60	63	95·2		
Govanhill	8	93	101	92·1	63	70
Kelvinside	12	238	250	95·2	79	63
Provanmill	17	106	123	86·2	78	85
Gloucester	180	31	211	14·7		
Gosforth, Northumberland	38	30	68	44·1		
Greenock, Renfrewshire	32	106	138	76·8		
Guildford, Surrey	137	18	155	11·6	15	42
Haddington, East Lothian	55	5	60	8·3		
Halifax, Yorkshire	20	96	116	82·8	42	110
Harrogate, Yorkshire	21	36	57	63·2	53	68
Haverfordwest, Pembrokeshire	31	0	31	0·0		
Hay-on-Wye, Herefordshire	14	1	15	6·7		
Henley in Arden, Warwickshire	333	21	354	5·9		
Hereford	39	10	49	20·4	16	28
Hexham, Northumberland	20	60	80	75·0		
Hopwood, Worcestershire	17	1	18	5·6		
Horsham, Sussex	150	4	154	2·6		
Hoylake, Cheshire	11	58	69	84·1	48	76
Ipswich, Suffolk	321	8	329	2·4	24	55
Kendal, Westmorland	10	101	111	91·0		
Kilmarnock, Ayrshire	65	90	155	58·1		
Kingston upon Hull, Yorkshire	119	4	123	3·3	53	60
Kirkcaldy, Fife	35	1	36	2·8		

TABLE 7.1—*continued*

Locality	Numbers			Pollutants		
				Per cent		
	Red	Black	Total	Black	Smoke	SO$_2$
Kirkintilloch, Dunbartonshire	40	146	186	78·5		
Knutsford, Cheshire	56	312	368	84·8		
Lancaster	24	73	97	75·3	37	71
Launceston, Cornwall	14	0	14	0·0		
Leeds, Yorkshire	19	109	128	85·2	52	154
Leek, Staffordshire	10	12	22	54·5		
Leicester	39	1	40	2·5	41	82
Leominster, Herefordshire	31	4	35	11·4		
Lichfield, Staffordshire	80	8	88	9·1		
Lincoln	31	0	31	0·0	34	56
Liverpool, Broad Green	2	70	72	97·2		
Princes Park	6	76	82	92·7	41	168
London, Banstead	46	0	46	0·0		
East Ham	55	0	55	0·0	24	90
Hyde Park	80	0	80	0·0	27	149
Mill Hill	50	0	50	0·0	18	80
Southall	24	1	25	4·0	22	87
Maidstone, Kent	16	0	16	0·0		
Malton, Yorkshire	40	2	42	4·8		
Manchester, Broughton	4	29	33	87·9	59	143
Heaton	5	34	39	87·2		
Mansfield Woodhouse, Nottinghamshire	45	1	46	2·2		
Marple, Cheshire	11	101	112	90·2		
Maw Green, Staffordshire	322	124	446	27·8	53	110
Middlesbrough, Yorkshire	8	30	38	78·9	62	71
Moffat, Dumfriesshire	30	1	31	3·2		
Morpeth, Northumberland	42	8	50	16·0		
Muirhead, Lanarkshire	6	156	162	96·3		
Musselburgh, Midlothian	55	107	162	66·1		
Nantwich, Cheshire	48	46	94	48·9		
Newcastle under Lyme	18	47	65	72·3	56	84
Newcastle upon Tyne	36	105	141	74·5	96	116
Newport, Monmouthshire	40	1	41	2·4	23	46
Northallerton, Yorkshire	62	61	123	49·6		
Northampton	23	0	23	0·0	23	37
Northwich, Cheshire	15	52	67	77·6		
Nottingham	82	8	90	8·9	33	88
Oldham, Lancashire	6	20	26	76·9	72	100
Oxford	210	6	216	2·8	19	32

6

TABLE 7.1—*continued*

Locality	Numbers			Pollutants		
				Per cent		
	Red	Black	Total	Black	Smoke	SO$_2$
Paisley, Renfrewshire	12	92	104	88·5	66	123
Penrith, Cumberland	80	47	127	37·0		
Perth	117	57	174	32·8		
Peterborough, Northamptonshire	100	0	100	0·0		
Pontefract, Yorkshire	55	6	61	9·8	67	133
Preston, Lancashire	15	94	109	86·2	96	120
Reading, Berkshire	221	6	227	2·6	19	34
Ripon, Yorkshire	42	34	76	44·7		
Rochdale, Lancashire	4	71	75	94·7	47	106
Rugby, Warwickshire	25	0	25	0·0	25	61
Scunthorpe, Lincolnshire	43	2	45	4·4	52	77
Selby, Yorkshire	63	3	66	4·5		
Sheffield, Yorkshire	49	77	126	61·1	55	111
Shipston on Stour, Warwickshire	44	0	44	0·0		
Shirley, Warwickshire	207	15	222	6·8		
Shrewsbury	70	13	83	15·7	29	50
Skipton, Yorkshire	82	85	167	50·9		
Stafford	153	18	171	10·5	39	62
Stirling	125	6	131	4·6	23	29
Stockport, Cheshire	9	91	100	91·0	54	99
Stourbridge, Worcestershire	79	8	87	9·2		
Swansea, Glamorgan	45	0	45	0·0	19	44
Uttoxeter, Staffordshire	36	6	42	14·3		
Wakefield, Yorkshire	63	91	154	59·1	65	122
Warrington, Lancashire	4	31	35	88·6	83	114
Water Orton, Warwickshire	199	47	246	19·1		
Whalley, Lancashire	17	25	42	59·5		
Whitchurch, Shropshire	19	22	41	53·7		
Wigan, Lancashire	22	75	97	77·3	125	102
Worcester	216	38	254	15·0		
York	31	15	46	32·6	45	62

Additional samples from southern Britain confirm the lower frequencies found in the earlier survey between Birmingham and Newcastle under Lyme, and the almost complete absence of melanics from eastern and southwestern England and south Wales. High frequencies persist to the north of Newcastle under Lyme into the Manchester area, but to

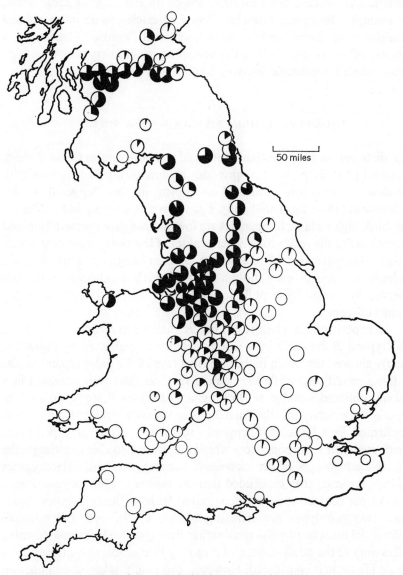

FIGURE 7.2. Map of Great Britain showing the frequency of the melanic forms of the Two-Spot Ladybird (black segments). The larger circles refer to larger sized samples.

the east of this industrial belt there is a very sharp decrease as one moves off the Pennines to the flatter country of Nottinghamshire.

All large areas with much heavy manufacturing industry have high melanic frequencies, but industrial towns on the edge of these areas, for example Doncaster, often have low frequencies, as do more isolated manufacturing towns such as Scunthorpe and Dundee. Conversely a number of towns with little in the way of heavy industry may have high frequencies: for example Bangor, Kendal and Perth.

INDUSTRIAL MELANISM IN OTHER SPECIES

The distribution of the melanic forms of *A. bipunctata* in Great Britain, as seen in the map (fig. 7.2) provides an interesting comparison with Kettlewell's data on industrial melanism in the Peppered Moth, *B. betularia* (1958, fig. 1 and 1965, fig. 1). In this species, and unlike the ladybird, high melanic frequencies are found throughout eastern England except near to the south coast where values of less than 50 per cent occur. About three-quarters of the moths in East Anglia are the extreme melanic, *carbonaria*, even though much of this area appears to be little affected by atmospheric pollution, and lichens are widespread on the trunks of trees.

An experiment involving marking, releasing and recapturing melanic and typical *B. betularia* in a wood in Dorset, unaffected by pollution, clearly showed the much higher survival rate of the paler typical moths in these conditions, while the relative survival rates were reversed in a heavily polluted wood near Birmingham (Kettlewell, 1955, 1956). The cryptic superiority of the melanics in heavily polluted areas was confirmed by Clarke & Sheppard (1966) in a series of experiments conducted near Liverpool, from which they were also able to estimate the selective advantages of the *carbonaria* homozygotes and heterozygotes and the typicals; they concluded that the *carbonaria* homozygotes were at a 15 per cent disadvantage compared with the heterozygotes. Since these two genotypes are indistinguishable to the eye, the selective differential must be physiological rather than visual. The estimated value refers only to the adult stage, and thus may be considerably altered if the whole life cycle is considered. However, this result, taken in conjunction with Ford's demonstration (1940) of the superiority under adverse conditions of the larvae of *Cleora repandata* destined to become melanics, when compared with their genetically non-melanic siblings, shows the existence of strong selection pressures associated with the phenomenon of industrial melanism, but not directly connected with the appearance of the insects. The occurrence of high frequencies of *carbonaria* in East Anglia, where the trees are not perceptibly darkened (Creed, Duckett & Lees, in prep.) suggests that such non-visual selection may be much more important in the maintenance of melanic polymorphisms than has been generally acknowledged hitherto.

The distribution of melanics in the Pale Brindled Beauty Moth, *Phigalia pedaria* (Lees, Chapter 8) shows a greater similarity to that in *A. bipunctata* than does the Peppered Moth. Frequencies of over 50 per cent are confined to areas with heavy industry, but, unlike the ladybird, the polymorphism is maintained at a higher frequency throughout all the non-industrial areas from which samples have been taken. As in the Peppered Moth the polymorphism must, in part at least, be maintained by visual selection which is independent of the effects of pollution, or by non-visual selection. In a multiple regression analysis of melanic frequencies Lees finds that, of the factors included, reflectance of the tree trunks is the most important, followed by the local rainfall and smoke pollution levels (though not sulphur dioxide). In this species, then, there is a pattern of industrial melanism superimposed upon one of non-industrial melanism.

SELECTION IN THE TWO-SPOT LADYBIRD

The most detailed data on the distribution of melanics in *A. bipunctata* obtained by Lusis (1961) were from Latvia and from the vicinity of Leningrad; for ease of comparison with the present survey they are represented in fig. 7.3. It will be seen that the melanic frequency falls on leaving the main urban areas of Riga and Leningrad. From this, and other data, Lusis concluded that the highest melanic frequencies are found in industrial areas; the present survey in Britain supports this, thus justifying the use of the descriptive term industrial melanism. Lusis also found relatively higher frequencies in places with a maritime, more humid climate. On first analysis some of the data presented here appear to be in agreement with this, in particular the very low frequencies found throughout much of eastern England. However, low frequencies are also found in unpolluted sites near to the much wetter west coast and in

FIGURE 7.3. Map of the eastern Baltic showing the frequency of the melanic forms of the Two-Spot Ladybird (black segments). The data are taken from Lusis (1961).

the more industrial parts of south Wales. Taken as a whole the present survey shows little correlation between melanic frequency and humidity in Great Britain and furthermore provides no support for my earlier suggestion that some unknown factor favours the melanics as one moves north.

Data from Scotland show a marked correlation with local smoke pollution levels, but little with the prevailing levels of sulphur dioxide (Creed 1969). This has now been tested in as many sites as possible throughout Britain. Measurements of pollution are made at over one thousand recording stations, and it is possible to pair some of these with ladybird collections. Great caution must be exercised since considerable local variation in pollution may occur. Annual summaries giving monthly, six monthly and annual mean values for smoke and sulphur dioxide are published, classifying each site according to the type of building, or sources of pollution in the immediate neighbourhood (Ministry of Technology, 1964–1969). To eliminate as much of the very local variation as possible, only sites in residential and rural areas have been included; those alongside factories or commercial centres are ignored. In table 7.1 are shown the most recent available mean summer pollution figures for the 69 sites that can be paired. A multiple regression of melanic frequency against smoke and sulphur dioxide shows that, even with the very imperfect method of estimating the general pollution levels (as opposed to the point sites at which measurements are taken), R^2, or the proportion of the observed variance in frequency that can be accounted for is 55 per cent; the partial regression on smoke is highly significant ($P < 0.001$) while that on sulphur dioxide is not ($0.3 > P > 0.2$). Ignoring sulphur dioxide, a simple regression against smoke gives a value of R^2 of 54 per cent. In the calculations the angular transformation of frequencies was

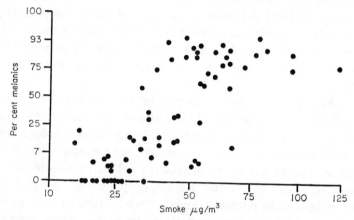

FIGURE 7.4. A scatter diagram showing the relationship between the melanic frequencies (using the angular transformation) and the square root of the mean summer smoke concentration.

used, though no weighting for sample size was introduced. Square roots of the pollution concentrations were used as this gives a more normal distribution. Figure 7.4 shows the melanic frequencies plotted against the local smoke concentrations.

A more extensive analysis, including many more independent variables, has been undertaken, though it covers a more limited geographical area. A total of 36 ladybird collecting sites, south of a line from Preston to York, can be paired with wooded areas in which nine measurements, comprising parameters of the epiphytes growing upon, and the mean reflectance of the bark of ten oak trees, were made (Creed, Duckett and Lees, in prep.). To these measurements were added various physical variables including distance north and east on the National Grid, altitude, rainfall, mean temperatures and local pollution levels. A multiple regression of melanic frequency on fifteen variables confirms that the most significant factor, taken singly, is the local smoke level. In this analysis the mean annual pollution concentrations were used; with all variables included, R^2 is estimated at 75 per cent and the regression on smoke alone gives a value of R^2 of 46 per cent. Perhaps more important is that none of the partial regressions on distance east or north, altitude, rainfall, sulphur dioxide concentration, or the reflectance of the bark of the trees, among others, were anywhere near formally significant.

The absence of any correlation with geographical grid co-ordinates effectively disposes of my earlier suggestion of a north–south cline that is independent of local pollution conditions. While the sulphur dioxide levels are almost as high in London and Birmingham as they are in the industrial parts of Lancashire and Yorkshire, the mean concentration of smoke shows a marked increase as one moves north in these three conurbations.

It is clear that differences in crypsis between mainly red and mainly black ladybirds can be of little significance so far as maintenance of the polymorphism is concerned. Red is certainly a warning coloration, and shiny black probably is also; furthermore they are equally visible when seen against a green leaf, even when heavily polluted with soot. When not on leaves, or out of sight hibernating in deep crevices, ladybirds would most often be seen by potential predators while crawling over, or resting on, the bark of trees or the stems of smaller plants. Again both colour patterns must be almost equally conspicuous. Although the measurements of reflectance mentioned previously were made at a height of 1·5 m on oak trees in woodland, one may suppose that they are proportional to the mean general darkness of the bark in that locality; and yet there was no significant partial regression of melanic frequency on reflectance. Thus the situation is quite different from that concerning *Ph. pedaria*.

Adalia bipunctata, in common with other ladybirds, is poisonous and actively rejected by many potential predators (Frazer & Rothschild 1960). However, no information is available as to possible qualitative or quantitative differences between the toxic or repellent substances of the various morphs. The red and the black insects both produce the characteristically scented secretion from the defensive glands on their legs that causes rejection by mice and bushbabies (Rothschild 1961); there is no discernible difference to the human nose.

Apart from the local smoke level, with its direct or indirect effect, gene flow from neighbouring populations may have a considerable influence on the frequency of melanics. Thus, a relatively small but very smoky town may have a low melanic frequency if it is surrounded by an extensive population not subject to great pollution; if, on the other hand the surrounding country does not provide a suitable habitat, such a town may have a high frequency. *A. bipunctata* is widespread and common over much of lowland Britain, but absent from upland moors and mountainous areas; here it is largely restricted to towns and sheltered valleys where the populations will be confined and gene flow limited. Towns situated in valleys, as they usually are in these mountainous and hilly areas, are also subject to high levels of pollution during temperature inversions when the smoke is held in by cold air above and by the walls of the valley at the sides. Thus, the ladybird population may be subject to exceptionally high local pollution levels while at the same time being relatively isolated. Examples are provided by Appleby, Kendal and Penrith; the centre of Bangor is similarly subject to extreme atmospheric pollution on occasions, being enclosed on three sides, though the fourth is open to the sea. Whether the smoke levels in such places are sufficient to account for the high melanic frequencies observed, and whether the mean annual pollution level, the mean during the active summer months or the few extreme values reached during certain climatic conditions are critical we cannot say; pollution figures in the National Survey are not available for any of these sites. By the same argument one might expect a high frequency of melanics in the valleys of south Wales, and yet at Pontypridd they represent less than six per cent of an admittedly small sample.

Samples with a higher frequency of melanics than might be predicted from the local smoke concentrations include Liverpool, Rochdale, Halifax, Edinburgh, Stockport and Leeds. These are all situated in extensive areas of high pollution, all neighbouring populations also having relatively high melanic frequencies, though to the east of Edinburgh and Leeds there is a rapid decline in frequency. The unexpectedly high frequencies in Hereford and Guildford are less easy to explain. In contrast with these populations, those in the large but relatively isolated industrial towns of Scunthorpe, Kingston upon Hull, Doncaster and Leicester, and in Pontefract, which lies on the eastern edge of the main Yorkshire industrial belt, have much lower melanic frequencies than would be predicted from smoke concentrations alone. These towns are all contiguous with, or surrounded by, areas inhabited by the insect and with relatively low smoke levels.

The steepness of the clines in morph frequency, even though it may be enhanced in some instances by discontinuities in the insects' range, shows the existence of strong selection pressures. Equally strong selection must have been responsible for the seasonal changes in melanic frequency observed in Berlin by Timofeeff-Ressovsky (1940a,b). He found that the melanic frequency rose each summer by up to thirty per cent and decreased again by a comparable amount during hibernation. Lusis (1961) suggested that this might be due to greater sexual activity of the black insects, resulting from their quicker response to changes of temperature in the diurnal cycle, and hence being able to make better use of the optimal temperatures during summer, and perhaps also of slightly lower temperatures; the black insects would heat up more quickly and would therefore be active earlier than the red ones. In winter the greater temperature changes experienced by the melanics would put them at a disadvantage, resulting in the higher mortality observed by Timofeeff-Ressovsky. However this cannot explain the high frequencies of melanics in industrial areas unless the prevailing smoke conditions reduce the sunlight to below a critical level at which selective disadvantages, associated with the alleles producing melanism, are counterbalanced by advantages resulting from the more rapid warming up period.

Alternatively, it may be argued that some constituent of the smoke is toxic, or in some other way harmful, to the ladybirds, and that the black insects are less susceptible than are the red ones. There is a measure of support for this view, though it is by no means conclusive. The frequency of melanics would be affected not only by the gross smoke pollution level, but also by its chemical composition, whereas the reduction in incident light and heat will be more nearly a function of the gross smoke level. Although the melanic frequencies in the Birmingham area, the Lancashire, Yorkshire and Durham industrial areas and the midland valley of Scotland all show a good correlation with the local smoke levels, the relative proportions of the two melanic types varies considerably. In Birmingham and Scotland *sexpustulata* is usually about twice as common as *quadrimaculata*, whereas in the industrial region centred on Manchester, and to a lesser extent in Durham, it is *quadrimaculata* that is the commoner by a factor of about two (Creed, unpublished). A geographical survey of smoke analysis would be required to verify this possible association between melanic type and a particular fraction of the pollution. However, in general *sexpustulata* predominates where the locally mined coal is of high volatility but only weakly caking, and *quadrimaculata* predominates where the coal is of high volatility and strongly caking. The principal source of low volatile coal in Britain is south Wales where (and possibly in consequence) melanic frequencies are negligible despite high local pollution and restricted gene flow.

A cline in gene frequencies in the Asiatic Ladybird, *Harmonia axyridis*, from the north to the south of Japan has been taken as evidence by Komai (1956) of the route of colonization of the islands by this species; the populations in the north resemble most the neighbouring mainland ones. Ford (1964) pointed out that the obviously strong climatic adaptation that is reflected in the morph frequencies invalidates the conclusion that the colonization path has had any lasting effect on local frequencies. However, Komai & Chino (1969) still feel that the difference in composition between the Korean and Kyushu populations is too great to assume colonization across the strait between Korea and Japan. Without in any way suggesting which route was taken, although multiple and repeated crossings seem most likely, the present survey of the Two-Spot Ladybird in Britain provides examples of much steeper changes in frequency in continuous populations than are observed in the Asiatic Ladybird between Korea and Japan. In the one species, and probably in both, strong selective forces are paramount in determining gene frequencies.

SUMMARY

1. The Two-Spot Ladybird has three common forms: red with two black spots, black with six red spots (*sexpustulata*) and black with four red spots (*quadrimaculata*).

2. Samples of the insect have been collected from nearly two hundred sites in Great Britain.

3. The black forms are most frequent in areas with much heavy industry or, as a result of other factors, subject to a high level of atmospheric pollution.

4. The frequency of black individuals is correlated with local smoke concentrations; it shows little correlation with sulphur dioxide.

5. Of the black forms, *sexpustulata* is the commoner around Birmingham and in the industrial midland valley of Scotland; *quadrimaculata* is

the commoner in Lancashire and Yorkshire. It is suggested that this balance may be influenced by differences in the nature of the smoke.

6. This species clearly shows Industrial Melanism determined by strong selective forces; however, camouflage and predation seem to be of little importance in the maintenance of the polymorphism.

ACKNOWLEDGMENTS

I should like to thank all those who have provided samples of ladybirds or who have helped to collect them, especially my wife and Dr C.J. Cadbury, Professor Th.Dobzhansky, Dr P.Harper, Dr D.A.Jones, Mr D.R.Lees, Professor K.G.McWhirter, Mr M.A.Palles-Clark and Mr J.M.S.Williams. I am also most grateful to Dr J.G.Duckett and Mr D.R. Lees for permission to quote our unpublished results on pollution and epiphytes.

REFERENCES

CLARKE C.A. & SHEPPARD P.M. (1966) A local survey of the distribution of industrial melanic forms in the moth *Biston betularia* and estimates of the selective values of these in an industrial environment. *Proc. R. Soc. B.* **165**, 424–439.

COOK L.M., ASKEW R.R. & BISHOP J.A. (1970) Increasing frequency of the typical form of the Peppered Moth in Manchester. *Nature, Lond.* **227**, 1155.

CREED E.R. (1966) Geographic variation in the Two Spot Ladybird in England and Wales. *Heredity* **21**, 57–72.

CREED E.R. (1969) Genetic Adaptation. In J.Rose (ed.), *Technological Injury.* Gordon & Breach.

CREED E.R. (1971) Industrial melanism in the Two Spot Ladybird and smoke abatement. *Evolution* **25**, in press.

FORD E.B. (1940) Genetic research in the Lepidoptera. *Ann. Eugen.* **10**, 227–252.

FORD E.B. (1964) *Ecological Genetics.* Methuen.

FRAZER J.F.D. & ROTHSCHILD M. (1960) Defence mechanisms in warningly-coloured moths and other insects. *Proc. 11th Int. Cong. Entomology* **3**, 249–256.

HALDANE J.B.S. (1956) The theory of selection for melanism in Lepidoptera. *Proc. R. Soc. B* **145**, 303–306.

HAWKES O.A.M. (1920) Observations on the life-history, biology and genetics of the ladybird beetle *Adalia bipunctata* (Mulsant). *Proc. zool. Soc. Lond.* **1920**, 475–490.

HAWKES O.A.M. (1927) The distribution of the ladybird *Adalia bipunctata* L. (Coleoptera). *Entomologist's mon. Mag.* **63**, 262–266.

KETTLEWELL H.B.D. (1955) Selection experiments on industrial melanism in the Lepidoptera. *Heredity* 9, 323–342.

KETTLEWELL H.B.D. (1956) Further selection experiments on industrial melanism in the Lepidoptera. *Heredity* 10, 287–301.

KETTLEWELL H.B.D. (1958) A survey of the frequencies of *Biston betularia* (L.) (Lep.) and its melanic forms in Great Britain. *Heredity* 12, 51–72.

KETTLEWELL H.B.D. (1965) A 12-year survey of the frequencies of *Biston betularia* (L.) (Lep.) and its melanic forms in Great Britain. *Ent. Rec.* 77, 195–218.

KOMAI T. (1956) Genetics of Ladybeetles. *Adv. Genet.* 8, 155–188.

KOMAI T. & CHINO M. (1969) Observations on geographic and temporal variations in the ladybeetle *Harmonia*. I. Elytral patterns. *Proc. Japan Acad.* 45, 284–288.

LUSIS J.J. (1961) On the biological meaning of colour polymorphism of ladybeetle *Adalia bipuncta* L. *Latvijas Entomologs* 4, 3–29.

MADER L. (1926–37) *Evidenz der paläarktischen Coccinelliden und ihrer Aberationen in Wort und Bild*. Vienna.

MINISTRY OF TECHNOLOGY 1964–1969. National survey annual summaries. Warren Spring Laboratory.

ROTHSCHILD M. (1961) Defensive odours and Müllerian mimicry among insects. *Trans. R. ent. Soc. Lond.* 113, 101–121.

SMITH B.C. (1958) Notes on relative abundance and variation in elytral patterns of some common coccinellids in the Belleville district. *Ann. Rept. Entomol. Soc. Ontario* 88, 59–60.

TIMOFEEFF-RESSOVSKY N.W. (1940a) Zur Analyse des Polymorphismus bei *Adalia bipunctata* L. *Biol. Zbl.* 60, 130–137.

TIMOFEEFF-RESSOVSKY N.W. (1940b) Mutations and geographical speciation. In Julian Huxley (ed.), *The New Systematics*. Oxford University Press.

8 ❀ The Distribution of Melanism in the Pale Brindled Beauty Moth, *Phigalia pedaria*, in Great Britain

D. R. LEES

INTRODUCTION

Melanism in moths is the phenomenon in which some individuals of a species are darker than the typical form, due to an increased deposition of melanin in their epidermal scales. For over a century this has received much attention from entomologists and more recently geneticists. In Britain early attempts to account for the occurrence of melanism in many different ecological situations stressed the importance of environmental factors acting directly. For example, White (1876) cites climatic factors as being the 'exciting cause' of melanism in the Highlands of Scotland. Tutt (1890, 1891) began by having similar views, correlating melanism with high rainfall and humidity. Later (1899) his emphasis changed and he argued that climatic factors were an indirect cause of melanism; in areas of high rainfall, particularly where smoke was also present in the atmosphere (e.g. Lancashire and Yorkshire) such surfaces as rocks and trees were darkened by the action of rain, or rain and soot. On such surfaces melanic moths were better concealed and 'natural selection' augmented by 'hereditary tendency' favoured these forms.

Harrison (1920) cast doubt on the importance of crypsis in maintaining melanism. He also pointed out that areas of high rainfall often have luxuriant growths of epiphytes that are far from dark. At the same time he noticed the paucity of epiphytic bryophytes and lichens in industrial areas. Harrison's explanation of melanism relied upon the presence of mutagens in the air pollution of industrial areas and in sea spray near coasts which, acting on the larvae, greatly increased the rate of production by mutation of those genes increasing melanin deposition. These melanics, he thought, were in their larval stage more resistant to the effects of ingesting pollution-covered foliage. He performed experiments to test this hypothesis (Harrison 1935), but the results obtained have been largely discredited on the grounds that the induced mutation rates

152

involved were far higher than any observed elsewhere, and the melanics were inherited as recessives whereas almost all industrial melanics are inherited as dominants (Fisher 1933; Ford 1937; Thomsen & Lemche 1933).

Ford (1937) argued that melanics have a superior viability and that in industrial areas, where this selective advantage is not counteracted by increased conspicuousness, the melanic form would spread. Evidence supporting his views followed when he showed that heterozygous melanic *Cleora repandata* L. are more resistant to starvation than the typical form (Ford 1940). In an extensive study of the geometrid moth *Biston betularia* L. Kettlewell (1955a, 1956) has shown the importance of crypsis on polluted and unpolluted backgrounds, and selective predation by birds, in maintaining a melanic polymorphism and accounting for the spread of industrial melanics in the last 120 years; in addition he demonstrated differences in the behaviour and physiology of the melanics (Kettlewell 1955b, 1957, 1961).

Surveys have shown how *carbonaria* and the other less extreme melanic *insularia*, in *B. betularia* are geographically distributed in Britain (Kettlewell 1958a, 1965; Clarke & Sheppard 1966). An association between urban areas (e.g. Lancashire, the Birmingham area, S.E. England and Glasgow) and high frequencies of *carbonaria* is apparent. Conversely such mainly rural areas as southwest England and northern Scotland are lacking *carbonaria*, or it occurs at low frequency (less than 5 per cent). The apparently anomalous frequencies of around 75 per cent in rural areas of East Anglia, Kettlewell attributes to pollution being carried large distances on prevailing winds from urban areas lying to the south and west.

The highest frequencies of *insularia* are found in areas where it is suggested that air pollution is having a noticeable, but not drastic, effect upon epiphytic plants (Kettlewell 1958a). Clarke & Sheppard (1966) also indicate that this form is at less of a disadvantage than *carbonaria* in rural areas of north Wales but at less of an advantage in the industrial areas of Lancashire. In this species the distribution of melanism has thus been described mainly in terms of air pollution and its indirect effect on the crypsis of the imaginal stage of the moth, the frequencies of the melanics being maintained by the selective predation of birds.

In contrast to the situation in *B. betularia* many species of moth are known in which melanic polymorphisms occur in places far removed from the influence of air pollution (see Chap. 9). The most extensively studied of these is the caradrinid *Amathes glareosa* Esp. in Shetland (Kettlewell & Berry 1961, 1969; Kettlewell, Berry, Cadbury & Phillips 1969). Here, as in *B. betularia*, selective predation by birds seems to be important in maintaining the polymorphism but other factors, for example differences in behaviour between populations, local adaptation and dominance modification also play a part.

The subject of this paper, *Phigalia pedaria* Fab. is widely distributed in the British Isles and its melanic forms are maintained at the present time as a polymorphism in the populations of both industrial and rural areas.

LIFE HISTORY AND DESCRIPTION OF FORMS

The adults of *Ph. pedaria* emerge during January, February and March. The males fly at night and rest during the day on tree trunks, fences, walls etc. The females are brachypterous and cannot fly, their wings being about 1 mm long. Eggs are laid in crevices in the bark of trees and these hatch in late April and early May. The larvae feed upon a variety of deciduous trees including oak, birch, hazel and sallow. In June they burrow into the ground to pupate and remain there until ready to emerge. The species occurs throughout Britain and continental Europe and it is especially common in dense deciduous woodland. A closely related species *Ph. titea* Cramer occurs in North America.

Typical male *pedaria* are variable in both the colour and patterning of the forewings. The general appearance is greyish, tinged with green, brown and yellow markings. Barrett (1901) contains coloured illustrations of this form (plate 302). Typical females also vary, from insects having a uniform dark brown dorsal surface of the abdomen and thorax, to those in which these parts are a much lighter brown with some dark markings; however the ventral surface of the abdomen is very light grey with a few dark scales.

In Britain two melanic forms are recognizable. One, *monacharia* Staud, has the typical patterning of the male completely replaced by uniform green-black or black on the forewings, with lighter grey-brown hindwings and a dark brown or black abdomen and thorax. The females are uniformly black except for some grey or white scales at the base of the antennae. The other melanic, referred to in this paper as intermediate, is less extreme than *monacharia* and some of the typical patterning on the

PLATE 8.1. Typical, intermediate and melanic *Phigalia pedaria*; top row males, bottom row females. The females are mounted to show the patterning of the ventral abdomen.

[to face p. 155

forewings of the male remains. The amount of patterning is variable but usually consists of black or dark brown patches across the costal margin, with some lighter scales flecking the remainder of the wing. The overall appearance is more green and brown than *monacharia*. Females of this form are indistinguishable from *monacharia* dorsally but they may be identified by their ventral surfaces where the black background is liberally flecked with grey and white scales (see plate 8.1 for males and females of all three morphs).

Ford (1937) reports crosses of *pedaria* which show that *monacharia* is inherited as a dominant to typical; this has been confirmed in the present study. The intermediate is controlled by an allele of the *monacharia* locus and is recessive to *monacharia* but dominant to typical (Lees, in prep.). This situation is similar to the genetic control of the melanics of the Peppered Moth *Biston betularia*, which are also allelomorphic, with *carbonaria* dominant to the range of intermediate forms collectively known as *insularia*, which are in turn dominant to typical (Clarke & Sheppard 1964, Lees 1968). The two species differ in that the intermediate form of *pedaria*, although slightly variable, seems to be controlled by a single allele whereas *insularia*, which is very variable, is probably controlled by several.

SAMPLING AND SCORING METHODS

The samples in this survey fall into three categories:

1   61 samples obtained by me using either assembly traps or mercury vapour light traps;
2   10 samples obtained by local entomologists at light traps, who sent their samples to me for scoring;
3   17 samples obtained and scored by local entomologists.

The 71 samples in categories 1 and 2 above have been scored for both the melanic forms and typical. Some misclassification of the two melanics, particularly when the specimens were worn, may have occurred. Most difficulty in separating these two forms was encountered in those samples where the overall melanic frequency was more than 50 per cent. There seems to be a tendency for intermediate to be darker in such samples. However this error is thought to have been small. The 17 samples in category 3 were scored for melanic and typical only; the two melanic

TABLE 8.1. Description of samples of *Phigalia pedaria*. Samples where the two melanic forms are separated.

Site	Locality	Year	Phenotype numbers				Phenotype (per cent)			Gene frequency		
			Typ	Int	Mel	Total	Typ	Int	Mel	Typ	Int	Mel
1	Ardross, Easter Ross	1968	10	0	0	10	100·0	0·0	0·0	100·0	0·0	0·0
2	Fort Augustus, Inverness-shire	1968	36	0	0	36	100·0	0·0	0·0	100·0	0·0	0·0
3	Killiecrankie, Perthshire	1968–69	340	65	38	443	76·7	14·7	8·6	87·6	8·0	4·4
4	Aberfoyle, Perthshire	1969	56	17	10	83	67·5	20·5	12·0	82·1	11·6	6·2
5	Rowardennan, Dumbartonshire	1968–69	21	3	4	28	75·0	10·7	14·3	86·6	6·0	7·4
6	Ince Blundell, Lancashire	1969	11	2	33	46	23·9	4·3	71·7	48·9	4·3	46·8
7	Wilmslow, Cheshire	1968–69	9	2	31	42	21·4	4·8	73·8	46·3	4·9	48·8
8	Sandbach, Cheshire	1969	11	5	25	41	26·8	12·2	61·0	51·8	10·7	37·5
9	Keele, Staffordshire	1969	4	3	13	20	20·0	15·0	65·0	44·7	14·4	40·8
10	Moddershall, Staffordshire	1969	17	6	31	54	31·5	11·1	57·4	56·1	9·2	34·7
11	Ellenhall, Staffordshire	1969	53	10	26	89	59·6	11·2	29·2	77·2	7·0	15·9
12	Shugborough, Staffordshire	1969	26	12	32	70	37·1	17·1	45·7	60·9	12·7	26·3
13	Gailey, Staffordshire	1969	45	7	16	68	66·2	10·3	23·5	81·3	6·1	12·6
14	Armitage, Staffordshire	1969	10	4	12	26	38·5	15·4	46·2	62·0	11·4	26·6
15	Hopwas, Staffordshire	1968–69	41	14	56	111	36·9	12·6	50·5	60·8	9·6	29·6
16	Walsall, Staffordshire	1969	14	8	36	58	24·1	13·8	62·1	49·1	12·5	38·4
17	Middleton, Warwickshire	1968	30	11	52	93	32·3	11·8	55·9	56·8	9·6	33·6
18	Himley, Worcestershire	1969	8	4	8	20	40·0	20·0	40·0	63·2	14·2	22·5
19	Coleshill, Warwickshire	1967–68	22	5	27	54	40·7	9·3	50·0	63·8	6·9	29·3
20	Lickey, Worcestershire	1967–68	101	8	47	156	64·7	5·1	30·1	80·5	3·1	16·4
21	Earlswood, Warwickshire	1968	44	5	18	67	65·7	7·5	26·9	81·0	4·5	14·5
22	Newland, Warwickshire	1968–69	53	8	15	76	69·7	10·5	19·7	83·5	6·1	10·4
23	Kenilworth, Warwickshire	1968	31	7	8	46	67·4	15·2	17·4	82·1	8·8	9·1

		Year									
24	Bretford, Warwickshire	1969	47	7	19	73	64·4	9·6	26·0	80·2	5·8	14·0
25	Chesterton, Warwickshire	1968	47	3	3	53	88·7	5·7	5·7	94·2	3·0	2·9
26	Alveston, Warwickshire	1968–69	65	3	4	72	90·3	4·2	5·6	95·0	2·2	2·8
27	Snitterfield, Warwickshire	1968	49	7	11	67	73·1	10·4	16·4	85·5	5·9	8·6
28	Alcester, Warwickshire	1969	24	2	4	30	80·0	6·7	13·3	89·4	3·7	6·9
29	Spetchley, Worcestershire	1969	23	1	2	26	88·5	3·8	7·7	94·1	2·0	3·9
30	Wormington, Gloucestershire	1969	45	2	1	48	93·8	4·2	2·1	96·8	2·1	1·0
31	Weston, Warwickshire	1968–69	54	5	0	59	91·5	8·5	0·0	95·7	4·3	0·0
32	Bruern, Oxfordshire	1969	46	0	2	48	95·8	0·0	4·2	97·9	0·0	2·1
33	Kiddington, Oxfordshire	1968–69	68	4	1	73	93·2	5·5	1·4	96·5	2·8	0·7
34	Steeple Barton, Oxfordshire	1968–69	58	0	2	60	96·7	0·0	3·3	98·3	0·0	1·7
35	Wytham, Berkshire	1967–69	312	5	9	326	95·7	1·5	2·8	97·8	0·8	1·4
36	Nuneham Courtenay, Oxfordshire	1968	36	2	0	38	94·7	5·3	0·0	97·3	2·7	0·0
37	Chalfont St. Peter, Buckinghamshire	1968	15	3	1	19	78·9	15·8	5·3	88·9	8·5	2·7
38	Shilton, Oxfordshire	1970	218	12	5	235	92·8	5·1	2·1	96·3	2·6	1·1
39	Purton, Wiltshire	1970	120	4	2	126	95·2	3·2	1·6	97·6	1·6	0·8
40	Chatcombe, Gloucestershire	1970	81	7	12	100	81·0	7·0	12·0	90·0	3·8	6·2
41	Wickwar, Gloucestershire	1970	122	9	6	137	89·1	6·6	4·4	94·4	3·4	2·2
42	Park End, Gloucestershire	1969	50	7	16	73	68·5	9·6	21·9	82·8	5·6	11·6
43	Hope Under Dinmore, Herefordshire	1970	18	1	2	21	85·7	4·8	9·5	92·6	2·5	4·9
44	Pontrilas, Herefordshire	1970	34	3	8	45	75·6	6·7	17·8	86·9	3·8	9·3
45	St Arvans, Monmouthshire	1970	309	41	56	406	76·1	10·1	13·8	87·2	5·6	7·2
46	Coldra, Monmouthshire	1970	42	18	13	73	57·5	24·7	17·8	75·9	14·8	9·3
47	Bettws Newydd, Monmouthshire	1970	55	10	22	87	63·2	11·5	25·3	79·5	6·9	13·6
48	Llangattock, Breckonshire	1970	126	17	26	169	74·6	10·1	15·4	86·3	5·6	8·0

TABLE 8.1.—continued

Site	Locality	Year	Phenotype numbers				Phenotype (per cent)			Gene frequency		
			Typ	Int	Mel	Total	Typ	Int	Mel	Typ	Int	Mel
49	Ebbw Vale, Monmouthshire	1970	15	6	27	48	31·3	12·5	56·3	55·9	10·2	33·9
50	Llanishen, Glamorgan	1970	48	6	25	79	60·8	7·6	31·6	77·9	4·7	17·3
51	Gelligaer, Glamorgan	1970	34	9	45	88	38·6	10·2	51·1	62·2	7·7	30·1
52	Llwyn Y Pia, Glamorgan	1970	27	7	21	55	49·1	12·7	38·2	70·1	8·6	21·4
53	St. Donats, Glamorgan	1970	27	1	4	32	84·4	3·1	12·5	91·9	1·7	6·5
54	Margam, Glamorgan	1970	15	4	26	45	33·3	8·9	57·8	57·7	7·2	35·0
55	Pont Rhyd Y Fen, Glamorgan	1969–70	18	7	23	48	37·5	14·6	47·9	61·2	10·9	27·8
56	Talybont, Breconshire	1969–70	127	24	70	221	57·5	10·9	31·7	75·8	6·9	17·3
57	Senny Bridge, Breconshire	1970	58	5	11	74	78·4	6·8	14·9	88·5	3·7	7·7
58	Llandovery, Carmarthenshire	1970	51	5	7	63	81·0	7·9	11·1	90·0	4·3	5·7
59	Clyne, Glamorgan	1970	21	8	41	70	30·0	11·4	58·6	54·8	9·6	35·6
60	Llanybri, Carmarthenshire	1970	24	2	2	28	85·7	7·1	7·1	92·6	3·8	3·6
61	Llangain, Carmarthenshire	1970	43	1	11	55	78·2	1·8	20·0	88·4	1·0	10·6
62	Cynwyl Elfed, Carmarthenshire	1970	53	4	6	63	84·1	6·3	9·5	91·7	3·4	4·9
63	Canaston, Pembrokeshire	1970	18	0	3	21	85·7	0·0	14·3	92·6	0·0	7·4
64	Llanybyther, Carmarthenshire	1970	83	6	13	102	81·4	5·9	12·7	90·2	3·2	6·6
65	Maentwrog, Merionethshire	1968–69	125	6	11	142	88·0	4·2	7·7	93·8	2·2	4·0
66	Madingley, Cambridgeshire	1968–70	32	3	1	36	88·9	8·3	2·8	94·3	4·3	1·4
67	Woolpit, Suffolk	1970	16	1	1	18	88·9	5·6	5·6	94·3	2·9	2·8
68	Tunstall, Suffolk	1970	110	9	5	124	88·7	7·3	4·0	94·2	3·8	2·0
69	Harleston, Norfolk	1970	41	8	2	51	80·4	15·7	3·9	89·7	8·4	2·0
70	Swanton Novers, Norfolk	1970	29	2	2	33	87·9	6·1	6·1	93·7	3·2	3·1
71	Brandon, Suffolk	1970	15	0	0	15	100·0	0·0	0·0	100·0	0·0	0·0

TABLE 8.2. Description of samples of *Phigalia pedaria*. Samples where the two melanic forms are combined.

Site	Locality	Year	Phenotype Numbers			Phenotype (per cent)		Gene frequency	
			Typ	Mel	Total	Typ	Mel	Typ	Mel
73	Rannoch, Perthshire	1968–69	29	5	34	85·3	14·7	92·3	7·7
74	Blairgowrie, Perthshire	1968	15	2	17	88·2	11·8	93·9	6·1
75	Penrith, Cumberland	1957–62	34	12	46	73·9	26·1	86·0	14·0
76	Milnthorpe, Westmorland	1968	26	20	46	56·5	43·5	75·2	24·8
77	Bebington, Cheshire	1957	7	22	29	24·1	75·9	49·1	50·9
78	Tregarth, Caernarvonshire	1963–66	34	1	35	97·1	2·9	98·5	1·5
79	Quorn, Leicestershire	1968	4	8	12	33·3	66·7	57·7	42·3
80	Ledbury, Herefordshire	1968	17	3	20	85·0	15·0	92·2	7·8
81	Radley, Berkshire	1968	20	2	22	90·9	9·1	95·3	4·7
82	Newbury, Berkshire	1967–68	30	2	32	93·8	6·3	96·8	3·2
83	Shipham, Somerset	1968–69	57	3	60	95·0	5·0	97·5	2·5
84	Ottershaw, Surrey	1946–59	96	5	101	95·0	5·0	97·5	2·5
85	Woking, Surrey	1968	33	10	43	76·7	23·3	87·6	12·4
86	Winchester, Hampshire	1968	13	1	14	92·9	7·1	96·3	3·7
87	Falmer, Sussex	1965	24	1	25	96·0	4·0	98·0	2·0
88	Minstead, Hampshire	1968–69	41	4	45	91·1	8·9	95·4	4·6
72	Newtonmore, Inverness-shire	1968	Melanics present in the population but details of the sample size are not known.						

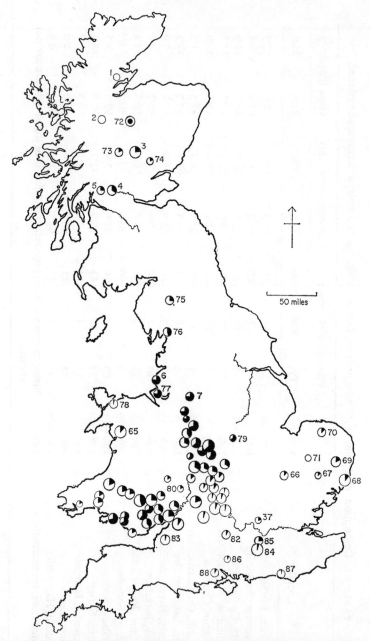

FIGURE 8.1. Distribution map showing the frequency of typical (white segments) and the combined frequency of intermediate and melanic (black segments) male *Phigalia pedaria* in England, Scotland and Wales. For key to the numbers see tables 8.1 and 8.2. The four sizes of circle refer to samples of 10–24, 25–49, 50–99 and 100 or more insects, in increasing order.

forms were not scored separately. Misclassification of typical in all three sampling categories is unlikely.

All the samples consist of male moths only, since females cannot fly and are thus not caught at light traps, while assembly traps utilize the female pheromone and also attract only males. Most of the samples were taken in the four years 1967–70.

DISTRIBUTION OF THE MELANICS

Table 8.1 contains phenotype and gene frequencies for those samples in categories 1 and 2 above, while table 8.2 contains frequencies for samples in category 3. The total melanic phenotype frequency (intermediate and *monacharia* combined) at each locality is shown pictorially in fig. 8.1. The survey is incomplete in several areas and samples are absent from southwest England, northeast England and southern Scotland, including the Glasgow and Edinburgh area. Conversely two areas have been studied in detail; the midlands of England and south Wales (figs. 8.2 and 8.3).

It is apparent that there is an association between high melanic frequencies and dense urban and industrial areas. For example, Lancashire and Cheshire (sites 7, 8 and 9), the Birmingham conurbation (sites 12, and 14 to 19), and the industrial areas of south Wales (sites 49 to 52, 54, 55 and 59), all have frequencies of 60 to 75 per cent total melanics. Conversely, some mainly rural areas have very low total melanic frequencies and this is most noticeable in central southern England (sites 35, 36, 82, 84, 86 to 88) where frequencies of from 5 to 10 per cent are found. However in three other completely rural areas higher frequencies than these do occur.

1 In East Anglia a frequency of 10 to 15 per cent is indicated by five samples (sites 66 to 70).

2 Samples obtained in south Wales up to 50 miles (85 km) away from the industrial valleys and separated from them by hills of over 2000 ft (610 m), the Brecon Beacons and Black Mountain, show 15 to 25 per cent melanics (sites 57, 58 and 61–4).

3 In central Scotland, north of the industrial Midland Valley, similar frequencies to those in 2 above are found (sites 3–5, 73 and 74).

In a total of 88 samples only three (sites 1, 2 and 71) contain no

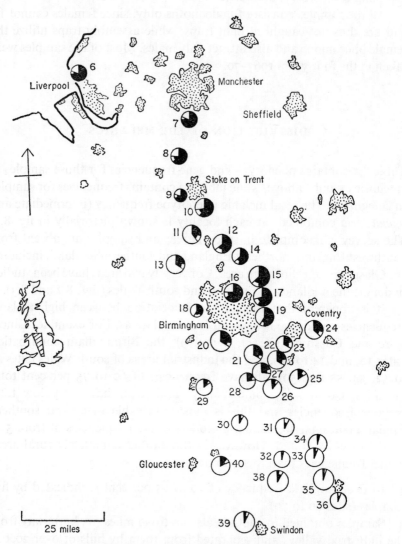

FIGURE 8.2. Distribution map showing the frequency of typical (white segments), intermediate (halved segments) and melanic (black segments) male *Phigalia pedaria* in the Midlands of England. For key to numbers see table 8.1. The four sizes of circle refer to samples of 10–24, 25–49, 50–99 and 100 or more insects, in increasing order. Large towns are represented by stippling. The inset shows the relationship of this map to the rest of Britain.

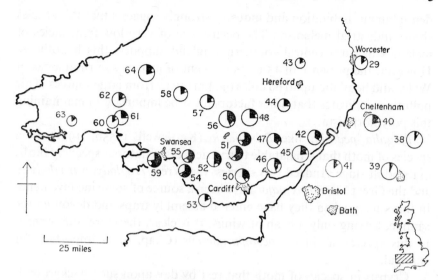

FIGURE 8.3. Distribution map showing the frequency of typical (white segments), intermediate (halved segments) and melanic (black segments) male *Phigalia pedaria* in south Wales. For key to numbers see table 8.1. The four sizes of circle refer to samples of 10–24, 25–49, 50–99 and 100 or more insects, in increasing order. Large towns are represented by stippling. The inset shows the relationship of this map to the rest of Britain.

melanics. The sizes of these are small (36, 10 and 16 respectively) and do not exclude the possibility that the real frequencies of melanics in these populations are 10 per cent or more. It is therefore possible that melanic *pedaria* occur throughout Britain at least at a low frequency.

The relative proportions of *monacharia* and intermediate within the melanic fraction of each sample varies with the total melanic frequency. Intermediate rarely represents more than a quarter of all melanics when their frequency is 50 per cent or more. However, at total frequencies of below 25 per cent, intermediate may account for between a third and a half of the melanics (see figs. 8.2 and 8.3). Where the melanics only constitute a very small proportion of the whole population, the number of them caught has always been too small to obtain a reliable estimate of the relative proportions of the two morphs.

DISCUSSION

The coincidence of high frequencies of melanics in *pedaria* with areas of dense human habitation and industry strongly suggests that this species shows industrial melanism. The occurrence of very low frequencies of melanics in rural central southern England supports this hypothesis. However, the presence of 15 to 35 per cent of melanics in rural areas of Wales and Scotland up to 60 miles (96 km) away from large sources of air pollution, indicates that other factors must be important in maintaining this polymorphism.

Phigalia pedaria, like *B. betularia* (Kettlewell 1955b, 1961) is a species of moth that is palatable to a variety of predators, especially birds, in its adult stage. Indeed two species, the Wren, *Troglodytes troglodytes* and the Great Tit, *Parus major* have been a source of some inconvenience in this survey when they have entered assembly traps and devoured the sample, leaving only a heap of wings. It is clear, therefore, that crypsis in this species, as in *Lasiocampa quercus* (Chap. 9), is important for survival.

Crypsis in species of moth that rest by day upon such backgrounds as tree trunks depends on several properties of the backgrounds (Cott 1940). The two most important of these are reflectance and the degree to which the surface is broken up by, for example, fissures in the bark and growths of epiphytic plants. Both of these properties are affected by two main components of air pollution from industrial and domestic sources in Britain; the gas sulphur dioxide and the solid fraction smoke.

It is well known that air pollution and the general acidification of the habitat that it causes, are toxic to a varying extent to all epiphytic plants (Gilbert 1965, 1968); the lichen and bryophyte 'deserts' around large towns are confirmation of this. The effect of this pollution is therefore to reduce the degree to which the background of the moth is broken up, thereby decreasing the effectiveness of the insect's disruptive patterning. Smoke has an additional effect, it consists mainly of carbonaceous material and its deposition on any surface tends to make that surface dark.

At 45 *Ph. pedaria* sampling sites in Great Britain, south of a line running from Preston to York, measurements have been made of nine parameters of the ecology of epiphytes on oak trees (*Quercus* sp.) in relation to air pollution (Creed, Duckett & Lees, in prep.). These include estimates, based on ten trees at each site, of percentile cover at 1·5 m of *Pleurococcus* sp., crustose and foliose lichens, bryophytes and bare wood. Measurements have also been made of the reflectance of the trunk at the same height. Information on six physical parameters, such as rainfall, temperature, altitude etc. has also been obtained. In addition values for the two atmospheric pollutants, sulphur dioxide and smoke, have been obtained either directly or by interpolation (Ministry of Technology, 1964–69).

It is therefore possible to analyse the geographic variation in the frequency of melanic *pedaria* at these 45 localities in terms of the 17 parameters mentioned above. To do this a series of multiple regressions have been computed of melanic frequency on varying combinations of the 17 possible independent variables. The number of these variables has been progressively reduced keeping the value of R^2 as high as possible (R^2 is the proportion of the variance of the dependent variable y that can be accounted for by the multiple regression, and is the square of the multiple correlation coefficient). By this method the number of parameters has been decreased in terms of which the variation in total melanic frequency of *Ph. pedaria* may be explained.

In this way a value for R^2 of 85 per cent is obtained when the multiple regression of total melanic frequency on all 17 independent variables is computed. However, the partial regressions for only three of these are

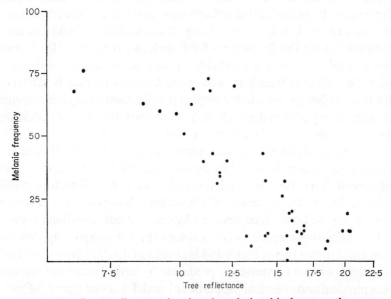

FIGURE 8.4. Scatter diagram showing the relationship between the combined frequency of intermediate and melanic *Phigalia pedaria* and the mean reflectance of oak trees at the same sites. Tree reflectance is expressed as percentage reflectance and plotted on a logarithmic scale.

formally significant: mean reflectance, mean annual rainfall and smoke. By reduction to these three variables R^2 only drops by 4, to 81 per cent. On further reduction, the highest value of R^2 obtained for a single independent variable is 64 per cent for mean log reflectance. It is thus possible to explain 64 per cent of the variation in melanic frequency in terms of this parameter. The scatter diagram of total melanic frequency plotted against the log reflectance of oak trees at the same site appears in fig. 8.4. The negative regression is highly significant ($t_{(43)} = -9\cdot14, P < 0\cdot001$).

Calculation of similar multiple regression analyses for the two melanic forms separately indicates that the extreme melanic, *monacharia*, contributes most to the values of R^2 presented above for total melanics; corresponding values for the intermediate are consistently much lower than that for *monacharia*. For example, R^2 values using all independent variables are 87 per cent for *monacharia* and 56 per cent for the intermediate. On reduction to the single independent variable mean reflectance, a value of 63 per cent is obtained for *monacharia* while that for the intermediate is only 22 per cent.

It seems, therefore, that variations in *monacharia* frequency may be largely explained in terms of the reflectance of oak trees, but the possibility cannot be excluded that reflectance is strongly correlated with some other unmeasured, or unknown, independent variable which has a causal relationship with the frequency of this melanic. It has not, for instance been possible to devise a workable method of assessing the degree to which the surface of trees is broken up, and reflectance may be correlated with this. Although reflectance may in part be used to explain variation in the frequency of the intermediate, it seems that there is some additional factor determining the distribution of this melanic form.

The choice of oak for the ecological survey of epiphytes (Creed *et al.* in prep.) was determined mainly by the fact that it is perhaps the most widespread large tree in England and Wales, and therefore readily available for study. In terms of *Ph. pedaria*, however, it is a realistic choice since the larvae feed upon its leaves, and oak woodland is one of the best habitats in which to catch the moth. The imago may therefore be assumed to rest on the trunk and branches of this tree during the hours of daylight, when it is exposed to predation by birds. Because of variation in the physical and chemical properties of bark from one species of tree to another (Skye 1968), the results of this survey are only directly applicable to oaks. However, studies on sycamores (*Acer pseudoplatanus*) indicate that although the two species of tree differ in the absolute values of the parameters measured, the same geographical trends are apparent in each.

This work indicates that tree (oak) reflectance is influenced to a large extent by air pollution, particularly the smoke fraction of it, but other factors, especially where pollution levels are low, may also be important. For instance, in areas of high rainfall and low pollution levels, where foliose lichens and bryophytes constitute a large proportion of the cover of the trunk, its surface tends to be very patchy, with areas of light and dark epiphytes and shadow intermingled. In such situations, although measurement of reflectance in the field is made difficult by the patchiness of the subject, it seems that the overall background available to a moth is darker than in slightly more polluted areas where trees have a much more uniform covering of crustose lichens. Nevertheless in areas where foliose lichens are absent it is possible to explain 57 per cent of the variation in reflectance of trees in terms of smoke.

The analysis of the distribution in England and Wales of melanics of the two-spot Ladybird, *Adalia bipunctata* (Creed 1966 and see Chap. 7) provides an interesting comparison with *Ph. pedaria*. This species is distasteful to predators and therefore, unlike *pedaria*, does not rely upon crypsis for survival. This is confirmed by the emergence of smoke in Creed's multiple regression analysis as the main independent variable in terms of which the melanic distribution of the ladybird may be explained; the corresponding parameter for *pedaria* is the reflectance of trees. It appears that the melanism of *A. bipunctata* is a response to a direct effect of smoke but in *Ph. pedaria* it is, at least in part, determined by the indirect effects of air pollution on crypsis in the imago.

A comparison of the *pedaria* melanic distribution with that of the moth *Biston betularia* L. (Kettlewell 1958a, 1965) is also profitable. It might be supposed that the melanics of these two species are subject to similar selection pressures; they belong to closely related genera, their habitats overlap, both species are palatable to avian predators and they have similar behaviour. On the other hand the two species differ in that sexual dimorphism is pronounced in *pedaria* but not in *betularia*, while the difference in appearance between the typical and melanic morphs is greater in *betularia* than in *pedaria*. In addition the imagines of the two species fly at different seasons of the year; *pedaria* during January, February and March and *betularia* during May, June and July.

The distributions of their melanics in Britain show one basic similarity: highest frequencies occur in dense urban and industrial areas. There are, however, several points of difference.

(a)   Melanics are absent in *betularia* from rural central Scotland where frequencies of around 25 per cent are found in *pedaria* populations.

(b)   In parts of East Anglia, little affected by air pollution, melanic *betularia* constitute 75 per cent of the population; the equivalent value for *pedaria* is less than 15 per cent.

(c)   In rural southwest Wales where *carbonaria* seems to be absent and *insularia* occurs at a frequency of less than 5 per cent (Lees, in prep.), *pedaria* melanic frequencies as high as 25 per cent are found.

(d)   Maximum melanic frequencies in highly polluted areas in *betularia* are around 98 per cent; in *pedaria* no sample has yet been obtained where melanics exceed 75 per cent of the population.

(e)   *carbonaria* were first recorded in an industrial area, Manchester, in 1848 (Edelston 1864), at which time melanics of *pedaria* were probably well established in the rural Highlands of Scotland; White (1876) describes *pedaria* in this area as being a species which is 'frequently melanochroic, but of which many individuals are not so.' It is likely, therefore, that the *pedaria* melanic polymorphism, at least in central Scotland, occurred before the growth of industry, and the consequent atmospheric pollution, in the second half of the nineteenth century. In contrast melanics in *betularia* were not recorded until this industrial growth had begun.

It is thus apparent that, although these two species show what has come to be known as 'industrial melanism', in the sense that a larger proportion of their populations in urban areas are melanic, they also show some very marked differences in the distribution of their melanics. These differences are not consistent from area to area and show no clear cut pattern; in some areas frequencies of melanic *betularia* are higher than those of *pedaria*, in others the reverse is true. It seems that simple differences between the species such as in degree of sexual dimorphism, crypsis or potential for migration cannot alone be invoked to explain such discrepancies in geographic variation. It is likely that the genes controlling the various melanics have differing pleiotropic effects, the selective values of which may vary from area to area.

Non-visual differences between melanic and typical morphs of moths are known. For instance, behavioural differences in background preference have been demonstrated for *B. betularia* (Kettlewell 1955b) and *Catocola ultronia* (Sargent 1966) but this is not the case with *Ph. titea*, the North American equivalent of *Ph. pedaria*, in which both morphs show a preference for light backgrounds (Sargent 1969).

Physiological differences between morphs have also been investigated. Heterozygous melanic *Cleora repandata* are more resistant to starvation than their typical sibs (Ford 1940). In *B. betularia, carbonaria* is more resistant to the effects of eating soot-covered foliage (Kettlewell 1958b). In addition, Clarke & Sheppard (1966) have demonstrated that the homozygote *carbonaria* is at a 15 per cent disadvantage compared with the heterozygote. The balanced state of the melanic polymorphism in this species in very polluted areas, with a maximum frequency of about 98 per cent *carbonaria*, is attributed to this disadvantage.

In *Ph. pedaria* the maximum frequency of about 75 per cent for all melanics in the heavily polluted areas around Birmingham, parts of Cheshire and Lancashire suggests that some non-visual attribute of the genes controlling these melanics is influencing their frequency. These maxima occur in the same areas as the highest melanic frequencies of *B. betularia*. It might be argued that migration of imagines of *pedaria* from neighbouring areas of lower melanic frequency is responsible for the 75 per cent maximum, but migration alone cannot account for an apparent 'plateau' at this frequency over large areas. In addition, *pedaria* is a smaller, less mobile species than *betularia* and only its males are able to fly; thus migration is likely to be a more important factor in *betularia* than in *pedaria*. But, as already indicated, the maximum frequency of the melanic phenotypes is 20 per cent higher in *betularia* than in *pedaria* in these areas.

It is thus clear that in *pedaria* the genes controlling melanism have some pleiotropic disadvantageous effects. Direct evidence of this comes from three broods which indicate that the homozygous intermediate genotype has a reduced viability. In these broods heterozygous intermediates have been crossed. The broods are homogeneous and segregations in the progeny do not differ from a 2:1 ratio in both sexes ($\chi^2_{(1)} = 0.29$; $0.9 > P > 0.8$), but they do differ significantly from the expected 3:1 ratio ($\chi^2_{(1)} = 8.49$; $0.005 > P > 0.001$). No such deficiency has been found in eight similar crosses made with *monacharia* heterozygotes. The apparent inviability of the intermediate homozygote may be part of the reason for its gene frequency never having been found to exceed 15 per cent (see table 8.1).

In addition to the polymorphic variation, which has been the subject of most of this chapter, *Ph. pedaria* also shows variation which is continuous and presumably under polygenic control: typical moths vary in the patterning and darkness of their forewings. In order to ascertain whether part of this variation, darkness, shows any geographic variation, samples of the forewings of males from 14 sites along a transect were measured for their reflectance. This transect ran roughly northwest for 96 miles (154 km) from the mainly rural Oxford area up to and through the heavily polluted Birmingham conurbation. Along the transect the total melanic frequency increases from 4 per cent at Wytham (site 35) to 68 per cent at Middleton (site 17), see fig. 8.2. The results of the measurement of wing reflectance show that typical males become darker towards Birmingham; there is a significant negative regression of the log reflectance of wings on distance from Oxford ($t_{(12)} = -5.26$; $P < 0.001$). That this darkening of typical wings is consequent upon a darkening of the backgrounds upon which the moth rests is suggested by the significant positive regression of log wing reflectance on log tree reflectance, at the 14 sites in the transect ($t_{(12)} = 3.06$; $0.01 > P > 0.002$), see scatter diagram fig. 8.5. Thus along the transect, in addition to increasing melanic frequency, selection favours increasing darkness of typical towards a heavily polluted, urban area; this corresponds with the darkening of the trees upon which the moth rests. E.R.Creed has pointed out to me that it may be most realistic to describe the darkness of a population of *pedaria* males in terms of the frequencies of the intermediate and *monacharia*, and the reflectance of typical individuals. In this way the mean reflectance

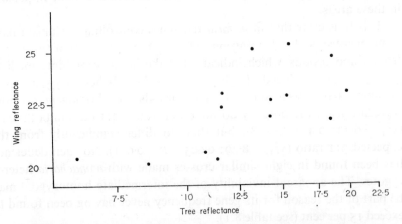

FIGURE 8.5. Scatter diagram showing the relationship between the mean reflectance of samples of the forewings of typical male *Phigalia pedaria* and the mean reflectance of oak trees at the same sites. Both tree and wing reflectances are expressed as percentage reflectance and plotted on logarithmic scales.

of all individuals in a sample would provide the best single measurement of melanism.

In having two methods of being dark, both polymorphic and polygenic, *pedaria* resembles *Gonodontis bidentata* in which the typical form is darkened polygenically in Birmingham, where the melanic form *nigra* is absent, while at Cannock Chase, 20 miles (32 km) north of Birmingham more than 50 per cent of the population are melanic (Kettlewell 1959).

The differences in the geographic variation of *Ph. pedaria*, *B. betularia* and the ladybird *A. bipunctata* emphasize the complexity of the phenomenon known as 'industrial melanism' and the dangers of generalizing, in this context, from one species to another. It is to be expected that differences exist between the two moth species, which rely upon crypsis for survival, and the ladybird which does not. However, differences in distribution between the two moth species are as great as those between them and the ladybird. In all three insects there is evidence of the importance of factors other than the effects of air pollution in maintaining their melanic polymorphisms. Some of these have been identified but it is likely that many others remain to be revealed. For instance in *Ph. pedaria* little is known at present of the selective forces acting on the brachypterous females.

SUMMARY

1. Two melanic forms of the moth *Ph. pedaria* occur throughout Britain. Alleles at the same locus control these two morphs; *monacharia*, the extreme melanic, is dominant to a slightly patterned melanic (intermediate) and both are dominant to typical. The effects of these alleles are also detectable in the females, which are brachypterous.

2. There is some evidence that insects homozygous for the intermediate allele are inviable.

3. Samples of males from 88 localities in England, Wales and Scotland have been obtained in the four seasons 1967–70. Highest melanic frequencies (around 75 per cent) have been found in dense urban areas of the English midlands, Lancashire and South Wales, and the lowest in rural areas of central southern England. In other rural areas, however, (in south Wales and central Scotland) frequencies as high as 25 per cent are found.

4. In a survey of the environmental factors affecting epiphytes of

7

oaks (*Quercus* sp.) in England and Wales, information for some or all of 17 parameters has been obtained (Creed *et al.* in prep.). Forty-five of these sites correspond with *pedaria* samples. In a multiple regression analysis of geographic variation in melanic frequency it is possible to explain 85 per cent of this variation in terms of all 17 independent variables, and, on reduction, 64 per cent in terms of the single variable, oak tree reflectance.

5. A tendency for the forewings of typical male *pedaria* to be darker in the polluted Birmingham area than in rural areas to the south is shown to be related to tree reflectance.

6. The distribution of melanic *Ph. pedaria* is compared with those of the ladybird *A. bipunctata* (Chap. 7) and the moth *B. betularia* (Kettlewell 1958a, 1965). It is concluded that *Ph. pedaria*, like *B. betularia*, shows industrial melanism but that in both species, at least in some areas, factors other than air pollution are responsible for their melanic polymorphisms.

ACKNOWLEDGMENTS

I am indebted to many landowners in England and Wales who gave me permission to use moth traps on their property, and to the following who collected samples of *Ph. pedaria* and kindly sent details to me:

Sir Eric Ansorge, J.Boorman, J.Briggs, D.L.Coates, M.D'Oyly, Dr J.G.Duckett, D.W.H.Ffennell, P.H.Gamble, J.C.Gladman, D.E.Hardy, Dr M.Harper, Dr P.Harper, Commander G.W.Harper, K.M.Harris, R.Leverton, L.W.Siggs, T.B.Silcocks, G.Sutherland, W.G.Vosper, D.R.J.Wallace, J.Thorpe, Rothamsted Experimental Station, Air Marshal Sir Robert Saundby, W.M.Webb, G.S.Woollatt.

In addition I should like to thank the many people who attempted to obtain samples but were unsuccessful.

Dr E.R.Creed has encouraged me throughout this study and I am additionally grateful to him, and to Mr P.T.Handford, for reading this chapter in typescript and making many helpful suggestions.

I acknowledge all those in the Department of Forestry, Oxford, who helped in computing the statistics of the multiple regression analysis.

Dr H.B.D.Kettlewell provided me with information on four early samples and without his stimulus the survey would not have been started; during it I have been privileged to be accommodated in Professor E.B.Ford's laboratory and I thank him for his generous help at all times.

REFERENCES

BARRETT C.G. (1901) The Lepidoptera of the British Islands. Vol. 7, pp. 135–139. London, Lovell Reeve.

CLARKE C.A. & SHEPPARD P.M. (1964) Genetic control of the melanic form *insularia* of the moth *Biston betularia* L. *Nature* 202, 215–216.

CLARKE C.A. & SHEPPARD P.M. (1966) A local survey of the distribution of industrial melanic forms in the moth *Biston betularia* and estimates of the selective values of these in an industrial environment. *Proc. R. Soc., B* 165, 424–439.

CREED E.R. (1966) Geographic variation in the Two-spot Ladybird in England and Wales. *Heredity* 21, 57–72.

COTT H.B. (1940) Adaptive coloration in animals. Methuen, London.

EDELSTON R.S. (1864) *Entomologist* 2, 150.

FISHER R.A. (1933) On the evidence against the chemical induction of melanism in Lepidoptera. *Proc. R. Soc., B* 112, 407–416.

FORD E.B. (1937) Problems of heredity in the Lepidoptera. *Biol. Rev.* 12, 461–503.

FORD E.B. (1940) Genetic research in the Lepidoptera. *Ann. Eugen., Lond.* 10, 227–252.

GILBERT O.L. (1965) Lichens as indicators of air pollution in the Tyne valley. In G.T.Goodman, R.W.Edwards and J.M.Lambert (eds.). *Ecology and the Industrial Society*, pp. 35–47. Blackwell Scientific Publications, Oxford.

GILBERT O.L. (1968) Bryophytes as indicators of air pollution in the Tyne valley. *New Phytol.* 67, 15–30.

HARRISON J.W.H. (1920) Genetical studies in the moths of the geometrid genus *Oporabia (Oporinia)* with a special consideration of melanism in the Lepidoptera. *J. Genet.* 9, 195–280.

HARRISON J.W.H. (1935) The experimental induction of melanism and other effects in the geometrid moth, *Selenia bilunaria* Esp., *Proc. R. Soc. B.* 117, 78–92.

KETTLEWELL H.B.D. (1955a) Selection experiments on industrial melanism in the Lepidoptera. *Heredity* 9, 323–342.

KETTLEWELL H.B.D. (1955b) Recognition of the appropriate backgrounds by pale and black phases of Lepidoptera. *Nature* 175, 943.

KETTLEWELL H.B.D. (1956) Further selection experiments on industrial melanism in the Lepidoptera. *Heredity* 10, 287–301.

KETTLEWELL H.B.D. (1957) Industrial melanism in the Lepidoptera and its contribution to our knowledge of evolution. *Proc. R. Instn. Gt. Br.* 36, 1–14.

KETTLEWELL H.B.D. (1958a) A survey of the frequencies of *Biston betularia* (L.) Lep. and its melanic forms in Great Britain. *Heredity* 12, 51–72.

KETTLEWELL H.B.D. (1958b) Industrial melanism in the Lepidoptera and its contribution to our knowledge of evolution. *Proc. 10th Int. Congr. Ent.* (1956) 2, 831–841.

KETTLEWELL H.B.D. (1959) New aspects of the genetic control of industrial melanism in the Lepidoptera. *Nature* **183**, 918–921.

KETTLEWELL H.B.D. (1961) The phenomenon of industrial melanism in the Lepidoptera. *A. Rev. Ent.* **6**, 245–262.

KETTLEWELL H.B.D. (1965) A twelve-year survey of the frequencies of *Biston betularia* (L.), and its melanic forms in Great Britain. *Ent. Rec.* **77**, 195–218.

KETTLEWELL H.B.D. & BERRY R.J. (1961) The study of a cline. *Heredity* **16**, 403–414.

KETTLEWELL H.B.D. & BERRY R.J. (1969) Gene flow in a cline. *Heredity* **24**, 1–14.

KETTLEWELL H.B.D., BERRY R.J., CADBURY C.J. & PHILLIPS G.C. (1969) Differences in behaviour, dominance and survival within a cline. *Heredity* **24**, 15–25.

LEES D.R. (1968) Genetic control of the melanic form *insularia* of the Peppered Moth *Biston betularia* (L.). *Nature* **220**, 1249–1250.

MINISTRY OF TECHNOLOGY (1964–69) Warren Spring Laboratory, 'The investigation of air pollution'. National Survey Annual Summaries, Warren Spring Laboratory.

SARGENT T.D. (1966) Background selections of geometrid and noctuid moths. *Science* **154**, 1674–1675.

SARGENT T.D. (1969) Background selections of the pale and melanic forms of the cryptic moth *Phigalia titea* (Cramer). *Nature* **222**, 585–586.

SKYE E. (1968) Lichens and air pollution. *Acta phytogeogr. suec.* **52**, 1–123.

THOMSEN M. & LEMCHE H. (1933) Experimente zur Erzeilung eines erblichen melanismus bei dem Spanner *Selenia bilunaria* Esp. *Biol. Zbl.* **53**, 541–560.

TUTT J.W. (1890) Melanism and melanochroism in British Lepidoptera. *Ent. Rec.* **I**, 5–325.

TUTT J.W. (1891) Melanism and melanochroism in British Lepidoptera. *Ent. Rec.* **2**, 3–149.

TUTT J.W. (1899) *A natural history of the British Lepidoptera.* Vol. I, pp. 63–66, Swann Sonnenschein, London.

WHITE D.F.B. (1876) On melanochroism and leucochroism. *Entomologist's mon. Mag.*, **13**, 172–179.

9 ❋ Recessive Melanism in the Moth *Lasiocampa quercus* L. in Industrial and Non-Industrial Areas

H. B. D. KETTLEWELL, C. J. CADBURY

AND D. R. LEES

INTRODUCTION

The Oak Eggar Moth, *Lasiocampa quercus*, Lasiocampidae, was selected for a special study in the field of ecological genetics because melanism in this species exhibits several unusual features. First, melanic polymorphisms were maintained among imagines in several populations which appeared to be isolated from one another. They exist both in regions influenced by industrial pollution and in others which are essentially unpolluted.

Secondly, though dark forms of the imago have been recorded from over one hundred species of Lepidoptera as industrial melanics, (Kettlewell 1961a) and from about forty in rural areas (Kettlewell, 1961b; Kettlewell & Cadbury 1963), *L. quercus* provides one of the few known examples of dark larval phases which are genetically controlled. Others have been quoted by Cockayne (1927–28) and Harrison (1932).

Thirdly, certain melanic forms of *L. quercus* were known to be recessive in contrast with the other melanics which have been investigated, both industrial and non-industrial, where the inheritance of the melanic is dominant to the typical form (Kettlewell 1961a,b, Ford 1955a). Recessive melanism is largely confined to rare, semilethal morphs of species which are distasteful and normally exhibit aposematic patterns. Instances are found among the Arctiidae, Hypsidae and Zygaenidae. It has been suggested that melanism may have conferred little advantage to the survival of these species in the past (Kettlewell 1965a).

A different situation is one in which melanism is disadvantageous and has become recessive but the gene controlling it confers heterozygous advantage. *L. quercus*, together with species of *Selenia* and *Ennomos*, Selidosemidae (Ford 1955b) and *Lycia hirtaria* Clerck, Selidosemidae (Cadbury 1969) provide examples of recessive dark forms of moths palatable to birds which, though scarce, occur in wild populations at a

175

higher frequency than could be maintained by recurrent mutation. Instances of recessive melanism among larvae appear to be even more restricted, for other than in *Lasiocampa quercus* it has been to our knowledge only recorded in *S. bilunaria* Esp. (Cockayne 1927–28).

LIFE HISTORY

In Britain, there are two races which are isolated from one another more by their life cycles and habitat than their geographical ranges. Race *quercus* which predominates in southern England, has a one-year cycle with the imagines emerging in July and early August. The larvae feed chiefly on woody shrubs such as *Rubus fruticosus* L. (agg.), *Crataegus monogyna* Jacq. and, on sand dunes, *Salix repens* L. In appearance there seems to be no constant phenotypic characters which distinguish *quercus* from race *callunae* Palmer which in Britain has a two-year cycle with the imagines emerging in June or early July. This race is found in moorland regions where the larvae feed mainly on *Calluna vulgaris* (L.) Hull. Not only does race *callunae* have a two-year life-history but populations throughout the British Isles are synchronized so that the great majority of imagines emerge in odd calendar years. Fully grown larvae are usually only found in even years.

Lasiocampa quercus pairs in the afternoon between 1400 and 1700 hours. At this time of day the males fly actively in the sunlight and are attracted by a pheromone which the female releases when resting among low vegetation. Since this scent may be detected by males up to a mile (1·7 km) downwind, 'assembling' traps containing virgin females provide a satisfactory method of sampling this species.

THE MELANIC FORMS AND THEIR GENETICS

IMAGINES

Sexual dimorphism is well marked in *L. quercus*. The typical males have reddish-brown wings which are conspicuously banded with yellow on both surfaces. The heavy bodied females are yellower in colour with a pale outer margin of the wings. There are three named, phenotypically distinct dark forms. In f.*olivaceo-fasciata* Cockerell (1889), the band across the wings is greenish rather than the normal yellow colour. Form

PLATE 9.1.
*Lasiocampa
quercus.*
Typical form
(Yorkshire)
   (i) Male
  (iii) Female
Melanic f.*olivacea*
Tutt (Caithness)
  (ii) Male
  (iii) Female
($\times$ 2/3)
Female f.*olivacea*
at rest, Caithness
($\times$ 2/3)

[to face p. 176

olivacea Tutt (1902) is a more extreme melanic in which not only the band but the wings and body as a whole are suffused with greenish black in both sexes. Form *lurida* Cockayne (1951) resembles *olivacea* but the ground colour of the wings is more red than green. In even the darkest forms the central white spot on the fore-wings usually remains conspicuous. Melanic and typical morphs are illustrated in plate 9.1.

Melanics phenotypically similar to *olivacea* occur in populations of race *quercus* from Denmark, the Netherlands and from England on the Cheshire coast and have been shown to be recessive to the typical form (Hoffmeyer 1960, Lempke 1951, Gordon Smith 1954). Form *olivacea* from certain Yorkshire Moors where race *callunae* exists are also recessive (Cadbury 1969).Both this melanic and the less extreme one, f.*olivaceo-fasciata* have occurred together in Cheshire and south Lancashire, and in Yorkshire. In these areas the observed frequency of *olivaceo-fasciata* is considerably lower than that expected if it be the heterozygote. Breeding experiments from Yorkshire stock have shown that only a small proportion of heterozygotes are recognizable from the normal form. Moreover, they indicated that, in this region, the two melanics, *olivacea* and *olivaceo-fasciata* are controlled by one gene, the expression of which is altered by modifiers (Cadbury 1969). Since many of the earliest melanic *L. quercus* recorded from England resembled the less extreme form it is possible that darkening has proceeded as a result of the selection of such modifiers.

In Caithness, a melanic which is phenotypically similar to f.*olivacea* from Yorkshire and elsewhere and an intermediate phase resembling f.*olivaceo-fasciata* occur with the typical form. Though the results of breeding experiments did not exclude the possibility that such intermediates are heterozygotes, a proportion of known heterozygotes were indistinguishable in appearance from the yellow-banded typical insects. Moreover, with few exceptions, among samples both of wild caught males and imagines reared from wild larvae, there was a significant deficiency of the intermediate form compared with the expected heterozygote frequency. It would appear therefore that another mutant may be responsible for the Caithness *olivaceo-fasciata*. If one accepts this, then data from five broods in which both parents were of Caithness origin indicate that f.*olivacea* in northeast Scotland is recessive to the typical form, as it is in other populations.

It has been possible to cross the melanic forms from Yorkshire and Caithness. If the same gene were responsible for f.*olivacea* in the two regions one might expect all the progeny of such a cross to be dark. They were, however, either normal (the majority of males) or intermediate in colour (females) with no *olivacea*. Melanism in the two regions appears, therefore, to be controlled by different genes which are complementary in that they supply the other's deficiency. It is possible that the darkening in the females was due to an additive effect of the two sets of genes. If the mutants responsible for melanism are indeed at different loci, independent segregation of the dark forms should occur in the F_2 generation. Since the melanics are similar in appearance and the double homozygote is doubtfully distinguishable, a 9:7 ratio is to be expected. In two such broods there was no significant departure from this (Cadbury 1969).

LARVAE

In the first two instars, the larvae have what appears to be a disruptive pattern which is composed of black, blue and yellow markings. In later instars, the segments of the body are usually covered by long, reddish-brown hairs. In several of the *L. quercus* populations in which a melanic polymorphism is maintained among the imagines, dark larvae also occur. These include two forms; one in which the hairs are dark chocolate-brown and another with black hairs of a silky texture. They appear to be genetically distinct.

Gordon Smith (1954) showed that the blackish form of larva from Cheshire was recessive to the typical form. The situation appears to be similar with both melanic forms of larva on certain Yorkshire Moors (Cadbury 1969). We have recently shown, however, that the genes controlling melanic larvae in Cheshire and Yorkshire are different.

Breeding experiments have shown that both melanic forms of larva from Yorkshire continue to be inherited as recessives when the genes which control them are introduced into the Caithness gene complex, which has no previous experience of dark larvae, except presumably as rare mutants. Though segregation occurred much as expected in three F_2 broods, the parents of which were heterozygous for the gene responsible for black hair, the melanic larvae were not all black. In one brood five were black and one was dark chocolate, as were all eighteen in a second brood. In the third, all eleven were black. A possible explanation for this is that the expression for black larval coloration is modified in the hybrid Yorkshire × Caithness gene complex.

It would seem from the data presented in table 9.1, that on Rombalds

TABLE 9.1. The relationship between melanic larvae and imagines from Rombalds Moor, Yorkshire.

	Typical	Olivacea	
		Numbers	Per cent
(a) Gordon Smith (1956) from wild larvae			
From dark larvae*	26	7	21·2
From normal larvae	477	9	1·9
$\chi^2_{(1)} = 32 \cdot 58\ P < 0 \cdot 001$			
(b) Kettlewell (unpublished) from wild larvae			
From dark larvae*	7	15	68·2
From normal larvae	359	7	1·9
$\chi^2_{(1)} = 158 \cdot 2\ P < 0 \cdot 001$			
(c) W. Collinson (unpublished) from wild larvae			
From black larvae		16	
From dark chocolate larvae	194	42	17·6
From normal larvae	1767	34	1·9
For dark chocolate and normal larvae			
$\chi^2_{(1)} = 142 \cdot 7\ P < 0 \cdot 001$			
(d) Cadbury and Lees (unpublished) from wild larvae			
From dark chocolate larvae	1 *oliv.-fasc.*	3	
From normal larvae	41 + 1 *oliv.-fasc.*	0	
(e) Cadbury (1969) from 4 broods			
From dark chocolate larvae	13 + 10 *oliv.-fasc.*	13	31·7
From normal larvae	35 + 3 *oliv.-fasc.*	2	4·7
For normal imagines: *olivacea* $\chi^2_{(1)} = 7 \cdot 698$; $0 \cdot 01 > P > 0 \cdot 001$			
For normal imagines: *olivacea* and intermediate $\chi^2_{(1)} = 19 \cdot 35$; $0 \cdot 01 > P > 0 \cdot 001$			

* These dark larvae may have included black ones.

Moor in Yorkshire, selection has favoured linkage between both forms of dark larvae and dark imagines to a highly significant degree. In the instances in which larvae of the dark chocolate phase have been distinguished, 34 in a sample of 194 wild dark larvae and twelve from 36 bred ones developed into melanic moths. The fact that normal caterpillars, both those from Rombalds Moor and from broods which segregated for larval coloration, also produced f.*olivacea* imagines (1·9 per cent in two wild samples; 5 per cent in laboratory-reared broods—see table 9.1) suggests that the genes responsible for the dark chocolate form of larva and the melanic imago are situated on the same chromosome but are loosely linked. Indeed, it has been possible (D.R.L.) to estimate a cross-over value of 8·9 per cent for one brood and 25·6 per cent for another when intermediate imagines were classified as *olivacea* or 10·1 and 27·8 per cent if they were included among the typical phase.

The association between the black form of larva and f.*olivacea* appears to be particularly close. With respect to the population on Rombalds Moor, the nineteen imagines that have been successfully reared from such extreme melanic caterpillars have all been dark. Moreover, in 15 F_2 broods reared by Gordon Smith (1954) from *L.quercus* stock which originated from the Wirral in Cheshire, 217 typical imagines developed from normal brown larvae, while all the *olivacea*, 77 in number, metamorphosed from blackish caterpillars. It might appear that melanism in the two stages is controlled by one gene. Among Lepidoptera, however, very few examples are known of a gene which influences colour in more than one phase in metamorphosis, although instances have been cited by Cockayne (1927–28), Ford (1937) and Kettlewell (1959). Alternatively it is possible that through selection favouring close linkage two genes, one responsible for black larval colour and the other for f.*loivacea*, have come to act as a single super-gene. The small number of the extreme melanic form of caterpillar which has been reared may explain why no cross-over classes have been recorded.

On the evidence available, it would seem that the two melanic forms of larvae which occur in Yorkshire are controlled by different, non-allelic genes which are situated on the same chromosome as the gene responsible for dark coloration in the imago.

There have been a number of independent records that melanic larvae from Cheshire, Yorkshire and Germany form dark cocoons (Frings 1905, Bell 1909, Niepelt 1911, Gordon Smith 1954, Kettlewell 1959) but typical larvae have been found, in some instances to produce cocoons of similar coloration.

DISTRIBUTION OF THE MELANIC POLYMORPHISMS

In Britain, occasional melanic imagines have been recorded from a number of widely scattered localities (Cockayne 1952). There are three regions where the melanics of *L.quercus* are more frequent and are clearly maintained as a polymorphism. These are (a) the Cheshire and south Lancashire coast (race *quercus*); (b) certain moors in the West Riding of

Yorkshire (race *callunae*) and (c) northeast Scotland together with Orkney (race *callunae*).

THE CHESHIRE AND SOUTH LANCASHIRE COAST

Both melanic larvae and imagines were recorded with some regularity on the sandy coast of the Wirral from the beginning of the century until fairly recently. Indeed, the moths reared from the dark caterpillars have without exception been f.*olivacea* (Bell 1909, Gordon Smith 1954). The Wirral population appears to have been much reduced in recent years, probably as a result of housing and golf course development. In 1959 only small samples were obtained and these included but one melanic (a single f.*olivaceo-fasciata* among 45 males taken at Caldy).

Across the Mersey in south Lancashire, *L.quercus* is still abundant among the extensive sand dune systems between Crosby, to the north of Liverpool, and Southport. The species is particularly associated with *Salix repens* which is the food plant of the larva. F.*olivacea* was recorded from this area in both 1880 and 1890 (Tutt 1902) and the polymorphism continues to be maintained. In a sample of 557 males obtained with 'assembling' traps in July 1959, there were seven f.*olivaceo-fasciata* and two *olivacea*. On the same set of dunes but two miles (3.4 km) to the south, P.Harper (unpublished report for the Nature Conservancy) took a sample of 457 male *quercus* of which six were *olivaceo-fasciata* and one was *olivacea*. In both years, the phenotype frequency of the two melanic forms combined was 1·6 per cent.

THE WEST RIDING OF YORKSHIRE

In the last 30 years of the nineteenth century there were several records of melanic *L.quercus* imagines from moorland close to the industrial regions of Huddersfield and Halifax (Tutt 1902). The areas in question have since been much reduced by building and the fragments of heather moor that remain are frequently burnt. In the 1920's a number of both *olivaceo-fasciata* and *olivacea* were reared from dark larvae on the moors near Penistone at the centre of the Huddersfield, Sheffield and Manchester triangle (Hewson 1953). However, a sample of 163 males obtained in 1959 from this area was all typical.

Form *olivacea* has persisted as a polymorphism on Rombalds Moor since it was first recorded there in 1893 (Pierce 1894). This moor, which lies northwest of the industrial region, is isolated from other heathery tracts by Airedale to the south and by Wharfedale to the north. Until 1955 there was no information available for the phenotype frequency of melanic *L. quercus* on Rombalds Moor (Gordon Smith 1956). In five subsequent 'moth' years (1957, 1959, 1961, 1963, 1967) the male population was sampled. In four of these seasons the daily populations of this sex were estimated over an area of approximately 1 square mile (250 ha) by means of mark, release and recapture techniques similar to those developed by Dowdeswell, Fisher & Ford (1940). From these data, the mean daily population of males was calculated over the main emergence period. The results, both for frequencies and population size, have been presented in detail by Cadbury (1969) but are summarized in table 9.2.

TABLE 9.2. Phenotype frequencies of *Lasiocampa quercus* imagines on Rombalds Moor, Yorkshire.

| Year | Sample size | Olivacea | | Olivaceo-faciata | | Mean daily population $\male\male$† |
		Numbers	Per cent	Observed	Expected*	
1955	519	16	3·1	?	151	Unknown
1957	493	23	4·7	?	168	661 ± 194
1959	463	6	1·3	0	94	392 ± 259
1961	142	2	1·4	3	30	151 ± 90
1963	313	3	1·0	1	55	247 ± 91
1967	214	1	0·5	0	28	Unknown

1955 from random sample of wild larvae (Gordon Smith 1956).

1957 from random sample of 'assembled' male imagines (Kettlewell 1959).

Other years from random samples of 'assembled' male imagines collected by the authors.

* The expected number of f.*olivaceo-fasciata* was calculated from the frequency of f.*olivacea*.

† The mean daily population was calculated over the peak emergence period.

The relatively high frequency of f.*olivacea* in 1955 and 1957 (3·1 and 4·7 per cent respectively) was associated with a large population. In 1959 and the following three 'moth' years, when the population of *L. quercus* imagines was low compared with the previous two, there was a significant drop ($P < 0.01$) in the frequency of f.*olivacea* to less than 1·5 per cent. Nevertheless, as table 9.2 shows, the frequency of the heterozygote would have remained fairly high.

Dark larvae have been known from Rombalds Moor since 1900 (Hewson 1953). W.Collinson (personal communication) has obtained a series of larval samples which, apart from the period 1940–44, provides a continuous record dating back to 1932 (see table 9.3). These data reflect both fluctuations in the population size and also show a general tendency for an increase in the frequency of dark larvae (both dark chocolate and black forms combined). Indeed there is a significant positive regression of melanic larval frequency on time ($t = 3\cdot429$, $0\cdot01 > P > 0\cdot001$). The majority of these larvae were of the chocolate-brown form. Though the black phase has been recorded in most years, its frequency is less than 1 per cent (see table 9.3).

Information on melanic *L. quercus* is available for few of the other moors on the northwest side of the industrial West Riding. Farnhill,

TABLE 9.3. Phenotype frequencies of larvae collected on Rombalds Moor, Yorkshire.

Year	Sample size	Dark-chocolate	Black	Per cent melanic
1932–4	280	0	0	
1936	*ca.* 2,000	0	0	
1938	*ca.* 3,000	12	1	0·4
1940–4	No samples			
1946	*ca*, 1,000	17	2	1·9
1948	*ca.* 700	15	1	2·2
1950a	*ca.* 700	0	0	0·3
1950b	50	2	0	
1952	*ca.* 700	13	3	2·3
1954a	*ca.* 6,000	30	4	0·6
1954b	681	48	0	7·1
1956a	*ca.* 14,000	157	32	1·3
1956b	*ca.* 2,000	82	0	4·2
1958	619	46	0	7·5
1960a	27	2	0	8·6
1960b	8	1	0	
1962	30	1	1	6·3
1964	106	5	2	6·9
1966	118	3	0	2·5

1950b (Hewson 1953), 1954b (Gordon Smith 1956), 1956b (Kettlewell 1959), 1960b (Cadbury, unpubl.), 1966 (Cadbury and Lees, unpubl.) All other samples, W. Collinson (personal communication).

seven miles up the Aire Valley from Rombalds Moor was sampled in
1967 when 101 males were caught, all of which were typical. However, on
another isolated moor, W.Collinson (personal communication) recorded
15 f.*olivacea* and 20 f.*olivaceo-fasciata* among the 145 males which he
obtained there between 1956 and 1964. Moreover, of the 86 larvae found
on this moor, six have been dark chocolate brown.

NORTHEAST SCOTLAND

The existence of melanic *L.quercus* imagines in Caithness has been known
since about 1919 when local inhabitants began sending cocoons south to
English collectors (Cockayne 1949). The situation, however, received
little attention until 1960 when the present investigation began.

It soon became apparent that the Caithness population differs in two
aspects from the populations on the Cheshire and Lancashire coast and
in Yorkshire. First, dark imagines occur at a high frequency in certain
areas and secondly melanic larvae are unknown. Indeed 431 larvae
collected in three years, 1960, 1962 and 1964 were all of the normal red-
brown colour. We have sampled the imagines both by using 'assembling'
traps and rearing them from wild-collected larvae. The proportions of
f.*olivacea* in the two sexes have been similar in samples obtained from the
second source.

Between 1960 and 1969, samples of over 30 insects were obtained from
17 localities in Caithness and three in neighbouring Sutherland to the
southwest and west. A further seven small samples were taken in Sutherland and one on the island of Hoy, Orkney. The phenotype frequencies
of f.*olivacea* are given in table 9.4 and shown on the map (fig. 9.1).
Frequencies of over 50 per cent of this melanic occurred in four localities
situated on the lowland moors below 400 ft (122 m) in northeast
Caithness. Along the extreme north coast of the county such high frequencies were not maintained and on Dunnet Head, an isolated headland,
only 11.5 per cent were f.*olivacea*. Similarly, both to the south and west,
the frequency of the melanic declines. In the higher moorland areas along
the Caithness/Sutherland border not only does *L.quercus* appear to be
scarce but the frequency of *olivacea* is below 20 per cent (see map).

An intermediate form which resembles f.*olivaceo-fasciata* was
present in nearly all but the smallest samples. As has already been pointed
out, its frequency in most instances is significantly lower than would be
expected if such insects were heterozygous for the *olivacea* gene; it will

be seen from table 9.4 that not only the highest frequencies (up to 35·8 per cent) of the intermediate form but also the lowest (below 10 per cent) occurred in areas where the heterozygotes were expected to represent 34 to 50 per cent of the population. If it is a correct assumption that the two dark forms in Caithness are under independent genetic control, then in the more extreme melanic, f.*olivacea*, the expression of the gene responsible for the 'intermediate' will tend to be obscured.

With worn male imagines caught in 'assembling' traps it was sometimes difficult to determine whether an insect was of the normal or intermediate form. For this reason the most battered and bleached individuals were excluded from the samples. However, by examination of the band on the underside of the hindwing, which usually retained both its colour and scales, most of the imagines could be classified with respect to their phenotype. At only one locality (Keiss in 1967) has it been possible to compare the proportion of typical and intermediate phases in 'assembled' samples of males with those reared in the same season from larvae. In this instance the difference was not significant $(\chi^2_{(1)} = 1·640; P = 0·2)$.

Form *olivacea* has been observed on Hoy in Orkney (E.B.Ford, personal communication) but there were no melanics in a sample of 24 males taken on the island in 1969. Elsewhere in Scotland only occasional specimens of this dark form have been recorded though *L. quercus* is both widespread and common, thus suggesting that the melanic occurs at low frequencies.

MELANIC *LASIOCAMPA QUERCUS* ON THE EUROPEAN CONTINENT

GERMANY

The only other region where both melanic larvae and imagines have been recorded as well established polymorphisms is a tract of sandy heathland along the River Havel near Brandenburg (Frings 1905, Niepelt 1911). The imagines, which were reared from blackish caterpillars, resembled f.*olivacea* except that even the white forewing spot was dusky. In this German locality, therefore, there appears to have been a close association between the gene controlling melanism in the two developmental stages, which is similar to the situation in the two English regions where black larvae occur.

TABLE 9.4. Phenotype frequencies of *Lasiocampa quercus* in northeast Scotland.

Locality*	Year	Sample†	Typical numbers	Olivaceo-fasciata		Olivacea	
				Observed	Expected‡	Numbers	Per cent
1. Dunnet Head	1967	61	49	5	27	7	11·5
2. Mey Hill	1967	50	32	3	25	15	30·0
3. Warth Hill	1967 & 1969	264	136	28	125	100	37·8
4. Keiss	1965	126	12	24	33	90	71·4
	1967	114 (♂ & ♀)	40	21	49	53	46·5
5. Black Hill	1967 & 1969	251	97	15	96	139	55·4
6. North Watten Moss	1967	37	6	7	10	24	68·6
7. Loch Olginey	1967	54	31	1	25	22	40·7
	1965	(10)	—	3	—	7	—
8. Mybster	1967	33	11	1	11	21	63·6
	1967 & 1969	109	33	11	38	65	59·6
9. Loch More	1969	37 (♂ & ♀)	14	—	12	23	62·2
10. Badlibster	1967	50	34	4	25	12	24·0
11. Tannach	1967	63	24	4	24	35	55·6
	1963	93 (♂ & ♀)	12	34	38	47	50·5
	1965	53 (♂ & ♀)	8	19	22	26	49·0
12. Camster	1965, 1967 & 1969	670	278	124	312	268	40·0
13. Rumster	1965 & 1967	78	27	21	37	30	38·4
14. Dunbeath	1967	51	16	10	21	25	49·0
15. Berriedale	1965	53	22	14	26	17	32·0
16. Golspie, Suth.	1965	33	24	4	16	5	15·1
17. Reay	1969	56	54	2	—	—	—
	1967	(26)	22	3	—	1	—

18. Trantlemore, Suth.	1967 & 1969	(32)	28	1	—	3	
19. Forsinard, Suth.	1967 & 1969	(19)	17	—	—	2	
20. Kinbrace	1969	(31)	31	—	—		
21. Kildonan, Suth.	1969	49	48	—	12	1	
22. Hoy, Orkney	1969	(24)	24				2·0

* Five samples were from localities in Sutherland, one in Orkney and the remainder in Caithness (see fig. 9.1).

† Unless indicated otherwise all samples were males captured in 'assembling' traps. The samples containing both sexes were reared from wild larvae. Where there is no heterogeneity between different years these samples have been combined.

‡ Expected number of heterozygotes calculated from frequency of f.*olivacea*.

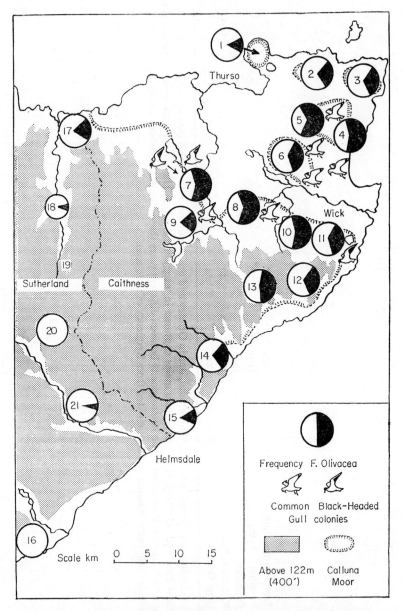

FIGURE 9.1 Phenotype frequencies of f.*olivacea* of *Lasiocampa quercus* L. with relation to gull colonies and topography in northeast Scotland. Reference is made to the numbered sample localities in table 9.4.

THE NETHERLANDS AND DENMARK

Dark imagines have been recorded on the coast of the Netherlands on several occasions (Lempke 1960) and more regularly from West Jutland in Denmark (Hoffmeyer 1960). Though the melanics occur mainly on the coastal sand dunes, as on the Cheshire and south Lancashire coast, they are also known from several inland localities.

With respect to the appearance of the melanics, the situation is comparable with that in Caithness. In both these regions an intermediate phase occurs in addition to f.*olivacea* and it is apparently frequent in West Jutland. Moreover the f.*olivacea* bred by Lempke (1951) in the Netherlands developed from normal larvae, and according to Hoffmeyer (1960) dark *quercus* larvae have not been recorded in Denmark.

OBSERVATIONS AND EXPERIMENTS ON PREDATION BY GULLS

After emerging from the cocoon, an *L.quercus* imago rests with wings folded tent-like over the body on the stems of a low shrub which on moorland is usually heather and on sand dunes is *Salix repens*. Except when 'assembling' to females for a fairly short period of the day males remain at rest. In overcast, cool weather they may be inactive all day. The females fly only briefly at dusk when scattering fertilized eggs. The crypsis of the insects, both at rest and on the wing, is therefore important for their survival since *L.quercus* is palatable to a variety of birds which include, the Merlin, *Falco columbarius* L., the Kestrel, *F. tinnunculus* L., the Swift, *Apus apus* L., the Skylark, *Alauda arvensis* L., the Meadow Pipit, *Anthus pratensis* L., the Wheatear, *Oenanthe oenanthe* L., the Common Gull, *Larus canus* L., and the Black-headed Gull, *L. ridibundus* L.

Of these predators the most important in the areas of Britain where *Lasiocampa quercus* maintains melanic polymorphisms are the Black-headed Gull and Common Gull. These two species have both increased the size of their breeding populations in Britain since the end of the nineteenth century (Parslow 1967) and the Black-headed Gull in particular has increased the number of its inland colonies in recent years. In Caithness both species breed commonly on the large tracts of poorly drained moorland while in the West Riding of Yorkshire and in Lancashire, Black-headed Gull colonies are often associated with water reservoirs on the high moors of the Pennines. Both gulls are largely insectivorous in summer (Sparck 1950), and appear to be adapted to take advantage of any local abundance of food which suddenly becomes available.

In 1955 and 1957, when *L. quercus* imagines were particularly abundant on Rombalds Moor, many Black-headed Gulls congregated to feed on the moths (Gordon Smith 1956, Kettlewell 1959). The majority of Gulls probably came from a breeding colony of 100 to 150 pairs on a nearby moorland tarn. As they quartered the moor they were observed constantly dropping into the heather to gorge themselves on the large female moths. Such was the level of predation that of the 44 normal yellow females and 2 f.*olivacea* which were released (by H.B.D.K.) on one occasion only a single melanic remained uneaten after 2 hours. These observations indicate the potential of gulls as selective agents in maintaining a melanic polymorphism among *L. quercus* imagines.

Further observations on the predation of imagines by gulls were made (by C.J.C. and H.B.D.K.) in Caithness in 1965 when, in three localities, up to eight Black-headed Gulls and two Common Gulls were observed on many occasions hawking male *L. quercus* as they flew upwind to virgin females in assembling traps. Of the two species, the Black-headed Gull was the more agile on the wing and therefore more successful at snapping up its fast flying prey. Even so a number of male *quercus* eluded capture. The number of instances in which the phenotype could be identified was too few to determine whether the aerial predation was selective, as has been suggested by Kettlewell (1957) with respect to *Alcis repandata* L., Selidosemidae.

It was not until Loch Olginey, with its breeding colony of *c.* 300 pairs of Black-headed Gulls and 12 pairs of Common Gulls, had been visited that gulls were observed picking up female *L. quercus* from the heather, as they had done on Rombalds Moor. This suggested that the area might be suitable for predation experiments.

In a series of tests equal numbers of typical and melanic f.*olivacea* were put out in natural resting positions among the heather. Moths of the intermediate form were excluded. Since males tend to flutter and even fly when released, recently killed specimens were used. Each of these was supported in an almost vertical position on the heather by a fine pin thrust through a stem so that the point impaled the moth from the under-side. Preliminary tests showed that freshly dead *quercus* were as accept-able to gulls as living ones. The females are more quiescent and it was therefore possible to include both living and dead specimens in the tests. The moths were put out 3 yards apart and 1 yard from a dark string line laid out among the heather. The two phenotypes were alternated, and the position of each typical *quercus* was marked by a red poultry ring on the string while a blue indicated f.*olivacea*. The rings were covered so as to be hidden from the gulls. It was possible to make observations from a car less than 30 yards away.

Up to 25 Black-headed and 8 Common Gulls were initially attracted to the site of the experiment by a small quantity of bait thrown out on the nearby grass (the heather was carefully avoided). These birds soon turned to quartering the 'release' area. The moths were in vertical resting positions with the upper surface of the wings exposed to the wind into which the gulls tended to fly when searching for their prey. In many instances therefore it would have been the underside of the moth which was visible to the predators. This may explain why there is a cryptic pattern on both surfaces of the *quercus* wings. In this situation, compared with aerial predation, the Common Gulls appeared to be more efficient at discovering the moths than the Black-headeds. With a longer neck they were not only able to look downwards but also backwards beneath the breast. Even though Common Gulls were out-numbered by three to one, on two occasions they picked up three out of the four moths which were seen to be taken.

Under these experimental conditions the rate of elimination of the typical form was higher than that of f.*olivacea* by 83·3 per cent in the male and 89·5 in the female (table 9.5). Both in the tests using male *quercus* and those involving females there was a highly significant difference ($P < 0.001$) in the proportion of the two forms taken. By comparison,

TABLE 9.5. Selective elimination of *L. quercus* imagines during experiments at Loch Olginey, Caithness. 24 June–6 July 1965.

	Male (all dead)		Female (live and dead)	
	Taken	Left	Taken	Left
Typical	19	3	17	2
Olivacea	4	18	0	20
	$\chi^2_{(1)} = 17.86, P < 0.001$		$\chi^2_{(1)} = 28.18, P < 0.001$	
Differential rate of elimination	83·3 per cent		89·5 per cent	

differential predation rates which favour the melanic form have been estimated as being over 70 per cent in *Biston betularia*, Selidosemidae, near Birmingham (Kettlewell 1956) and in the order of 13–14 per cent in *Lycia hirtaria*, Selidosemidae, in London (Cadbury 1969). In *B. betularia* near Liverpool, Clarke & Sheppard (1966) obtained similar results to Kettlewell's Birmingham figures.

In three seasons (1965, 1967 and 1969) only small samples of imagines were obtained near the site of the predation experiment at Loch Olginey but in a total of 45 for the three years, 30 were f.*olivacea* (66·7 per cent); this is one of the highest frequencies of the melanic in Caithness.

Although gulls seem to be the most important predators of *L. quercus* it has been possible to observe the effects of predation by one other predator, on Rombalds Moor, Yorkshire. There, in 1957, a pair of Merlins was observed capturing large numbers of male imagines (Kettlewell 1959). These falcons removed the appendages from their prey on certain boulders. At the plucking sites 60 wings of *L. quercus* were found, and these were all of the typical form. No record was kept of the number of imagines from which these wings came but, on the assumption that they came from a minimum of 30 and a maximum of 60 imagines, it is possible to calculate, for these two extreme possibilities, whether the absence of dark wings is due to chance. $\chi^2_{(1)}$ values of 1·46 ($P > 0·1$) and 2·92 ($0·1 > P > 0·05$) are then obtained. Both of these are below the formal level of significant departure from chance.

The highly toxic hairs of *L. quercus* larvae provide protection against the attacks of most birds with the exception of the Cuckoo, *Cuculus canorus* L. Though the caterpillars in the last three instars usually feed in the late afternoon and after dark, they may expose themselves at other times after rain. When not feeding the larvae are frequently to be found stretched along the woody stems of their food plant. It is possible that melanism may be advantageous to larvae exposed to predation by Cuckoos in situations where the heather (or *Salix repens* on sand dunes) is blackened by soot. In Caithness however, where dark caterpillars are unknown, the physiological advantages associated with the gene responsible for melanism at this stage may outweigh selection pressures which favour dark coloration. On Rombalds Moor, three ichneumonids and a tachinid fly have been recorded as parasites of the larva (Cadbury 1969) but there is no evidence that these influence the polymorphism of either the larva, or the imago.

DISCUSSION

In addition to *L. quercus*, there are a few other species of Lepidoptera in Britain which have phenotypically similar melanics existing in widely separated and dissimilar environments. The chocolate-brown f.*alopecurus* Esp. of *Apamea crenata* Hufn., Caradrinidae, which occurs at a high frequency in industrial regions, is also found in Caithness (at a frequency of about 50 per cent—J. Rosie, personal communication), and Shetland (9 in a sample of 28). *Entephria caesiata* Schiff. f.*atrata* Lange, Hydriomenidae, not only occurs in the Scottish Highlands and in Shetland, but also in the polluted environment around Paisley and Glasgow (Kettlewell & Cadbury 1963). *Phigalia pedaria* Fab. Selidosemidae, provides another example of an industrial melanic, f.*monacharia* Staud, which has a phenotypically similar counterpart in rural areas including the Scottish Highlands and west Wales (see Chap. 8). In this species it is likely that the same gene controls the melanics in both environments.

In many parts of England which are influenced by industrial pollution, *Alcis repandata* f.*nigra* Tutt has largely replaced the pale phase. A similar form f.*nigricata* Fuchs, also occurs very locally at a frequency of up to 10 per cent in relict pine woodland in Central Scotland. It has been shown to be dominant to the pale form of that region (Williams 1949, Kettlewell 1957, Cadbury 1969). It is not known, however, whether the same gene is responsible for this non-industrial melanic and the industrial one which is also inherited as a dominant (Walther 1927).

In contrast with the examples cited above, the melanic forms of the imago of *L. quercus* are inherited as recessives, and in the very different environments of Yorkshire and Caithness they have been shown to be controlled by different genes. The fact that the melanic forms of the imago are recessive suggests that dark coloration may not have been advantageous until fairly recently. It might be expected that if crypsis is important in the survival of this species and *olivacea* is more cryptic than the typical form, at least in some circumstances, then there would be selection for any modifying genes darkening the heterozygote. The time taken to achieve dominance may largely depend upon the selection pressures involved. However, the adjustment of the gene responsible for *olivacea* in the complex may be hindered by adverse physiological effects of the modifiers. Nevertheless, the frequencies of the genes responsible for melanism in Yorkshire populations have been as high as 27 per cent for the melanic larvae and up to 20 per cent for the melanic imagines, while in Caithness the frequency of the *olivacea* gene exceeds 80 per cent in some areas. It would seem therefore that there has been considerable selection for such genes. Moreover, the selective advantage of inter-acting genes has resulted in linkage between those controlling dark coloration in two stages of metamorphosis, not only in Yorkshire, but apparently also on the Cheshire and south Lancashire coast and in Germany.

The broad band of yellow across the dark brown forewings of the normal male *L. quercus* provides a disruptive pattern which tends to render this sex less conspicuous to the human eye than the yellow females. This has been tested under natural conditions by placing moths in their resting position on the vegetation of Rombalds Moor, and in Caithness; the distances at which they ceased to be visible to the human eye have then been recorded (see table 9.6). The crypsis of f.*olivacea* is enhanced, at least in some areas, in Caithness where the heather is growing on a blanket of dark peat and on such Yorkshire moors as Rombalds where the heather stems are to some extent blackened by aerial pollution from adjacent towns. However, on Rombalds Moor, and on the dunes of

TABLE 9.6. The visibility of *Lasiocampa quercus* at rest on heather. The observations were made to test visibility against certain backgrounds. The distances given are the maximum at which the moth was still visible.

Sex	Typical (m)	Intermediate (from Caithness) (m)	f.*olivacea* (m)	Observer
Rombalds Moor, Yorkshire—Polluted heather, clear sky				
Male	49		15	H.B.D.K.
			11	
			10	
	30		11	C.J.C.
	26		11	
Female	57		15	H.B.D.K.
	51			
	52	31	21	C.J.C.
	50	26	18	
Caithness—Unpolluted heather on damp peat, overcast sky				
Male	11		5	H.B.D.K.
	9		5	C.J.C.
Female	40	21	14	

Lancashire and Cheshire which are also subject to air pollution, the phenotype frequency of f.*olivacea* has not risen above 5 per cent. It seems therefore that, unless selection of the melanic homozygote is frequency dependent in these areas, and there is some advantage in being rare, then the commoner heterozygote, which is phenotypically normal, is being selected for some non-visual attribute. The gene frequency must have reached a relatively high level, due to other selective agencies, before there were enough homozygotes for visual selection to be effective.

In contrast, the melanic forms of about 40 other species of Lepidoptera occur at high frequencies in the vicinity of Rombalds Moor (Mansbridge 1893, Porritt 1906, Kettlewell 1956b). Among them are several moorland species which normally rest upon the ground or upon rocks, and walls. The inheritance of these melanics, where known, is dominant, while in *L.quercus* it is recessive; but this alone cannot account for the frequency of f.*olivacea* never having to our knowledge, exceeded 5 per cent in these areas.

In comparison with the low frequencies of *olivacea* in Yorkshire and Lancashire, phenotype frequencies as high as 50–70 per cent are found in several localities in the rural and unpolluted far northeast of Scotland. There, in one locality, at Loch Olginey, experiments on predation by gulls have shown the degree of the selective elimination of the typical form of *quercus*.

High *olivacea* frequencies are found in Caithness at localities where the heather is mainly confined to growing on the banks of peat cuttings. It is thought that these cuttings provide a dark background against which the moths rest. This, together with the survival values of the melanic form estimated from the experiments at Loch Olginey, suggest that selective predation is an important factor in maintaining the polymorphism. However, rapid declines in the percentage of *olivacea* westwards and southwards of the areas of highest frequency (see fig. 9.1) do occur, and the balance of selective factors and gene flow that maintain them are of the greatest interest.

An example of such a decline is the 50 per cent drop in frequency from about 60 in the Keiss/Black Hill area (sites 4 and 5) to 11 at Dunnet Head (site 1) some 12 miles to the northwest. It seems unlikely that large changes such as these, over short distances, can be entirely attributable to changes in predation intensity or in the crypsis of the imagines at rest on their background of heather and peat. However, information is so far lacking on the relative survival of the two phenotypes in areas of northeast Scotland where the frequency of *olivacea* is below 10 per cent. General observations indicate that the darkness of the background soils and the vegetation of the moorland in Caithness and Sutherland do not vary a great deal, although cutting of the peat for fuel is much commoner in the more densely populated parts of central Caithness, and this may make backgrounds darker in such areas.

Predation of *quercus* by gulls may certainly be heavy in at least some of the localities of low *olivacea* frequency in Caithness. For instance, in the neighbourhood of Golspie in Sutherland, where no *olivacea* and two intermediates were found in a sample of 56, there are several large Black-headed Gull colonies. Heavy predation of imagines, both at rest and in flight, was observed there on two occasions in 1969 (D.R.L.), but the phenotypes of the moths taken could not be seen.

The effect of gene flow on the steepness of clines in *olivacea* frequency is similarly uncertain. Male imagines are fast, strong fliers and will 'assemble' to females over distances as great as 1 mile (1·7 km) in an afternoon. However, our observations indicate that they will not fly over arable or pasture land; they show a distinct preference for heather moorland and will turn back in flight at its boundary. The large areas of farmland in Caithness, which break up the moorland into comparatively small and isolated tracts, may thus constitute considerable barriers to gene flow and therefore enhance the steepness of the clines. However, Kettlewell & Berry (1969) have shown in the caradrinid moth *Amathes glareosa* in Shetland, that it may be that on either side of a barrier to gene flow in a cline, similar frequencies are found. In the areas of low *olivacea* frequency to the south and west of the region studied in Caithness, the moorland is more continuous and little interrupted by farmland. It seems that gene flow there may be relatively unimpeded.

It is unlikely, therefore, that the melanic polymorphism of *L. quercus* in northeast Scotland is maintained by selective predation alone, although this is clearly important in some situations. In addition, any cryptic advantage that *olivacea* may have is only applicable to a short phase of three to four weeks in a total life cycle of two years. Other counteracting factors probably influence the selective advantage of one genotype over another when the complete cycle is taken into account. Nevertheless, the magnitude of the selective disadvantage of the typical imagines when at rest, as shown by the predation experiment near the gull colony at Loch Olginey, indicates that few may survive long enough to reproduce, at least in some localities. In species other than *L. quercus* it has been shown that physiological and behavioural differences have played a part in the maintenance of melanic polymorphisms in the Lepidoptera (Ford 1940, Kettlewell 1957, 1958) and in ladybirds (Lusis 1961). We have evidence that this is also true of *L. quercus*.

Selection pressures may vary considerably from year to year and their influence on different genotypes may extend to more than one phase in the life cycle. Differences in the frequency of f.*olivacea* from one season to the next could reflect such changes. It is possible that homozygous *olivacea* are physiologically less viable than the normal homozygote and the heterozygote in the various developmental stages. Under optimum conditions, which for example could have existed on Rombalds Moor in Yorkshire in 1955 and 1957, when the population of *L. quercus* imagines was high (see table 2), the selective balance might be expected to swing in favour of the melanic as a result of its superior crypsis. With the onset of less favourable conditions and increasing intensity of non-visual selection, however, the disadvantageous component in the cumulative effects of the *olivacea* gene may have been accentuated. This could provide an explanation for the significant decrease ($\chi^2_{(1)} = 8 \cdot 106$; $0 \cdot 01 > P > 0 \cdot 001$) in the phenotype frequency between 1957 ($4 \cdot 7 \pm 1 \cdot 0$ per cent) and 1959 ($1 \cdot 3 \pm 0 \cdot 5$ per cent), see table 9.2.

In Caithness, among the samples of male imagines taken by means of 'assembling' traps, it has been possible to compare the phenotype frequencies from one 'moth' year to the next at five localities (see table 9.4). There was neither a significant difference between those at Tannach over three seasons (1965, 1967 and 1969), nor over two at Camster (1965 and 1967) and at Mybster, Warth Hill and Keiss (1967 and 1969). At Keiss, however, the frequency of *olivacea* in 1965 ($71 \cdot 4$ per cent) was significantly different from that in both 1967 ($57 \cdot 9$ per cent) and 1969 ($52 \cdot 2$ per cent) $- \chi^2_{(1)} = 4 \cdot 745$, $0 \cdot 05 > P > 0 \cdot 02$; and $\chi^2_{(1)} = 8 \cdot 448$, $0 \cdot 01 > P > 0 \cdot 001$, respectively.

At three sites in Caithness (Tannach and Keiss in 1967, and Mybster in 1969) it was also possible to compare phenotype frequencies in samples of male imagines with those of imagines reared from larvae collected when nearly full grown in the wild state (see table 9.4). Since the differences were not significant there was no evidence from these data of differential survival (or a behavioural difference) between homozygous *olivacea* and the other genotypes (typical and intermediate phases were combined for the purpose of contingency tests) in the later developmental stages.

Although *olivacea* in northeast Scotland and in Yorkshire are controlled by different genes, they are phenotypically similar. However, there is no reason to assume that these genes are similar in their other effects; for example in their influence on viability, rate of development, mating and 'assembling' preferences.

In an attempt to determine whether a behavioural difference exists between the *olivacea* and typical phenotypes in Caithness, a mark, release and recapture experiment was conducted in an area where the two phenotypes were in approximate equality (Tannach site 11). In this experiment, although there was a deficiency in the number of *olivacea* recaptured, it was not significant ($\chi^2_{(1)} = 3 \cdot 553$, $0 \cdot 10 > P > 0 \cdot 05$). Previously, differences in flight activity have been shown between those populations containing high frequencies of melanics and those containing few melanics in *Amathes glareosa*, in Shetland (Kettlewell 1961c, Kettlewell, Berry, Cadbury & Phillips 1969). However, these flight habit differences were not between melanic and typical forms as such, but rather between populations.

The melanic polymorphisms of *L. quercus* have many points of interest. The recessive inheritance of both larval and imaginal melanics is unusual inasmuch as these are maintained as polymorphisms in several quite distinct environments; melanic imagines occur, for instance, in polluted industrial areas of Yorkshire and, at a much higher frequency, in nonpolluted environments in northeast Scotland. Melanism in the larva and in the imago is controlled by separate genes which are linked and, in the future, a profitable field of study will be the interaction of these genes. In addition it is hoped to investigate some of their non-visual attributes.

SUMMARY

1. Larval and imaginal melanic polymorphisms occur in the moth *L.quercus* L., Lep., Lasiocampidae.

2. In Britain, *olivacea*, the imaginal melanic, is established in
(a) northeast Scotland,
(b) the West Riding of Yorkshire,
(c) the Cheshire and south Lancashire coast.
In (b) and (c), both regions influenced by air pollution, phenotype frequencies have not exceeded 5 per cent; in (a), a non-industrial area, frequencies of more than 50 per cent are recorded. Larval melanics occur in (b) and (c) but not in (a). On the continent of Europe *olivacea*

occurs in Denmark, the Netherlands and in Germany, where larval melanics also occur.

3. Form *olivacea* and both larval melanics are recessive to typical. Yorkshire intermediate imagines appear to be modifications of the expression of the allele controlling *olivacea*; in Caithness it seems that they are controlled by another gene.

4. The gene for dark chocolate larva is loosely linked to the *olivacea* locus while that for the black larva has come to act as a super-gene with it. The genes for dark larvae in Lancashire and Yorkshire have been shown to be different.

5. A significant increase in the frequency of dark larvae on Rombalds Moor, Yorkshire occurred between 1933 and 1966.

6. Selective predation is shown to be important in maintaining the *olivacea* polymorphism in northeast Scotland but on Rombalds Moor, Yorkshire, where melanic homozygotes occur at low frequency, selection of the heterozygotes maintains the polymorphism, but at certain times selection of the homozygote may also be important.

ACKNOWLEDGEMENTS

We wish to thank the following for their collaboration: Dr R.J.Berry, Dr P.Harper, Dr G.Howard and Mr A.Shapiro, who all helped so much with our field work in Caithness and Lancashire; Mr W.Collinson helped us in the field in Yorkshire and kindly allowed us to use some of his data.

We are particularly grateful to Miss G.Brooks for her care and skill with breeding stock in the laboratory.

We acknowledge valuable help and suggestions, at various stages in the course of this work, from Professor E.B.Ford F.R.S., Professor P.M. Sheppard F.R.S., Dr L.M.Cook, Dr E.R.Creed and Dr J.R.G.Turner.

We received financial support from the Nuffield Foundation and the Science Research Council.

REFERENCES

BELL W. (1909) Parallel variation in larvae and imagines of *Lasiocampa quercus*. *Ent. Rec.* **21**, 45.

CADBURY C.J. (1969) Melanism in moths with special reference to selective predation by birds. D.Phil. thesis, Oxford.

CLARKE C.A. & SHEPPARD P.M. (1966) A local survey of the distribution of industrial melanic forms of the moth *Biston betularia* and estimates of the selective value of these in an industrial environment. *Proc. R. Soc. B* **165**, 424–439.

COCKAYNE E.A. (1927–8) Annual Address—Larval variation. *Proc. R. ent. Soc. Lond.* **7**, 51–52.

COCKAYNE E.A. (1949) Leonard Woods Newman. *Ent. Rec.* **61**, 80–81.

COCKAYNE E.A. (1951) Aberrations of British Macrolepidoptera. *Entomologist* **84**, 241–245.

COCKAYNE E.A. (1953) The problem of *Lasiocampa quercus* Linnaeus ab. *olivaceo-fasciata* Cockerell ab. *olivacea* Tutt and melanic larvae. *Ent. Rec.* **64**, 306–309.

COCKERELL T.D.A. (1889) On the variation of insects. *Entomologist* **22**, 1–6.

DOWDESWELL W.H., FISHER R.A. & FORD E.B. (1940) The quantitative study of populations of the Lepidoptera. *Ann. Eugen. Lond.* **10**, 123–136.

FORD E.B. (1937) Problems of heredity in the Lepidoptera. *Biol. Rev.* **12**, 461–503.

FORD E.B. (1940) Genetic Research in Lepidoptera. *Ann. Eugen. Lond.* **10**, 227–252.

FORD E.B. (1955a) Polymorphism and taxonomy. *Heredity* **9**, 255–264.

FORD E.B. (1955b) *Moths.* Collins, London.

FRINGS C. (1905) *Las.quercus* L. ab. nov. *paradoxa* Frgs. *Soc. ent.* **20**, 89–90.

GORDON SMITH S. (1954) Experiments with a strain of *Lasiocampa quercus* ab, *olivaceo-fasciata* Cockerell with descriptions of two new aberrations. *Entomologist* **87**, 225–228.

GORDON SMITH S. (1956) Experiments with a strain of *Lasiocampa quercus* ab. *olivaceo-fasciata*. Cockerell from Cheshire and *Lasiocampa quercus* race *callunae*. Palmer from Yorkshire. (Lep. Lasiocampidae). *Entomologist* **89**, 137–138.

HARRISON J.W.H. (1932) The recent development of melanism in the larvae of certain species of Lepidoptera, with an account of its inheritance in *Selenia bilunaria* Esp. *Proc. R. Soc. B* **III**, 188–200.

HEWSON F. (1953) 'Black' larvae of *Lasiocampa quercus* L. in Yorkshire. *Ent. Rec.* **65**, 1–2.

HOFFMEYER S. (1960) De Danska Spindera. Universitets forlaget. Aarhus.

KETTLEWELL H.B.D. (1956) Further selection experiments on industrial melanism in Lepidoptera. *Heredity* **10**, 323–342.

KETTLEWELL H.B.D. (1957) Industrial melanism in the Lepidoptera and its contribution to our knowledge of evolution. *Proc. R. Instn. Gt. Br.* **36**, 1–14.

KETTLEWELL H.B.D. (1958) A survey of the frequencies of *Biston betularia* (L.) (Lep.) and its melanic forms in Great Britain. *Heredity*, **12**, 51–72.

KETTLEWELL H.B.D. (1959) New aspects of the genetic control of industrial melanism in the Lepidoptera. *Nature* **183**, 918–921.

KETTLEWELL H.B.D. (1961a) The phenomenon of industrial melanism in Lepidoptera. *Ann. Rev. Ent.* **6**, 245–262.

KETTLEWELL H.B.D. (1961b) Geographical melanism in the Lepidoptera of Shetland. *Heredity* **16**, 393–402.

KETTLEWELL H.B.D. (1961c) Selection experiments in *Amathes glareosa* Esp. (Lepidoptera). *Heredity* **16**, 415–434.

KETTLEWELL H.B.D. (1965a) Insect survival and selection for pattern. *Science* **148**, 1290–1296.

KETTLEWELL H.B.D. (1965b) A 12 year survey of the frequencies of *Biston betularia* L. (Lep.) and its melanic forms in Great Britain. *Ent. Rec.* **77**, 195–218.

KETTLEWELL H.B.D. & BERRY R.J. Gene flow in a cline. *Heredity* **24**, 1–14.

KETTLEWELL H.B.D., BERRY R.J., CADBURY C.J. & PHILLIPS G.C. (1969) Differences in behaviour, dominance and survival within a cline. *Heredity* **24**, 15–25.

KETTLEWELL H.B.D. & CADBURY C.J. (1963) Investigations on the origins of non-industrial melanism. *Ent. Rec.* **75**, 149–160.

LEMPKE B.J. (1951) A contribution to the genetics of *Lasiocampa quercus* L. *Ent. Rec.* **63**, 200–203.

LEMPKE B.J. (1960) Catalogus der Nederlanse Macrolepidoptera. *Tisdschr. Ent.* **103**, 145–215.

LUSIS J.J. (1961) On the biological meaning of colour polymorphism of Lady-beetle *Adalia bipunctata. Latvijas Entomologs* **4**, 3–29.

MANSBRIDGE W. (1893) Melanism in Yorkshire Lepidoptera. *Ent. Rec.* **4**, 110–111.

NIEPELT W. (1911) Zur Biologie von *Lasiocampa quercus* ab *olivaceo-fasciata* Cock. *Int. ent. Seit.* **23**, 185.

PARSLOW J.L.F. (1967) Changes in status among breeding birds in Britain and Ireland (part 3). *Br. Birds* **60**, 177–202.

PIERCE F.N. (1894) Lancashire and Cheshire Entomological Society. *Entomologist* **27**, 359.

PORRITT G.T. (1906) Melanism in Yorkshire Lepidoptera. *Rep. Br. Ass. Advmt Sci.* 316–325.

SPARCK R. (1950) Food of the north European gulls. *Proc. X International Ornithological Congress*, pp. 588–591.

TUTT J.W. (1902) *The natural history of the British Lepidoptera.* Vol. 3. Swann Sonnenschein, London.

WALTHER H. (1927) Veber Melanismus, *Dt. ent. Z. Iris (Dresden)* **41**, 32–49.

WILLIAMS H.B. (1949) Melanic *Boarmia repandata* at Rannoch, Perthshire. *Ent. Rec.* **61**, 5.

10 ❀ Speculations about Mimicry with Henry Ford

MIRIAM ROTHSCHILD

> One is apt to forget where one's ideas originally came from.
> Karl Jordan
> (1906 *Novit. Zool.* **13**, 431)

INTRODUCTION

When I first moved to Oxford I was a flea specialist and according to the *Daily Express* had published half a million excruciatingly boring words devoted to this subject. Mercifully E.B.Ford does not read newspapers, so he was probably unaware of this fact. On the first occasion on which he shared a meal with my family, I noticed that he took it for granted that Charles Lane, then aged 7, kept live moths in his hair, and he pointed out —a fact which had escaped me until that moment—that the Frosted Orange (*Gortyna flavago* Scheff.) which had just surfaced among these carroty curls was clearly preadapted to this environment and, indeed, in that particular situation provided a perfect example of crypsis. After the meal it was suggested that catching the Meadow Browns (*Maniola jurtina* L.) disporting themselves in the field-like lawn in front of the house might prove a distracting occupation for both children and adults. E.B.Ford showed no surprise when seven orange cats appeared as if by magic in the garden and deployed themselves silently along the edge of the lawn and, advancing into the long grass, assisted us by flushing out the butterflies. By the end of the afternoon the fleas hopped into the background of my mind and I began to think about warning coloration, butterflies, mimicry and evolution. I had been catching insects in that field accompanied by copper-haired children and copper coloured cats all that summer, and ever since I could remember, Karl Jordan, backed up by a collection of $2\frac{1}{2}$ million set specimens and an encyclopaedic fund of knowledge and information, had talked to me about the Lepidoptera. But up to that moment the subject had not seemed to me particularly interesting. Now quite suddenly all this was changed.

The born teacher is the man who can catch the imagination and stimulate ideas. The art is subtle and difficult to analyse. All that may be required is the appropriate phrase, possibly only a raised eyebrow and the single word 'Indeed' . . . At other times the perfect lecture, the marshalling of difficult ideas and a matchless clarity of expression, overwhelms the listener who leaves the lecture hall filled with excitement and the feeling that at last—at last—he understands genetics! But much as I appreciate these two qualities in E.B.Ford I think I am most grateful to him for his ability to make his pupil or his companion feel intelligent and amusing. This is a gift which engenders discovery at every level. We have all—but particularly the self-taught amateur such as myself—suffered from the cold douche thrown over us by the specialist in the field, the leaden weight of encyclopaedic knowledge, the not-so-surreptitious glance at the wrist watch which convinces you—who, anyhow, needs so little convincing!—that what you are about to say is nonsense, and long-winded nonsense at that. Natural selection has seen to it that we cannot apprehend the thoughts actually passing through our companions' minds. This is probably fortunate. But whether I appeared in E.B.Ford's laboratory accompanied by a female child reputed to have been stung in the leg by a Garden Tiger, or with a moth which I asserted—then on insufficient evidence—exhaled prussic acid, or with a long involved story about the sex ratio of fleas found on migrant wheatears in Fair Isle, I always left with the conviction that the idea was of even greater interest than I had originally realized. What a miraculous gift to be able to knock some scientific sense into a pupil and at the same time increase their zest and pleasure in the problem to be investigated!

E.B.Ford (1945) once remarked about collecting: 'The accomplished naturalist. . . reaches his conclusions by a synthesis, subconscious as well as conscious, of the varied characteristics of the spot weighed up with great experience. But this is a work of art rather than science.' This is the type of thinking E.B.Ford encourages in his associates—the blossoming of intuitive ideas engendered by wide reading, coupled with observations in the field, which lend themselves to fruitful experiments.

The quotation at the beginning of this chapter was a thinly veiled and, perhaps, rather severe rap over the knuckles administered by Karl Jordan to his collaborator Walter Lionel, 2nd Baron Rothschild. And it has come echoing down the years to impress another generation. Recently Rettenmeyer (1970) has published an excellent review on insect mimicry and among the two hundred odd references I counted at least thirty—

8

perhaps sixty—of which E.B.Ford should really have been joint author. It is now very fashionable to append a string of names to the title of a paper; authorship is eagerly claimed by those who, during the course of the investigation in question, caught or reared a few butterflies, cleaned the glassware, or looked up an entry or two in the *Zoological Record*. But these writers are less willing to acknowledge the source of their ideas, particularly if they have been culled from books rather than short papers, or developed during conversations. Perhaps, as Karl Jordan suggested, it *is* difficult to remember from where we got our inspiration, especially those occasions when some comment of ours has been transformed by a companion into what we thereafter cherish as one of our most brilliant and creative ideas!

I am not suggesting that E.B.Ford is responsible, or even partly responsible for the speculations and observations outlined in the following paragraphs, but it is with great pleasure and immense gratitude that I record the fact that almost all these ideas and deductions, were patiently heard and discussed and developed and embellished, either strolling on the lawn of All Souls or in E.B.Ford's room in the Department of Genetics—a room so neat and tidy that it verges on the impossible. Furthermore these discussions and ideas would certainly never have been projected on to paper but for Henry's encouragement and powers of persuasion.

'Really my dear Mrs Lane, you must write a book* about this!' and with an enigmatical smile he pours you out another glass of white Cinzano. And once again you almost believe it was all your own idea ...

SOME PROBLEMS CONNECTED WITH THE RELATIVE NUMBERS OF MIMIC AND MODEL

The gift of drawing valid generalizations from a few penetrating observations is one with which only a few of us are endowed, and it has always seemed to me well nigh miraculous that Bates (1862) on the evidence then available was able to formulate the theory of warning coloration and mimicry. On the other hand, the opposition to this theory seemed a little

---

* By persuading us to write books, E.B.Ford not only forces us to organize and clarify our thoughts, but also keeps us safely boxed up in the Radcliffe Library for a few months.

strange, once he had pointed out the resemblance of insects to inanimate objects such as leaves, sticks, or grass stems—a suggestion which was easily accepted by his contemporaries. As I have said elsewhere (Kellett & Rothschild 1971) the critics of the theory of mimicry at one time over-loaded the literature with proof that birds did not prey on butterflies and then subsequently overloaded it with proof that they ate *all* butterflies—models and mimics alike. For those of us who accepted the theory without much difficulty, attention has recently been centred on rather a different angle of the problem, namely the mechanisms by which the successful mimics evolve and maintain their *modus vivendi*. It was proposed by Bates and generally accepted by subsequent writers (Carpenter & Ford 1933) that in order for Batesian mimicry to succeed, the numbers of the model must considerably exceed those of the mimic. At the same time a species so situated that scarcity of numbers *is selected for*, must find itself in an extremely precarious situation under the constant threat of extinc-tion. All Batesian mimics are therefore under pressure to evolve specializ-ations whereby they can increase their own numbers *vis-à-vis* the model, without interfering with, or seriously altering, the learning pattern of the predators. It is to be expected that different situations call forth different examples of such specialization. Holling's mathematical predictions (1961, 1963), based on the important concept of alternative prey (Marshall 1902) which greatly increases the effectiveness of mimicry, suggest that 30 per cent models in a population can protect 70 per cent mimics. By means of some ingenious and well planned experiments Jane Brower (1960) showed that the memory and experience of a distasteful species, in the case of starlings (and probably other birds such as tits (*Parus* sp.) which have long memories for detail) protect a much larger relative number of mimics than was previously believed possible (even a proportion of 10 per cent models with 90 per cent mimics protected 17 per cent of the mimics). Both these situations, for example, would enable the Batesian mimic or feeble Müllerian mimic to shelter behind *an early emergence date of the model*—thus ensuring that the predators initial experience of the colour pattern in question is of the more noxious species (also at a time when alternative prey is less abundant). This will allow the mimic to rely on long bird-memory and to be slightly more numerous than would otherwise be possible. Very few emergence dates have been studied carefully but, for instance, the Ctenuchid moths *Amata* spp. which serve as models for the Burnet *Z. ephialtes* (L.) are always on the wing before the mimic (Bullini, Sbordoni & Ragazzini 1969 and personal communication) and the White Ermine (*Spilosoma lubricipeda* (L.)) consistently ahead of the Buff Ermine (*S.lutea* (Huf.)) (Rothschild 1963). The earliest strikingly aposematic moth seen on the wing in Britain, is the Cinnabar (*Tyria jacobaeae* (L.)) one of the most distasteful warningly-coloured moths found in this country (Frazer & Rothschild 1962) and it emerges when alternative prey is relatively scarce. Selection in this situation will tend to push the emergence dates of the most distasteful species forward, and hold back or delay the less distasteful. However, great scarcity of alternative prey in the early spring in temperate climates will prevent this gap widening, and will tend to keep the mimics and models on the wing in the period of maximum insect activity. In a brilliant paper Waldbauer & Sheldon (1971) have shown, however, that this is not always so, since the time of the fledgling period of young birds plays an important role in determining the relative numbers of model or mimic on the wing simultaneously. Poulton (1908) was the first to suggest that migratory predators could also influence the course of mimicry and this is confirmed by these authors.

Various observers including Swynnerton (1915), Morton Jones (1932) and Lane (1963) have stressed the fact that other things being equal, birds, like children, are inclined to choose the largest piece of food available at any given time. If, therefore, two similar butterflies are on the wing together, birds may be expected to try for the larger of the pair (Rothschild 1967). This suggests that in cases of Batesian mimicry (or again where a Müllerian mimic is more distasteful or powerfully protected than another) it would be advantageous for the model to be larger than the mimic and for the mimic to be smaller than the model. Where rather isolated pairs, such as the Monarch and the Viceroy are concerned, this discrepancy in size is obvious, and it is probable that the smaller species can thus, so to speak, shelter under the larger wing of the model. In the case of the African Papilios the reverse is true and the Batesian mimics such as *Papilio dardanus* and their various morphs are consistently larger than their models. The explanation of this apparent contradiction may be found in a different 'system' the butterfly has evolved to protect itself against the hazards of too great a numerical inferiority. This is achieved by means of polymorphism (the mimicry of several different well protected models) and *the limitation of mimicry to the female sex.* By this latter specialization the females which are in greater need of protection than the males (Carpenter & Ford 1933, p. 122) and tend to be more variable than the opposite sex (Fisher & Ford 1928) are enabled to be twice as numerous *vis-à-vis* the model than they could afford to be if both sexes were of similar appearance. This extremely important facet of sex-limited polymorphism may perhaps be regarded as a 'side effect' of the enduring and stable pre-mimetic pattern of the male, just as—in Ford's own words—heterozygous advantage is a necessary 'side effect' of the genetic control of Batesian mimicry. Sheppard (1961) stresses the fact that selection for female Batesian mimicry is fundamentally different from that controlling male stability. His experiments showed that, unlike the mimetic female patterns, the male pattern of *P. dardanus* has not evolved gradually, since it does not break down in hybrids, not even in crosses with the monomorphic Madagascan race. This suggests that in this instance male constancy is palaeogenic, whereas female mimetic polymorphism is neogenic in origin.

Magnus (1955, 1958a,b, 1963) pointed out that in all butterflies where mating behaviour has been examined, optical stimuli appear to play a large part, and, in all probability are a fundamental component of their mating behaviour. The male is always the active partner. He has also pointed out that the male butterflies have a tendency to respond to 'super normal' models, an observation corroborated by Crane (1955, 1957). Possibly this predilection for size accounts for the apparent contradiction noted above, that is to say where sex-limited mimicry is concerned the female mimics tend to be larger than their models, the possible disadvantages of the situation being more than compensated for by the response of the males of *P. dardanus* to this 'super model' among the confusing plethora of warningly coloured models and Müllerian mimics with which the females associate.

The importance of sex-limited mimicry is further emphasized in those cases in which the male pattern is not only constant but also mimetic. The same effect of doubling of female numbers *vis-à-vis* the model is still achieved, although in a different manner. Thus the male may mimic another species altogether, or the opposite sex ($\female$) of the same model. Even in those rare cases in which both sexes are similar mimics (for example a form of *Pseudacraea eurytus* (L.)), there are two distinct female morphs, thereby allowing one of them to increase its numbers to a higher and safer level, in relation to its model. One can perhaps surmise that in these groups of butterflies the pattern of the male will break down in crosses between different races, showing that in this case it is, like the female pattern, neogenic rather than palaeogenic, in origin. The chief point is, of course, that despite the spread of mimicry to the male sex a situation is maintained which minimizes the threat inherent in a life-style which demands selection for low numbers.

In the case of Müllerian mimics, and even in those butterflies or diurnal moths which merely adopt the warning life-style and lack mimics altogether, exactly the reverse situation can be expected. The males in such cases will *closely resemble* the females in appearance and consort with them in the field; in certain circumstances it may be an advantage for them to out-number the females, especially in very large congregations of butterflies, a fact originally noted by Bates (1862) for certain flocks of Heliconids in Brazil. It is extremely impressive that in the only instance in which a member of the genus *Papilio* (see below, p. 210) is known to be truly toxic, not only are the sexes alike, both in colour and ground pattern, but the female is very considerably smaller than the male. The apparent enormous disparity in the number of the sexes may be due in the case of *P. antimachus* (Drury) to the different habits of males and females, which result in phenomenally low catches of the latter, and not to a true excess of males. Nevertheless, it is significant, that in the same genus you can find examples of the two life-styles.

The question of sex-limited polymorphism and its variations is such an important aspect of mimicry that it has without doubt been evolved more than once and it is not necessary, as we have explained, to assume that in every case exactly the same set of propitious circumstances have either led to its development or perfected the details. Here again one must examine the special situation against wider general phenomena.

SOME PROBLEMS CONNECTED WITH AVIAN VISION AND THE MINUTE RESEMBLANCES BETWEEN MODELS AND THEIR MIMICS

E.B.Ford (Carpenter & Ford 1933) stated that as a rule mimics occupy the same areas as their models and 'follow' geographical races, and this is as true today as it was 35 years ago when he penned these lines. He went on to say 'The similarity between mimic and model cannot be explained as due to the effects of the common environment in which they both live.'

The opponents of the theory of mimicry, however, were quick to point out that in widely separated parts of the world many models did in fact look rather alike. If Acraeines of Africa and Heliconids of Brazil bore such a strong superficial resemblance to one another, why should not the mimic and its model, subjected to presumably very similar environmental pressures, resemble one another much more closely? The rather contradictory elements in this argument can be reconciled if the right importance is accorded in both instances to the selective pressures imposed by *birds' eyesight*, coupled with their learning capabilities. These determine the broad parallel evolution of many models (and aposematic insects which lack mimics) and the finer resemblances of the mimics. Furthermore the psychology of birds also determines certain details by which the mimic differs from the model—for instance in relative size and in behaviour.

In order to appreciate the role of avian eyesight in determining aposematic form and colouring you have to live, like Swynnerton, with birds *as companions* rather than experimental animals. This is a slight weakness I shared with this superb naturalist* and I can claim, like he did, to have watched our captive birds eat not hundreds but thousands of insects. We both developed a profound respect for their adroitness, lightning movements, diversity of feeding habit and their type of memory and ability to generalize, coupled with intense acuity of vision. My tame Shama (*Kittacincla malabarica* (Scop.)) for example, could *watch* a flea jumping, and jump after it before it landed.

We, not unnaturally—since it reaches an acceleration of 135 gravities during its leap—lose track of the flea at the moment of take off. Judging from the result of crop analysis, various Martins (*Progne* spp.) etc. can distinguish a drone from a worker bee when both parties are on the wing! Not only the birds' vision, but the intense scrutiny to which their prey is subjected after capture is of importance, and whether they grasp food in their claws and pick it to pieces, or chatter it through their beaks, or drum it on the ground or whether they capture their prey in full flight or are ground feeders, or bark searchers, or flower haunters—all these details of behaviour play a part in determining the shape, size and defence mechanisms of the winged model and its mimic. While many entomologists have stressed the fact that plants and insects are contemporaneously evolving organisms (Ehrlich & Raven 1965, Dethier 1970) they seem scarcely aware that birds and insects are likewise exerting heavy selection

---

* Swynnerton (1915) was the first to record vomiting in birds following the ingestion of Danaids and noted the different effect on those with a full or empty stomach. He was also the first to record warning signals by experienced birds when they suddenly caught sight of aposematic butterflies, and the attempts on the part of parent birds to dissuade their young (p. 106) from eating aposematic prey.

pressures on one another. It seems to me likely that once an insect has succeeded in establishing itself on a toxic plant and has evolved a method of storing its poisonous properties, rather than metabolizing or excreting them, it is frequently *destined* by the nature of the avian eye to develop along broadly similar lines. The part of the environment which selects the model's orange and black, or red and black, or white and black pattern, the rows of white spots, the colour 'flashes' and so forth are the bird predators present, for example, in Brazil and Africa. In turn it is the predelictions of the *species* of bird predators which determine the close resemblance between mimic and model.

It is striking that in those areas where a model is very scarce, or lacking altogether, the morphs of a mimetic species such as *Papilio dardanus* or *Pseudacraea eurytus* often tend to be more variable (Carpenter 1949; Ford 1964, p. 229) than when a model is present. In other words it is avian vision which checks variation within the gene complex.

Recently we have come across an interesting example of a toxic butterfly which may well illustrate the point in question. In the genus *Papilio* (Ehrlich 1970) there is an outstandingly aberrant species *P. antimachus* (Drury). This butterfly superficially resembles a giant *Acraea* —so much so that Trimen (1887) suggested it mimicked some huge Acraeine model, now extinct. At the time this was a reasonable supposition because in Africa there were apparently no 'model' *Papilios*—in fact the so-called Pharmacophagous species are lacking on the mainland of the African continent. We (Rothschild *et al.* 1970b) have examined the chemistry of one male specimen of *P. antimachus* and despite the fact that it was captured over 70 years ago it was possible to identify considerable quantities of an unknown cardenolide* in its dessicated body tissues. This aberrant member of the genus *Papilio* is therefore what it appears to be—an aposematic butterfly, and, if by chance, it ever became involved in a mimicry situation it would serve as a model or a Müllerian mimic. Finding a 'blue print' of a genuine warningly coloured species in this genus is not perhaps so surprising as the finer details which have been

* The specimen contained five times as much cardenolides as a single Monarch (Reichstein *et al.* 1968). von Euw compared this substance with pure samples of the following cardenolides but it differed from them all: Digistroside, Oleandrin, Andynerin, Desacetylanhydro-cryptogr. A, Odoroside A and B, Coroglaucigenin, Corotoxigenin Afroside, Calactin, Calatropin, Intermedioside, Sarveroside, Sarmentocymarin, Cymarin, Somalin and Hongheloside A. However, we have since examined further specimens of this species which contained no cardenolides, suggesting that, as in the case of *D. plexippus*, some of its food plants lack these substances.

imposed on the general ground plan. It must surely be the predelictions of the bird predators (of which the smaller hawks are an extremely important element, as I have seen for myself in Israel) and their reactions to certain colours, shapes, sizes, and patterns which exert the necessary parallel selection pressures to produce a basically similar aposematic insect in *Danaus, Papilio* (note that *P. antimachus* lacks tails) *Acraea* and even *Heliconius*, and it is surely the needle-sharp Avian eye that decrees that all four genera have developed fake bleeding spots on the thorax. The significance of these small spots and bands of colour on the bodies of aposematic insects, which emphasize and draw attention to the weakened areas in the cuticle where the toxin-laden haemolymph escapes easily on to the surface, can only be appreciated if one has observed the minute attention and careful scrutiny bestowed by birds on these signs and signals on their captive prey. Certain species of *Acraea* have bright yellow palps and tips to the antennae, which mark bleeding points and simulate drops of haemolymph. In the case of the toxic grasshopper, *Phymateus viridipes* (Stal.) the area on the thorax which becomes covered with foam where the poison gland content is frothed out, is raised in a series of lumps and bumps which catch the light, thus giving the impression that the flow of toxic material extends over a larger area than is in fact the case—or that the bubbles are present long after the supply is, in reality exhausted. On the thorax a belt of scarlet hair often marks the aperture of a defensive or repellent gland (as in *Arctia caja* (L.)). It has frequently been noted that the mimics seem unable to 'copy' these markings on the body, and apparently the 'best they can achieve' is a red flash or 'epaulette' on the wing near its junction with the thorax. This is of course greatly underestimating the inherent possibilities in the mimic. It is important that a predator which has previously caught a model and become acquainted with the repellent and nauseous qualities underlined by the red or yellow bleeding areas, should notice these marks on the mimic *from a safe distance*. Unlike the model, with its tough cuticle and other disagreeable qualities, the mimic cannot afford to be examined at close quarters. Thus the placing of the red marks on the surface of the wings is a further tribute to the keen eye-sight of birds.

It is extraordinary how, within the general overall aposematic pattern, different 'methods' have been evolved to deal with particular facets of the situation. Thus in some cases one can see how the mimic has taken advantage of the birds' predeliction for the super-model by remaining distinctly smaller than the model (see above p. 207), while another species has been forced to the opposite extreme, and has to exploit the male butterfly's predeliction for large females.

E.B.Ford drew my attention to the fact that the flight behaviour of the female *Papilio polytes* (L.) is very different from that of its models *Pachlioptera hector* (L.) and *P. aristolochiae* (Fabr.) Punnett (1915) and various other sceptics regarded this as evidence that this species was not really a mimic at all, for surely such poor resemblance could instantly be detected by any bird! While it is important for the flight of the model to be so slow that it can be easily caught by the inexperienced predator and on all subsequent occasions viewed at leisure by the birds, it is important that the mimic is not caught—important incidentally, for the model as well. Consequently there is every reason why there should be a slight but definite divergence in behaviour on the wing by mimic and model. Birds must see the mimic well enough to be *reminded* of the model but its flight must be such that it escapes without stimulating a chase or falling victim to an attack. There is, however, a variety of refinements in the behaviour pattern of mimics and some depend on last minute 'avoiding action' rather than a subtle difference in speed and wing beat.

Basically it is the development of secondary plant substances which has launched various primarily and secondarily day flying Lepidoptera into the aposematic way of life and it is the avian predators which have largely determined the type of warning *ensemble* of both the moth and its mimics. The birds in turn have evolved varying degrees of generalized immunity to poisonous compounds and have modified their feeding habits accordingly. *Danaus chrysippus* was described by Poulton as the commonest butterfly in the world. Such success makes it inevitable that certain bird predators have become Danaid 'specialists' and choose to feed on their poison-loaded tissues rather than those of more innocuous species.

PROBLEMS CONNECTED WITH ZYGAENIDS

E.B.Ford once described to me how the red-haired daughter of a distinguished entomologist was chased across a sand dune by a swarm of sex-starved male Burnet moths. This rekindled my interest in this wholly delightful group of insects, but it is with regret that I must record that neither of my red-haired children—despite countless opportunities— have kindled even a tiny flame in the heart of a male Burnet. If the stimulus which emanated from the young lady in question was scent, it is not linked, I regret to say, to the type of red hair enjoyed by the Lanes.

It was certainly thanks to E.B.Ford that we discovered that the Burnet when crushed (at all stages of its life-cycle) released HCN. At the beginning of these investigations he suggested that Courtenay Phillips—a pioneer in the field of gas chromatography—should 'exercise his engine' on the defensive secretions of the Lepidoptera, and thus the Burnet was probably the first moth to be introduced on to an apiezon column. This convinced me that before any further theories were introduced it was necesssary to prove Haase's mimicry hypothesis (1896), Swynnerton's deductions (1915) and Morton Jones' observations (1932) by means of some pharmacology and chemistry, which resulted in our identification of heart poisons in the body tissues of four aposematic grasshoppers,* three butterflies, two moths, an Aphid and a Lygaeid bug, all apparently derived from the food plants on which the insects were feeding. (Rothschild & Parsons 1962, Rothschild *et al.* 1970a,b, 1971). One of the most stimulating aspects of the 'Ford approach' is the alacrity with which he seizes upon a new technique—quite unperturbed by the fact that many of us have not yet mastered the old ones—and turns it to good effect in the course of the wider issues of ecological genetics—whether it is the computer or gel electrophoresis, or copper coloured cats.

A striking attribute of these moths is the bright yellow colour of the haemolymph to which Ford has drawn attention (Carpenter & Ford 1933, p. 94). This is a specialization found in both aposematic moths and butterflies, and is obviously of great importance to this life-style. The yellow colour is very varied in tint, being deep yellow in the Cinnabar and Burnet, light yellow in the Jersey Tiger, pale straw coloured in *Arctia virgo* (L.) etc., etc. The yellow haemolymph of Danaids and the Acraeinae are too well known to call for comment, but it is worth noting that aposematic diurnal geometers such as *Obeidia tigrata* (Guen.), *Cartaletis libyssa* (Hoff.) *Euschema transversa* (Walk.) etc. also have bright yellow haemolymph. In the case of the Burnets the watery, colourless, warning scent secretion, so characteristic of this species, is exuded

* An extremely interesting situation is developing in the case of one of these grasshopper *Poekilocerus pictus* (Fab.). This aposematic species up till relatively recent times was like *P. bufonius* only associated with Asclepiad plants. Lately it has developed catholic tastes and is rapidly becoming a pest, in India, of a wide variety of cultivated crops. How closely is its aposematic colouring linked with its capacity to store cardenolides? Has it sufficient inherent protective qualities of taste, smell or self-secreted toxins to maintain the aposematic life-style, or will it now lose its red and black pattern and return to the basic brown or green colouring of the cryptic species? Or will a polymorphism appear, or even speciation occur?

from the mouth, but if the insect is roughly handled the haemolymph squirts out of weakly sclerotized areas on the thorax, and mixes with this secretion from the repellent glands, on the body of the insect. In the case of *Arctia caja* (L.) the cervical glands first secrete a colourless odoriferous fluid, but subsequently haemolymph is pumped out through the same aperture (Rothschild & Haskell 1966) and the secretion becomes yellow in colour. In the related species, *Utethesia bella* (L.) (plate, fig. 3) the secretion which also contains blood cells is yellow when it first exudes from the gland. The pigments responsible for the golden hue of the haemolymph are not all similar but we (Aplin & Rothschild, personal observations) have found that in the Cinnabar—as in Coccinellids—they are chiefly carotenoids. It seems to me probable that in aposematic Lepidoptera these pigments serve a dual purpose. First of all they enhance

PLATE 10.1

1   The Lappet Moth (*Gastropacha quericifolia* L.). A great tribute to the acuity of avian vision.
2   Caterpillar of the Alder Moth showing both cryptic and aposematic stages.

EXPLANATION OF PLATE 10.1

There are two further types of mimicry which are both extremely impressive and exploit different traits in the bird predator. One is characterized by the extreme perfection of detail which is well illustrated, for example, by the Lappet Moth (*Gastropacha quericifolia* L.). Note for example the stalk-like palps and the antennae which are held in such a position that they supply the anterior end of the 'mid-rib' of the leaf (fig. 1). In the case of another cryptic moth, *Cricula trifenestrata* Helfer, not only are there transparent areas in the wing simulating holes in dried foliage, but, if disturbed, the flight of the moth exactly emulates the erratic path of a dead leaf whisked around by the breeze. This fantastic perfection of detail is also demonstrated by certain cryptic caterpillars when a 'shine' is developed on the surface of the body to simulate, for example, the glistening, sugary exudate on the flowers of the Mullein on which the full grown larvae of *Cucullia verbasci* L. feeds. A similar effect is seen on the surface of the caterpillar of the Alder Moth (*Apatele almi* L.) which clearly resembles shiny wet, fresh bird droppings (fig. 2). It should be noted that after the third moult this latter species changes from an extreme cryptic into an aposematic form, complete with yellow and black stripes and long clubbed hairs. It is not known whether a chemical change accompanies the switch to another life-style, but it is likely to be distasteful throughout all its larval stages but assumes warning coloration when it is too big to pass itself off as a bird dropping.

1

2

3

4

5

the general aposematic display or appearance of the insect if it is molested by the predator. Portschinsky (1897) realized that certain marks on the body of warningly coloured insects actually mimic these drops of haemolymph, so that they continue to exert a warning effect on the predator before or after they are actually secreted. The second function of these pigments is, I surmise, to enhance the odour of the repellent secretion associated with aposematic Lepidoptera, that is to say they amplify the smell by increasing the sensitivity of the predator. Recently it has been suggested (Rosenberg, Misra & Switzer 1968) that the mechanism for olfactory transduction depends on the formation of a weak bound complex between an absorbed gas molecule and certain carotenoid pigments, such as those which are found in the nasal cleft of mammals. Professor R.H.Wright informs me (1970, personal communication, and see also McCartney 1968) that injections of carotenoids into patients who are suffering from anosmia may partially restore their sense of

PLATE IO.I *continued*

3 *Utethesia bella* (L.) frothing out toxic secretion ($\beta\beta$-dimethylalcrylylcholine) mixed with yellow haemolymph from one of its cervical glands.
4 The Muslin Moth ($\male$) showing the pointed bee-like face, an effect produced by the position in which the legs are held.
5 Buff Ermine (*Spilosoma lutea* (L.)) feigning death, in a bee-like attitude.

EXPLANATION OF PLATE IO.I *continued*

The second type of mimicry depends not on accuracy of detail, but a 'snap' overall impression. There is no real similarity at all between mimic or model but some subtle action produces a generalized resemblance to a harmful or distasteful species. The male Muslin Moth, *Cyncnia mendica* Clerck, if touched, drops to the ground, curls round the abdomen and draws up its legs suddenly, thus irresistably reminding the observer of a small worker bee with a pointed face and well-filled yellow pollen bags (fig. 4). A bee-like general effect (fig. 5) is also achieved by the Buff Ermine (*Spilosoma lutea* L.) which 'feigns death' if seized or roughly handled. In the first type of mimicry (personal observation) even the experienced avian predator entirely overlooks the moth or the caterpillar: in the second it is startled by the sudden 'metamorphosis' and often responds by flying off without examining the prey at all.

The great contrast between the two basic life-styles is well illustrated by comparing figs. 1, 2, 4 and 5 with fig. 3 showing the aposematic *Utethesia bella* (L.) frothing out toxic secretion ($\beta\beta$-dimethylalcrylylcholine) mixed with yellow haemolymph from one of its cervical glands. Its right wing has been caught between forceps and it responds by secreting from the gland on that side of the body.

smell, thus demonstrating that the pigments in question need not be originally present in the cells of the olfactory receptors themselves in order to exert an effect. It is of considerable interest that the defensive secretions of many Lepidoptera are so often produced in the form of a froth (plate, fig. 3) since when these bubbles burst the secretion forms an aerosol of spray droplets (not a vapour), which would facilitate and favour the distribution of a substance of high molecular weight. Searching through the old literature one is continually impressed by the linking of the 'yellow juices' of the Lepidoptera with nauseous or penetrating or evocative smells, and the different ways in which the mixing of the haemolymph with the odoriferous secretions is achieved suggests that it is a phenomenon of great importance, and has been independently evolved on several occasions.

Another feature of the Burnet moth, characteristic of the majority of aposematic insects, is what the older writers describe as 'tenacity of life', and the ability to withstand injury. In the case of the Monarch the cuticle is extremely tough and 'leathery' and Carpenter & Ford (1933) have drawn attention to the fact that their wings 'when bent in the net do not break as is often the case with other butterflies'. The Burnets achieve the same ends in a slightly different manner. Far from being hard or tough their bodies are excessively soft, pliable and elastic, giving the superficial impression that they are chiefly composed of intersegmental membrane rather than tanned cuticle. Furthermore they appear to have unusual powers of wound healing. It is possible to inflict severe injuries upon the Burnet and to find that the tears in the integument have been sealed off like a puncture in a self-sealing tyre and the insect is moving about apparently none the worse for its injuries. This is a facet of the Burnet's specializations well worth investigating. It suggests an unusually efficient clotting mechanism in conjunction with an elastic type of cuticle.

Another aspect of the Burnet's tenacity of life is its resistance to HCN which has no doubt been developed in connection with its ability to secrete this toxic substance. Be that as it may, walking around in the killing bottle, *how does the Burnet respire*? This is a question to which none of us have so far yet produced an entirely satisfactory answer. It is the sort of problem one should think about in traffic jams ...

The Zygaenids are not the only Lepidoptera which have developed resistance to HCN, although so far the only known group of secretors. Thus there are three records from Arctiids, and a whole group of cryptic geometers, *Asthena* Hubner (Portschinsky, 1897; Rothschild, 1961–2 and

Szent-Ivany, personal communication) which survive for unusually long periods in the 'killing bottle'. The older literature abounds with gruesome stories of supposedly dead and luckless Danaids or Acraeines, found in some cabinet drawer 'flying on their pins'. It is by no means necessary to postulate a similar mechanism for all these different species, just as it is unnecessary to suppose that all insects which feed on cyanogenic plants cope with these poisons in the same fashion. *Zygaena*, which itself secretes HCN is immune to the substance; the Common Blue (*Polyommatus icarus* (Rott.)) turns HCN into harmless thyocyanate by means of the enzyme rhodanese; the caterpillar of the Lackey Moth (*Malacosoma neustria* (L.)) feeds with impunity on laurel leaves, rich in HCN, yet it neither stores cyanide at any stage of its life-cycle nor does it contain rhodanese. This species has evidently evolved a different method, not yet elucidated, for dealing with the protective devices of the plant (Parsons & Rothschild 1964). Although so far not investigated from this angle, the aposematic Heliconids and Acraeines which feed on *Passiflora* —plants which themselves produce HCN—are almost certain to be secretors. One can imagine that the former species after moving on to *Passiflora* evolved immunity and secretion in parallel, whereas *Acraea* which feeds on a great variety of plants was able to take advantage of a protected habitat—like the Burnet—by the fact that it already secreted this substance. It is of course well known that quite a wide range of animals which secrete or store specific poisons are themselves insensitive to that particular group of substances. Thus the toad which secretes Bufotalin is immune to the effect of digitalis whereas the frog is particularly susceptible; the grasshopper *Poekilocerus bufonius* Klug. which stores cardenolides in its body tissues (von Euw *et al.* 1967) is 300 times less sensitive to digitalis than the desert locust, while various snakes and the Gila Monster are immune from the effects of their own venom. Nevertheless this does not explain how Zygaenids survive in a regulation killing bottle for an hour or more.

The first possibility which might be investigated is the selective impermeability of certain cell membranes or the inner or outer mitochondrial membranes. A second possibility is the use by the insect of other cytochromes, such as those which are found in bacteria, yeasts and plants. But this appears less likely, since resistant insects seem to be relatively immune to other quite unrelated toxic substances and one therefore imagines a less specific mechanism. Furthermore, animals on the whole appear rather conservative in their use of 'standard' cytochromes. It is also possible that extremely rapid inactivation occurs in the tissues by binding to some other substance, in the same way that plants attach sugars to CN; thus in the laurel it appears as a glucoside, amygdalin. Detoxification in either the haemolymph or tissues seems unlikely since it would scarcely have time to function where tracheal respiration is concerned. Nor is the enzyme rhodanese, present in the body of Zygaenids (Jones, Parsons & Rothschild 1962). It is also possible to envisage an active pump for removing this poison from within a cell or mitochondrion, acting in conjunction with reduced membrane permeability. Finally, E.B.Ford has suggested that a highly active glycolytic pathway might enable the Burnet to live for some time under effectively anaerobic conditions in the presence of HCN. All this is pure speculation—but at least it indicates that the mechanism by which the Burnets may dispose of CN or prevent it reaching sensitive body tissues could prove of considerable biochemical interest.

It should be stressed that there is no evidence that the HCN is present as free cyanide in the Burnet's intact cells. Certain insects have developed very complicated methods for avoiding this situation and protecting their tissues from contact with toxic chemicals which they store. Thus for example in reactor glands (Eisner 1970, p. 167) the stored products mandelonitrile and an enzyme which catalyses it into HCN and benzaldehyde are liberated simultaneously when the glands are discharged. Up to that moment they are stored separately as the chemical precursers of the active secretion. An Arctiid moth (Teas 1967) achieves a somewhat similar effect by different means. The caterpillar ingests a toxic alkylating agent with its Cycad food-plant—re-synthesizing it in the gut and storing it in specific tissues as the beta-glycoside, cycasin. It leaves the predator to reconvert this relatively harmless substance into its toxic aglycone.

It is perhaps rather peculiar that some of these problems can equally arise in connection with the predators which in turn consume toxic insects. Thus, two ground feeders the Quail (*Coturnix japonica* (Tem.)) and the hedgehog (*Erinaceus europaeus* (L.)) are insensitive to large oral doses of digitalis and to cantharadin; and Swynnerton's Wood Hoopoes (*Irrisor erythrorhynchus* (Lath.)) have become specialists (p. 100) in eating Danaids, Melöeid beetles etc., etc. One is left with the impression that we are dealing with some rather broad principle which applies to both prey and predator. A general insensitivity to toxins is thus achieved, but the mechanism involved has, somehow, eluded us.

I have already drawn attention to the obvious fact that odours play an important part in defence mechanisms of insects, but it is interesting that so many species, particularly the Lepidoptera, share them in common. This is an example of Wickler's (1968) 'standardization of signals'. There is an infinite variety of scent patterns in nature just as we find an infinite variety of colour patterns. Yet animals ranging from frogs to frog-hoppers, sea slugs to cinnabars, who wish to advertise their presence, resort to the colour red or orange with which to put their message across. Nor are the major scent signals confined to one family or order, for like colour signals they even span the Phyla. Thus it comes as no great surprise to us to learn that both alligators (Lederer 1950) and certain Danaid butterflies share a musk-like scent which serves as a sex attractant. But perhaps because we are animals which hunt by sight rather than smell, the scent patterns of aposematic insects are more difficult for us to appreciate than the stripes and the spots of their warning colour schemes. Whose nose, for instance, has decreed that the unmistakable warning isobutyl-quinolene-like scent of the ladybirds (Coccinellids) should be common to certain Arctiids, Danaids, Acraeas, Syntomids, Heliconids, Lymantrids, and even one Sphingid? It is a sad reflection on our lack of techniques that so far we have been unable to identify this highly evocative and by no means unpleasant poppy-like aroma. It is, probably, its memory provoking rather than its repellent qualities which are important in this particular aposematic pattern, a fact which has confused collectors who are looking for nauseous effects such as those evoked by Necrophorus beetles and Pentatomid bugs. When trying to interpret the scents or 'evocatives' shared by those particular Lepidoptera one should probably think in terms of stimulation of the frontal lobes or hippocampus rather than its 'nice' or 'nasty' effect on the taste or the nasal organ of predators. Unfortunately one cannot study scent or taste patterns in preserved specimens as one can study aposematic colours. E.B.Ford realized this difficulty and in the past has been an intrepid masticator of Lepidoptera in the field— pointing out at the same time that although he uses those insects for the purpose of research he does not love butterflies but rather dislikes them... Our discovery of HCN, lethal proteins and heart poisons in the adult insect has introduced a certain note of caution into this angle of our studies. I was lamenting the difficulty of procuring *Papilio antimachus* alive and suggesting that if we ever did so, and he had the temerity to taste it, no doubt we should find it had yellow, bitter-flavoured haemolymph with a persistent smell of ladybirds. On this occasion E.B.Ford considered the proposal with slightly raised eyebrows:

'When I was in Texas,' he remarked, 'I used to eat rattlesnakes and I recall they tasted of cold scrambled eggs. Indeed my dear Mrs Lane, I think you really *should* write a book about all this.'

ACKNOWLEDGEMENTS

I have to thank Dr J.von Euw and Professor T.Reichstein for the footnote on p. 210, and Dr John Parsons and Dr Peter Miller for their help with the paragraphs relating to *Zygaena*, and Professor Philip Sheppard for a most useful discussion on polymorphism in female *P.dardanus*. My special thanks are due to Dr B.Kettlewell for providing the larvae illustrated on Plate 10.1, fig. 2.

REFERENCES

BATES H.W. (1862) Contribution to an insect fauna of the Amazon Valley. Lepidoptera: Heliconidae. *Trans. Linn. Soc. Lond. Zool.* 23, 495–566.

BROWER J.V.Z. (1960) Experimental studies of mimicry. IV: The reactions of starlings to different proportions of models and mimics. *Am. Nat.* 94, 271–282.

BULLINI L., SBORDONI V. & RAGAZZINI P. (1969) Mimetismo Mülleriano in populazioni Italiane di *Zygaena ephialtes* (L.) (Lepidoptera: Zygaenidae). *Arch. zool. ital.* 54, 181–213.

CARPENTER G.D.H. & FORD E.B. (1933) *Mimicry.* Methuen, London.

CARPENTER G.D.H. (1949) *Pseudacraea eurytus* (L.) (Lep. Nymphalidae): a study of a polymorphic mimic in various degrees of speciation. *Trans. R. ent. Soc. Lond.* 100, 71–133.

CRANE J. (1955) Imaginal behavior of a Trinidad butterfly, *Heliconius erato hydara* Hewitson, with special reference to the social use of color. *Zoologica, N.Y.* 40, 167–196.

CRANE J. (1957) Imaginal behavior of butterflies of the family Heliconiidae: Changing social patterns and irrelevant actions. *Zoologica, N.Y.* 42, 135–145.

DETHIER V.G. (1970) Chemical interactions between plants and insects. In E.Sondheimer & J.B.Simeone (eds.) *Chemical Ecology*, Chap 5. Academic Press, New York.

EHRLICH P.R. & RAVEN P.H. (1965) Butterflies and Plants: A study in coevolution. *Evolution, Lancaster Pa.* 18, 586–608.

EHRLICH P.R. (1970) A note on the systematic position of *Papilio antimachus* Drury. *J. Lep. Soc. New Haven, Conn.* 24 (3), 224–225.

EISNER T. (1970) Chemical defence against predation in arthropods. In E.Sondheimer & J.B.Simeone (eds.) *Chemical Ecology*, Chap 8. Academic Press, New York.

VON EUW J., FISHELSON L., PARSONS J.A., REICHSTEIN T. & ROTHSCHILD M. (1967) Cardenolides (heart poisons) in a grasshopper feeding on milkweeds. *Nature (Lond.)* 214, 35–39.

VON EUW J., REICHSTEIN T. & ROTHSCHILD M. (1971) Heart poisons (cardenolides) in the Oleander bug, *Caenocoris nerii*. *J. Insect Physiol.* 17 (in press).

FISHER R.A. & FORD E.B. (1928) The variability of species in the Lepidoptera with reference to abundance and sex, with an appendix by Professor E.B.Poulton. *Trans. ent. Soc. Lond.* 76, 367–384.

FORD E.B. (1945) *Butterflies*. Collins, London.

FORD E.B. (1964) *Ecological Genetics*. Methuen, London.

FOX H.M. & VEVERS G. (1960) *The nature of animal colours*. Sidgwick & Jackson, London.

FRAZER J.F.D. & ROTHSCHILD M. (1962) Defence mechanisms in warningly coloured moths and other insects. *Proc. XI int. Congr. Ent. Wien*, 1960, Bd. 3, 249–256.

HAASE E. (1893) Untersuchungen über die Mimikry auf Grundlage eines natürlichen systems der Papilioniden. 2. Untersuchungen über die Mimikry. Nägele, Stuttgart.

HAASE E. (1896) Mimicry in butterflies and moths. Researches on mimicry on the basis of a natural classification of the Papilionidae. Pt. 2. Researches on mimicry, translated by C.M.Child, pp. 154. Nägele, Stuttgart: Baillière, Tindall & Cox, London.

HOLLING C.S. (1961) Principles of insect predation. *A. Rev. Ent.* 6, 163–182.

HOLLING C.S. (1963) Mimicry and predator behaviour. *Proc. int. Congr. Zool*, 16, Washington 1963, 4, 166–172.

JONES D.A., PARSONS J. & ROTHSCHILD M. (1962). Release of Hydrocyanic acid from crushed tissues of all stages in the life-cycle of species of the Zygaeninae (Lepidoptera). *Nature (Lond.)* 193, 52–53.

JONES F.M. (1932) Insect coloration and the relative acceptability of insects to birds. *Trans. ent. Soc. Lond.* 80, 345–386.

KELLETT D.N. & ROTHSCHILD M. (1971) Notes on the reactions of various predators to insects storing heart poisons (cardiac glycosides) in their body tissues. *J. ent. A* (in press).

LANE C. (1963) Round the blue lamp. *Entomologist's mon. Mag.* 99, 189–195.

LEDERER E. (1950) Odeurs et parfums des Animaux. *Fortschr. Chem. org. NatStoffe* 6, 87–153.

McCARTNEY W. (1968) *Olfaction and odours: an osphiésiològical essay*. Springer-Verlag, Berlin.

MAGNUS D. (1955) Zum problem der 'überoptimalen' Schlusselreize. *Verh. dt. zool. Ges.* 1954, 317–325.

MAGNUS D.B.E. (1958a) Experimental analysis of some 'overoptimal' sign-stimuli in the mating-behaviour of the Fritillary butterfly *Argynnis paphia* L. (Lepidoptera, Nymphalidae). *Proc. X int. Congr. Ent. Montreal* 1956, 2, 405–418.

MAGNUS D. (1958b) Experimentelle Untersuchungen zur Bionomie und Ethologie der Kaiser-mantels *Argynnis paphia* L. (Lep, Nymph.). I. Über optische Auslöser von Anfliegereaktionen und ihre Bedeutung für das Sichfinden der Geschlechter. *Z. Tierpsychol.* 15, 397–426.

MAGNUS D.B.E. (1963) Sex limited mimicry. II. Visual selection in the male choice of butterflies. *Proc. XVI int. Congr. Zool.*, Washington 1963, **4**, 179–183.

MARSHALL G.A.K. (1902) Five years' observations and experiments (1896–1901) on the bionomics of South African insects, chiefly directed to the investigation of mimicry and warning colours. *Trans. ent. Soc. Lond.* **1902**, 292–584.

PARSONS J. & ROTHSCHILD M. (1964) Rhodanese in the larva and pupa of the common blue butterfly (*Polyommatus icarus* (Rott.)) (Lepidoptera). *Entomologist's Gaz.* **15**, 58–59.

PORTSCHINSKY J. (1897) (Caterpillars and moths of St. Petersburg Province. Biological observations and investigations.) Lepidopterorum Rossiae biologia. V. Coloration marquante et taches oscelléses, leur origine et développement. (In Russian). *Horae Soc. ent. Ross.* **30** (1895–1896), 358–428.

POULTON E.B. (1908) *Essays on evolution*, 1889–1907. Clarendon Press, Oxford.

PUNNETT R.C. (1915) *Mimicry in butterflies.* Cambridge University Press.

REICHSTEIN T., VON EUW J., PARSONS J.A. & ROTHSCHILD M. (1968) Heart poisons in the Monarch butterfly. *Science, N.Y.* **161**, 861–866.

RETTENMEYER C.W. (1970) Insect mimicry. *A. Rev. Ent.* **15**, 43–74.

ROCCI U. (1916). Sur une substance vénéuse contenue dans les Zygènes. *Archs ital. Biol.* **66**, 73–96.

ROSENBERG B., MISRA T.N. & SWITZER R. (1968) Mechanism of olfactory transduction. *Nature (Lond.)* **217**, 423–427.

ROTHSCHILD M. (1961–62) A female of the Crimson Speckled Footman (*Utethesia pulchella* L.) captured at Ashton Wold. *Proc. R. ent. Soc. Lond.* C. **26**, 35–36.

ROTHSCHILD M. & PARSONS J. (1962) Pharmacology of the poison gland of the locust *Poekilocerus bufonius* Klug. *Proc. R. ent. Soc. Lond.* C. **27**, 21–22.

ROTHSCHILD M. (1963) Is the Buff Ermine (*Spilosoma lutea* (Huf.)) a mimic of the White Ermine (*Spilosoma lubricipeda* (L.))? *Proc. R. ent. Soc. Lond.* A. **38**, 159–164.

ROTHSCHILD M. & HASKELL P.T. (1966) Stridulation of the Garden Tiger Moth, *Arctia caja* L. audible to the human ear. *Proc. R. ent. Soc. Lond.* A. **41**, 167–170.

ROTHSCHILD M. (1967) Mimicry: the deceptive way of life. *Nat. Hist. N.Y.* **76** (2), 44–51.

ROTHSCHILD M., VON EUW J. & REICHSTEIN T. (1970a) Cardiac glycosides in the oleander aphid, *Aphis nerii. J. Insect Physiol.* **16**, 1141–1145.

ROTHSCHILD M., REICHSTEIN T., VON EUW J., APLIN R. & HARMAN R.R.M. (1970b) Toxic Lepidoptera. *Toxicon* **8**, 293–299.

ROTHSCHILD M. & ALPIN R.T. (1971) Toxins in Tiger Moths (Arctiidae: Lepidoptera). *Proceedings 2nd International Pesticide Congress, Tel-Aviv* (in press).

SHEPPARD P.M. (1961) Recent genetical work on polymorphic mimetic Papilios. In J.S.Kennedy (ed.) *Insect Polymorphism*, pp. 20–29. Symposia No. 1. Royal Entomological Society of London.

SWYNNERTON C.F. (1915) Birds in relation to their prey: experiments on Wood-Hoopoes, small Hornbills and a babbler. *Jl. S. Afr. Orn. Un.* **11**, 32–108.

SWYNNERTON C.F.M. (in CARPENTER G.D.H.) (1942) Observations and experiments in Africa by the late C.F.M.Swynnerton on wild birds eating butterflies and the preference shown. *Proc. Linn. Soc. Lond.* **1941–42**, 10–46.

TEAS H.J. (1967) Cycasin synthesis in *Seirarctia echo* (Lepidoptera) larvae fed Methylazoxymethanol. *Biochem. biophys. Res. Commun.* **26**, 686–690.

TRIMEN R. (1887) *South African Butterflies: a monograph of the extra tropical species*, Vols. I–III. London, 1887–1889.

WALDBAUER G.P. & SHELDON J.K. (1971) Phenological relationships of some aculeate Hymenoptera, their dipteran mimics, and insectivorous birds. *Evolution Lancaster, Pa.* (in press).

WICKLER W. (1968) *Mimicry in plants and animals*. Translated from the German by R.D.Martin. Weidenfeld & Nicholson, London.

Note added in proof: A recent paper by Owen (*Oikos* **21**, 333, 1970) confirms the suggestion that members of the Acraeinae secrete their own toxins (Reichstein *et al.* 1968, p. 862) and in particular HCN (see above, p. 217).

11 ❋ Studies of Müllerian Mimicry and its Evolution in Burnet Moths and Heliconid Butterflies

J. R. G. TURNER

Müllerian mimicry does not of itself lead to variation of pattern or colour within a species. The function of Müllerian mimicry is to protect its possessors, all of them unpalatable species, by their mutual resemblance. This it does first by reducing the number of individuals of each species which predators kill before they learn to avoid them (because only one pattern needs to be learned) and second, one assumes, by what the experimental psychologist calls secondary reinforcement: the predator's avoidance of the unpleasant species can be strengthened merely by seeing its pattern, without the predator's actually tasting it. The cornerstone of Müllerian mimicry is thus uniformity of pattern, and Müllerian mimics are not likely to be polymorphic, unlike Batesian mimics (palatable species resembling unpalatable ones), in which polymorphism is favoured by natural selection (Ford 1953). (For a general review of mimicry, see Ford 1964 or Wickler 1968.)

This chapter discusses some recent research on two groups of Lepidoptera which show both Müllerian mimicry and genetic variation, and thus allow us to examine the evolution of this kind of mimicry.

ZYGAENA EPHIALTES

The four forms of the day-flying burnet moth *Zygaena ephialtes* (fig. 11.1) are produced by variation in two characters, colour and pattern. The moth may have either red or yellow pigment on the wings and abdominal band. The overall pattern may be of coloured (red or yellow) forewing spots, and a coloured hindwing; or all the forewing spots except the two basal ones may be white, the hindwings being black with a white spot. Bovey (1941) calls these two patterns peucedanoid and ephialtoid; in this chapter they will be P-form and E-form for short. In addition, the four

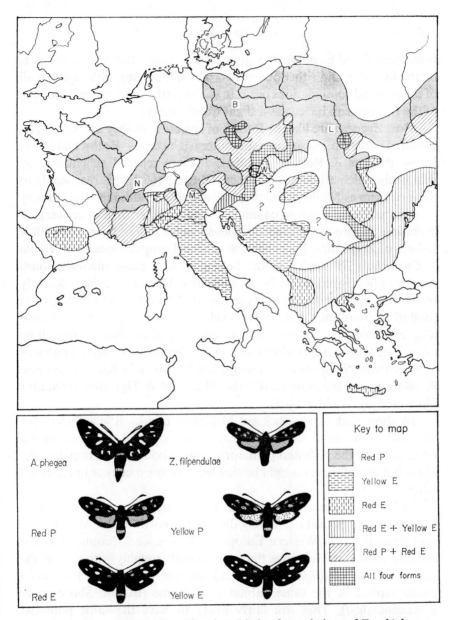

FIGURE 11.1. Distribution of various kinds of populations of *Z. ephialtes* in western Europe, redrawn from Bovey (1941). Inset: *A. phegea* (left) and *Z. filipendulae* (right), with the four forms of *Z. ephialtes* (red P, yellow P, red E, yellow E); the red E-form is shown with one hindwing spot, the yellow E-form with two (about ½ natural size). "?" indicates that the area is probably occupied by yellow E populations. Locations (origins of breeding stocks): B—Berlin, L—Lublin, M—Merano, N—Nyon, W—Wien, Z—Zaleshchiki. The circle contains Niederösterreich (see fig. 11.2).

forms (red P, red E, yellow P, yellow E) may have either five or six spots on the forewing and either one or two on the hindwing, although not all of the possible combinations occur in the wild (the spots on the hindwing of the P-forms can be seen as a slight transparency).

Most *Zygaena* are like the P-form of *ephialtes*; as they are warningly-coloured, contain poisonous substances (Frazer & Rothschild 1960, Jones, Parsons & Rothschild 1962), and are rejected vehemently by wild birds which attempt to eat them (Burton 1962), it is reasonable to think of them as Müllerian mimics one of another, although the mimicry is presumably the result of common ancestry rather than of convergent evolution. Figure 11.1 shows the widespread *Z. filipendulae*. The yellow E-form of *Z. ephialtes* looks like the various European species of *Amata* (= *Ctenucha* = *Syntomis*), moths belonging to a predominantly South American family renowned for its warning coloration (Beebe & Kenedy 1957), and we may think that these too are Müllerian mimics (fig. 11.1) (Bullini & Sbordoni 1970). The two yellow spots at the base of the fore-wing of *Z. ephialtes*, where *A. phegea* has only white spots, may well be mimicking the anterior yellow band on the abdomen of *phegea*; mimicry of body-marks by spots on the wings is well known in Batesian mimics, *Papilio memnon* for example (Clarke, Sheppard & Thornton 1968, also illustrated by Wickler 1968).

To show that two species are Müllerian mimics it is sufficient for practical purposes to show that they share potential predators in the wild, and that the potential predators, finding both species unpalatable, will reject one of them on sight having learned by experience to reject the other.

Kirby (1898) describes *A. phegea* as 'common throughout Southern Europe (except Spain)..., widely distributed north of the Alps, but... very sporadic... in Western Europe... In France... confined to the extreme southeast.' Its range thus corresponds roughly with that of the yellow E-form of *ephialtes* (fig. 11.1) and the two species have been photographed in the same habitat near Rome (Bullini, Sbordoni & Ragazzini 1969). They are fairly likely to have the same potential predators in the wild.

The mimicry of *A. phegea* and *Z. ephialtes* has been investigated by a series of short controlled experiments by Bullini *et al.* (1969), using caged blackbirds, starlings, nightingales and Oriental Nightingales (*Liothrix lutea*). The eight birds used rapidly learned not to touch *A. phegea*, while still eating various palatable satyrid butterflies; a ninth similarly learned to avoid *Z.filipendulae*. Four of the birds trained to avoid *A.phegea*, were presented with a palatable butterfly, one *A.phegea* and one *Z.filipendulae* in rapid succession; all twelve of the *phegea* presented in three replications were untouched, but the birds pecked at ten of the twelve *filipendulae*. In the terms of this experiment at least, *filipendulae* is not a mimic of *phegea*. These four birds, when presented with a palatable butterfly, one *A.phegea* and one *Z.ephialtes* (yellow E), left untouched both of the warningly coloured species, showing that *ephialtes* is an effective mimic of *phegea*, as the birds, having learned to avoid *phegea* then avoid *ephialtes* on sight. A third group of birds vomited or gave signs of considerable discomfort, according to the species, after eating *Z.ephialtes*; the species is therefore distasteful like other *Zygaena*, and is a Müllerian rather than Batesian mimic.

The species thus has two mimetic forms, the red P-form, mimicking other *Zygaena*, and the yellow E-form, mimicking various *Amata*, and two apparently non-mimetic forms (yellow P and red E). Bovey (1941, 1942, 1948, 1950, 1966), Dryja (1959) and Povolný & Pijáček (1949) have thoroughly worked out their genetics. A single locus (P, p) controls the pattern, the P-form being dominant, and another (R, r) controls colour, red being dominant to yellow. Bovey's and Dryja's backcross broods (table 11.1) show that the loci are not linked, there being no departure from the expected $1:1:1:1$ ratios in broods of any geographical origin (Ford 1964, has pointed out that Bovey's broods alone do not exclude loose linkage). The number of spots on the hindwing is controlled by another locus, with two spots dominant to one; an F_2 of 94 individuals (Bovey 1941, brood 61) shows that this locus is unlinked to P, and another of 104 moths (Bovey 1966, brood 158) shows that it is unlinked to R. Bovey (1941) reports that the number of spots on the forewing (five or six) is difficult to score, as the sixth spot is variable in appearance; it seems likely that it is controlled by several genetic factors, including the sex of the individual. Dryja (1959) attributed the character to a single locus, but as he does not give any numerical data, I am reluctant to accept his conclusion.

North of an irregular line running from southern France, through Switzerland, Austria, Czechoslovakia, southern Poland and the Ukraine, populations of *ephialtes* are of the red P-form (fig. 11.1). South of this line they are of the E-form, and may be pure yellow, as in Italy, Jugoslavia and, probably, Hungary, or pure red, as in southwestern France and parts of Greece and the Balkans, or polymorphic red and yellow, as in much of

TABLE 11.1. Segregation of broods of *Zygaena ephialtes*, tested for agreement to the ratios 1:1:1:1 and 9:3:3:1.

Cross	Brood				Deviation			Heterogeneity		
	Red P	Yellow P	Red E	Yellow E	χ^2	d.f.	P	χ^2	d.f.	P
BACKCROSS (1:1:1:1)										
Merano × Lublin[a]	696	672	653	663	1·51	3	0·68	42·93	33	0·12
Merano × Berlin[b]	390	375	381	367	0·75	3	0·86	5·01	33	1·00
Wien × Nyon[c]	48	39	31	31	5·28	3	0·15	5·60	3	0·13
Total	1134	1086	1065	1061	3·10	3	0·38	4·44	6	0·62
F$_2$ (9:3:3:1)										
Merano × Lublin[d]	290	115	93	29	3·60	3	0·31	16·44	12	0·17
Merano × Berlin[e]	1381	507	487	173	5·06	3	0·17	52·60	39	0·07
Wien × Nyon[f]	56	5	19	2	11·88*	3	0·01	3·72	6	0·71
Opava × Napajedla[g]	172	54	60	19	0·32	3	0·96	6·67	9	0·67
Total	1899	681	659	223	—	—	—	17·74*	9	0·04

* Significant.

[a] Dryja's broods 7,9,17,25–27,29,30,33,35,36,38.

[b] Dryja's broods 61,65–72, 75–77.

[c] Bovey's broods 70, 81.

[d] Dryja's broods 12, 15, 21, 32, 34.

[e] Dryja's broods 51–54, 56–60, 62, 63, 74, 89, 96.

[f] Bovey's broods 69, 79, 80.

[g] Broods 14, 14a, 14b, 112 of Povolný & Pijáček. Both localities are in Moravia.

the Balkans and the southern Ukraine. Along the line where the P- and E-forms meet there is a series of polymorphic populations, including those which are pure red but polymorphic for the two patterns, and those which are polymorphic for both colour and pattern, and therefore contain all four forms. With a few gaps, which may result from lack of information, these populations follow the line of the mountains of central Europe, from the Alps in the west, through the Tirol and Steiermark, north into Czechoslovakia, east through the Tatry and down the Dnestr valley on the northern side of the Carpathians, into the mountains of central Romania (compare fig. 11.1 with any physiographic map).

A few reservations must be made about the construction of fig. 11.1: the insect forms isolated colonies, and is not found continuously in the shaded areas; it is always a matter of opinion, given a few locality records, how these are joined up to give a shaded area, the results being approximate only, especially in the eastern third of the map, where records are sparse; in some of the zones shown as occupied by polymorphic populations, some colonies may contain less than the maximum number of forms (say only two forms in an area shown as having all four), and some of these zones are probably much narrower than is indicated. With these reservations, it seems as though a northern race, entirely red P, meets a southern race which is predominantly yellow E (but which is sometimes red E, or polymorphic red and yellow), producing at the meeting place along the mountains, a zone of polymorphic populations. This is particularly clear in eastern Austria (W in fig. 11.1) where a yellow E population which probably occupies the plains of Hungary (a pair of ?-marks on the map, there being no precise records), meets the northern red P race, giving a band of populations which are polymorphic in various ways; the p allele continues at a low frequency for a considerable distance into the populations shown as monomorphic red P-form, so that E-forms appear as rarities as far north as Berlin (Bovey 1941).

Bovey points out that there are no monomorphic populations of the yellow P-form, which appears only in localities where yellow E and red P-forms are found together. He concludes that the yellow P-form is at a selective disadvantage, a suggestion upheld by other *Zygaena* species, in which yellow forms occur only as rare mutations; Ford (1937) has shown them to be recessive in *Z. filipendulae*. Reichl (1958) has studied in detail the populations of *ephialtes* in the Austrian province of Niederöstereich (fig. 11.2, lower half; circle in fig. 11.1). The populations are

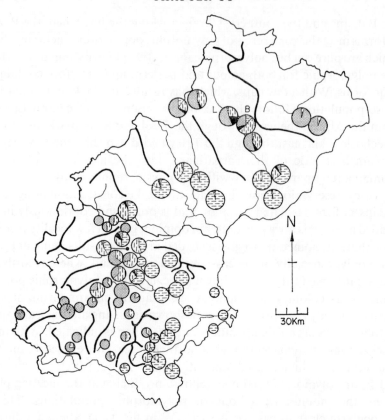

FIGURE II.2. Composition of populations of *Z. ephialtes* in (lower half) Niederösterreich (lower Austria), redrawn from Reichl (1958), and (upper half) Moravia, compiled from Povolný & Gregor (1946). Some small samples have been combined with neighbouring populations. Shading as in fig. II.I; yellow P-forms indicated by black. The thick black bars indicate watersheds. In Niederösterreich small circles are samples of less than 10 individuals, large circles of 10 to 57 individuals; exact sample sizes are not known for Moravia, except at Litovel (L) and Bělkovice (B) (see text), and percentages are estimates based on inspection of the populations and of entomological collections (Povolný, personal communication).

not exactly those expected from the intergradation of two races with the genotypes *PPRR* and *pprr*. Examining the valleys, working clockwise from the northeast of fig. II.2, we see populations which are pure yellow E, then a group of valleys in which the moths are all of the E-form, but polymorphic for colour, then a series of valleys (including the Donau, which flows across the centre of the province from west to east) in which there is a polymorphism for the P- and E-forms, but in which the moths are all red. The valley running into the Donau from the northwest, and the valley sloping northwards into Moravia contain similar populations. Only in the low hills and valleys to the north of the Donau in the centre of the province are there populations polymorphic for both pattern and colour, and only in these did the yellow P-forms appear. Putting it another way the change of colour occurs, on the whole, 10 kilometres east of the change of pattern (Reichl 1958). It looks very much as though polymorphisms for colour and for pattern tend to be mutually exclusive; selection against the yellow P phenotype might well be the reason. In Moravia likewise there are too few areas with high frequencies of yellow P (fig. 11.2, upper half).

The breeding experiments give little information about the disadvantage of the yellow P-form. Bovey's F_2 broods (table 11.1) show a significant deficiency of them, and all those surviving were heterozygotes, distinguishable because there is a slight lack of dominance in this cross: the homozygous *PPrr* individuals may well have died. On the other hand the broods of Dryja and Povolný & Pijáček show no such deficiency, the difference between these broods and those of Bovey being statistically significant.

Z. ephialtes may take one or two years to complete its life-cycle; in F_2 and backcross broods within the Merano × Berlin stock there is no association of phenotype with year of emergence (table 11.2). In the backcross within the Merano × Lublin stock there is marked heterogeneity between broods ($\chi^2_{(33)} = 64 \cdot 29$; $P < 0 \cdot 001$); in some the red E-forms emerge predominantly after two years and the yellow E after only one year, in other broods the red E-forms tend to emerge after one year. Two of the extreme examples are shown in table 11.2. It seems that the forms can differ in length of life-cycle, but the direction of the difference, and even its presence, is dependent upon the external or the genetic environment.

It is not worth commenting on the geographical distribution of the various numbers of spots, as little is known (Bovey 1941, gives the available information), but two genetical complications are worthy of note, although we do not understand their significance. The gene for yellow occurs as two iso-alleles, r^j and r^J, r^j being completely recessive to red, r^J giving an orange heterozygote (Bovey 1966). The distribution of these alleles is poorly known; r^j is known from the region of Wien (Bovey 1941) and Merano (Dryja 1959); r^J is known from the Valais

TABLE 11.2. Emergence of phenotypes of Z. ephialtes after one year, or two or more years; data from Dryja (1959).

Cross	Brood				Heterogeneity					
					Within broods			Between broods		
	Red P	Yellow P	Red E	Yellow E	χ^2	d.f.	P	χ^2	d.f.	P
BACKCROSS: Merano × Berlin										
Total Year 1	240	234	249	225	1·69	3	0·64	7·24	33	1·00
Year 2	150	141	132	142						
F₂: Merano × Berlin										
Total Year 1	1074	401	383	128	2·13	3	0·54	38·15	39	0·51
Year 2	307	106	104	45						
BACKCROSS: Merano × Lublin										
Total Year 1	621	615	586	607	—			64·29*	33	0·0009
Year 2	74	57	67	56						
Brood 17 Year 1	36	35	30	46	9·52*	3	0·02	—		
Year 2	6	5	11	2						
Brood 36 Year 1	53	33	53	42	10·10*	3	0·02	—		
Year 2	15	4	2	11						

* Significant.

(Switzerland) (Bovey 1966) and Zaleshchiki (Dnestr valley, Ukrainian SSR) (Dryja 1959). Orange forms are known in natural populations from the southern Ukraine (Kharkhov, Nikolayev), and from the band of polymorphism in Austria, Czechoslovakia and Poland. It is likely that these alleles show a widespread polymorphism. A sample from Litovel in Moravia, which is probably tolerably representative of the population from which it came, contained 98 red, 17 orange and 25 yellow (Povolný & Gregor 1946); this gives gene frequencies of 0·58 (R), 0·10 (r^j) and 0·32 (r^j), so that r^j in this population comprises about one-quarter of the r alleles.

In some crosses (Warszawa × Zaleshchiki, Merano × Lublin, Nyon × Wien) the gene P is incompletely dominant over p; in the cross Merano × Berlin, Dryja reports that dominance is complete. It is not clear whether this results from differences in genetic background or the presence of iso-alleles of P or p. Samples from Litovel and Bělkovice in Moravia contained P-forms, E-forms and intermediates in the numbers 53, 37 and 50, and 6, 46 and 19 respectively (Povolný & Gregor 1946); the deficiency of intermediates compared with the Hardy–Weinberg law, significant in the first sample, might be due to bias or mis-scoring, or variable expression, among other causes.

In summary then, the northern populations of *ephialtes* resemble other *Zygaena* species and are probably, by retention of common ancestral patterns, Müllerian mimics of them (red P-form). Some southern populations mimic *Amata* species, rather than *Zygaena* which are found in the same areas, the change to this pattern (yellow E) being produced by alleles at two unlinked loci. In addition some southern populations contain a third form (red E) which has the pattern of the *Amata*-mimic but the colour of the northern forms. We do not understand the significance of this; it would be very interesting if the third form occurred where *Amata* species are rare. A fourth form (yellow P) is produced only in populations where the southern and northern populations meet in the mountains of central Europe; there is evidence that this form may be at a selective disadvantage. Reichl (1959) has in addition shown that the distribution of yellow forms in central Europe seldom extends beyond the 9°C isotherm (actual), this being shown very strikingly by the populations containing all four forms in the region of Prague (northernmost such populations in fig. 11.1) which are isolated from other yellow populations, but which are contained approximately in an isolated ring of the 9°C isotherm.

MIMICRY

The mimetic relations of butterflies of the subfamily Heliconiinae, found predominantly in South and Central America, both with each other and with other butterflies, are so complex that it will be possible to discuss only some of the better studied examples. Most of these are in the genus *Heliconius*, which have elongated wings and, usually, bright colours, but some related genera, containing orange-brown butterflies with less elongated wings (resembling the north-temperate fritillaries or checkerspots), and called collectively the 'colaenids', will be mentioned from time to time.

To avoid a profusion of subspecific names, the geographical races of two of the species will be named by the colour-code used in plate 11.1, so that we have the *red, pink, grey* to *orange* races. The systematics of this group is particularly difficult; classification and names in the following account are based in part upon work on museum collections, in part on the following papers: Alexander (1961a,b), Beebe, Crane & Fleming (1960), Brown (1972*), Brown & Mielke (1971*), Emsley (1963*, 1965*), Lichy (1960), Michener (1942), and Turner (1966, 1967a,b,c,d, 1968b) (asterisk denotes major revision of the group). Geographical distributions are drawn from data obtained from public and private collections, and from literature records, particularly those of Biezanko (1938, 1949), Biezanko, Ruffinelli & Carbonell (1957), Ebert (1965, 1970) and Masters (1969).

There is one important change of name: the Trinidadian form of *H. ethilla* has in most recent papers been wrongly attributed to *H. numata*; I here follow what I believe to be the correct usage, applied by Emsley (1965) and Brown & Mielke (1971).

Of the many patterns displayed by the genus *Heliconius* we may note the following:

Red and yellow (stripes): for example the *red, pink, grey* and *orange* subspecies of *melpomene* and *erato*. Three species, as in table 11.3, plus *heurippa*.
Red and yellow (rays): for example the *green, blue* and *dark-blue* subspecies of *melpomene* and *erato*. The eleven species listed in table 11.3, and in addition *xanthocles* and some forms of *eanes*.
Red and yellow (blotches): yellow on the forewing, red on the hindwing,

TABLE 11.3. Mimicry within the genus *Heliconius*: a selection of species classified by pattern and morphology.

Group	Red and yellow (stripes)	Red and yellow (rays)	Red and yellow (blotches)	Tiger	Blue and yellow or black and white	Ithomiine or Acraeine	Non-mimetic
Subgenus *Laparus*	—	*doris* (red, S. America)*†	*doris* (red, C. America)†	—	*doris* (blue)*†	—	*doris* (green)*†
Subgenus *Eueides*	—	*tales**; *eanes* (form)	—	*isabella**	*eanes* (form)	*pavana*; *vibilia*	*lybia*
Subgenus *Heliconius*							
Group IA	*melpomene**† (extra Amazonian)	*melpomene*† (Amazonian); *timareta* (form)	*timareta* (form)	*ethilla**†; *nattereri* ♀†	*timareta* (form)	—	*nattereri* ♂*
Group IB	*besckei*†	*elevatus*†	—	—	*luciana*	—	—
Group II	*erato**† (extra Amazonian)	*erato*† (Amazonian); *demeter*	*clysonymus*†	—	*sara**†; *antiochus*; *congener*	*charitonia* ssp.; *peruviana*	*charitonia**†
Group III	—	*burneyi**†; *egeria*; *astraea*	*ricini**†	—	*wallacei**	—	—
Group IV	—	*aoede**†	—	—	*metharme*	*godmani*	—

* Known from early stages (pupa at least, usually all stages) (Beebe *et al.* 1960, Turner 1968b, Brown 1970).
† Illustrated in this chapter.

as the three species in fig. 11.7. The four listed in table 11.3, and in addition *hortense*, *himera* and *hierax*.

Tiger: orange-brown and yellow stripes and blotches on a black ground, for example *ethilla* (fig. 11.3). These patterns are like those of many

FIGURE 11.3. Some patterns in the genus *Heliconius*: (a) red and yellow (rays)—*H. burneyi*; (b) tiger—*H. ethilla*, yellow form; (c) blue and yellow—*H. sara*; (d) non-mimetic—*H. charitonia*. About ½ natural size.

species of Ithomiidae, and are shared by butterflies of the families Nymphalidae, Papilionidae, Riodinidae and Pieridae. The systematics of these *Heliconius* is difficult; the latest revision (Brown 1972) tentatively gives seven tiger species in the subgenus *Heliconius* (*ismenius*, *silvana*, *numata*, *ethilla*, *pardalinus*, *hecale*, and *aulicus*), in addition to *H.* (*Eueides*) *isabella* and the female of *H.*(*H.*) *nattereri*.

Blue and yellow, or *black and white*: iridescent blue with yellow forewing blotches (e.g. *sara* in fig. 11.3), or black with white (sometimes yellow) blotches and stripes. The species are listed in table 11.3.

Ithomiine or *Acraeine*: some species resemble ithomiids (other than tigers), or Acraeids (genus *Actinote*). Two *Heliconius* (*Eueides*) (*pavana* and *vibilia*) mimic *Actinote*; members of the subgenus *Heliconius* mimicking ithomiids include *charitonia* in western Peru (subspecies *peruviana*), *atthis*, *godmani*, and probably *hecuba*, but the list cannot be completed until both ithomiids and heliconids are studied together.

Non-mimetic: some species are not known to mimic any other, among them *charitonia* (fig. 11.3), *hermathena*, which is like *charitonia* with the addition of a red blotch in the forewing, the male of *H. nattereri*, like a *charitonia* with much widened yellow marks, and the green form of *doris* (fig. 11.6).

Although Latin names of exotic butterflies can mean little to anyone who has not studied the insects themselves, the above list, which is not exhaustive, with the illustrations, should give the reader an idea of the wealth of patterns and of mutually mimetic species within the subfamily.

When the species are grouped according to their morphology, preferably of early stages as well as of imagines, it can be seen (table 11.3) that morphologically related species may have markedly different patterns, and species difficult to distinguish from their patterns may have quite different early stages. *H. burneyi* and *H. wallacei* for instance have virtually indistinguishable pupae, but totally different patterns, and *burneyi* and the red form of *doris*, often difficult to separate without a close examination, are not at all similar as pupae. The pupae of *ethilla* (tiger) and *melpomene* (red and yellow) are almost identical. If, as seems likely, the morphological groups represent phyletically related species, it becomes difficult to imagine that the mimetic resemblances result simply from retained ancestral patterns (as with *Z. ephialtes* and other *Zygaena*); they must result from a complicated history of convergence and divergence of patterns.

That the mimicry is Müllerian is shown by an experiment (Brower, Brower & Collins 1963) in which Silverbeak Tanagers learned to reject Trinidadian heliconids on sight while still eating cryptically coloured satyrid butterflies (genus *Euptychia*). Mimicry was shown between *melpomene* and *erato* (the *pink* subspecies of plate 11.1), *sara* and *doris* (blue form), *ethilla* and *isabella* (both tigers) and two colaenid butterflies, *Agraulis vanillae* and *Dryas iulia*. All the mimetic pairs except *ethilla* and *isabella* were used reciprocally, so that, for example, birds trained to avoid *sara* avoided *doris*, and those trained to avoid *doris* avoided *sara*. A lower grade of mimicry was shown between rather dissimilar species, in that birds trained to avoid *iulia* or *vanillae* showed some aversion to *doris* when it was presented to them for the first time.

GEOGRAPHICAL VARIATION

It is likely that mimicry is not confined to experiments, but is effective in the wild. Mimetic heliconids certainly fly in the same areas (although it would take careful fieldwork to prove that, in the complex communities of the rain forest, they shared the same 'habitat'), and beak-marked specimens show that birds sometimes attack them in the wild (Collenette & Talbot 1930, Turner 1968b). The tanagers used by Brower *et al.* were caught in the wild in Trinidad, and one-third of them rejected the first heliconid (or colaenid) presented to them. As the species most rejected were those which were commonest in Trinidad at the time (and those which belonged to the commonest mimicry groups), this indicated that the birds had experienced, and learned to avoid, heliconid butterflies in nature. Moreover, certain species show parallel geographical variation which is difficult to explain as due to anything other than mimicry. Thus Moulton (1909) describes how the tiger butterflies, heliconids,

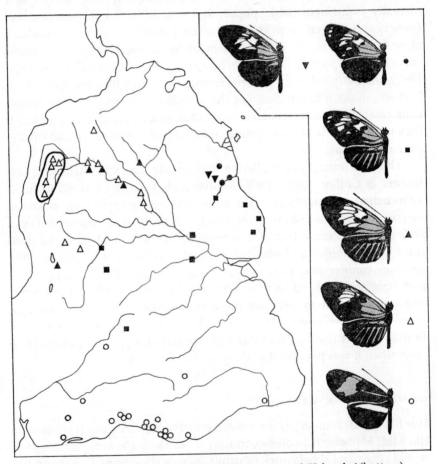

FIGURE 11.4. Distribution of races of *H. elevatus* and *H. besckei* (bottom). Compare with plate 11.1. In the region inside the black line the yellow bar on the forewing is very narrow. Original data from museums, with a few points from the literature and from K.S.Brown (personal communication).

ithomiids and others, all share the same coloration in the Guianas, a different pattern in Peru, Central America, Brazil and so on, although this work has yet to be examined in the light of the latest taxonomic revisions.

The red and yellow heliconids show this parallel variation even more strikingly. Plate 11.1 shows that *melpomene* and *erato* not only both have the red and yellow rayed pattern in the Amazon basin and the red and yellow striped pattern in extra-Amazonian areas, but that within both these large areas both species show a remarkable parallel variation, similar patterns having closely similar distributions. Other species in the red and yellow group share this parallel variation and distribution. *H. elevatus* (fig. 11.4) copies all the Amazonian races of *melpomene* (*green, purple-blue, blue* and *dark-blue*) with such exactness that the species have only recently been properly separated, and its relative (or perhaps subspecies) *besckei* (fig. 11.4) mimics the *orange* subspecies.

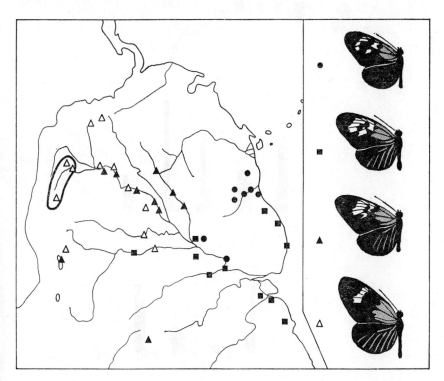

FIGURE 11.5. Distribution of races of *H. aoede*. Compare with plate 11.1; for the black line see fig. 11.4. Original data.

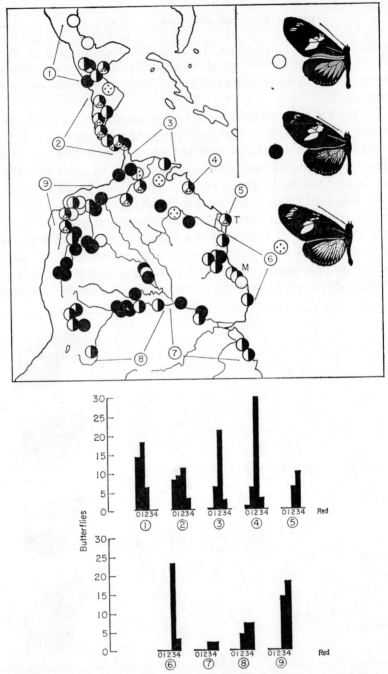

FIGURE 11.6. Distribution of the red, blue and green forms of *H. doris* (discs indicate only presence or absence, not frequencies), with histograms showing the number of butterflies of the red form with differing amounts of red on the forewing. Data from museum collections. M—Moengo, T— Trinidad (see table 11.4). Compare with fig. 11.3a,c.

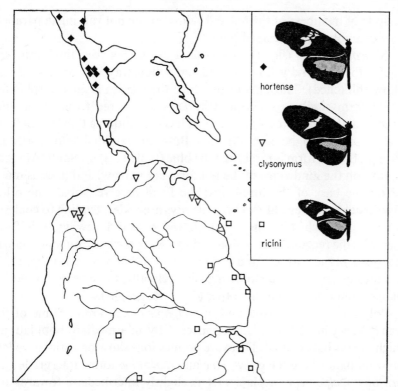

FIGURE II.7. Distribution of the three red and yellow (blotched) species, *H.hortense*, *H.clysonymus* and *H.ricini*. Original data from museum collections; *ricini* in central Amazonia from data communicated by K.S.Brown, jr.

The species *aoede* (fig. 11.5), *xanthocles*, *demeter* and *tales* also copy both the pattern and distribution of Amazonian subspecies; *burneyi*, *astraea* and *egeria* also do this, with a little less exactness. In the valley of the Rio Huallaga in northern Peru, and in some other valleys near the headwaters of that river, *elevatus*, *aoede* and *xanthocles* are like the *blue* form of *melpomene* and *erato* but have a narrower yellow band in the forewing. The area occupied by these forms is ringed in figs. 11.4 and 11.5, and it will be seen by comparison with plate 11.1 that in this valley *melpomene* and *erato* are both of an extra-Amazonian type, the *red* form, very similar in appearance to the forms found in Central America (although at the lower end of the valley, *melpomene* at least does produce a form resembling the other species). This is the most notable example of members of the red and yellow group not varying in parallel; other examples are discussed later.

Heliconius doris (fig. 11.6) acts as a bridge between three mimicry groups, the blue and yellow, the red and yellow (rayed), and the red and yellow (blotched). Throughout its range the species is polymorphic for a blue form similar to *sara* and *wallacei* and a red form resembling *burneyi*; in the northwest of its range it has in addition a third green form which appears to be non-mimetic. Both in Trinidad and northeast Surinam the red form is rarer than the blue (table 11.4). In South America the rays on the hindwings of the red form are narrow, and there is much red on the base of the forewings; the form here resembles the other rayed species. In Central America the rays are wider, tending to coalesce into a red blotch, and there is little red on the forewing; this form resembles the red and yellow blotched species, which are found in Central America and the northern and western parts of South America (fig. 11.7). The amount of red on the forewing varies clinally, there being no sudden change from one form to the other (fig. 11.6, graphs).

Polymorphism is also found in *melpomene* and *erato*. Some of the forms shown in plate 11.1 are connected by intermediate populations, which are probably clinal. There are populations showing variable yellow forewing bars where the *green, blue* and *dark-blue* forms intergrade; the *dark-blue melpomene* in plate 11.1 is slightly intermediate towards the *green* form. There are also populations on the west coast of Colombia intermediate between the *grey* and *red* forms, in that they are black (not blue) on the upper-side, with no marks but the red band, but have the yellow bar on the underside of the hindwing (the *grey* forms also have this bar on the underside, not shown in the plate). In addition the species show rampant polymorphism in areas of transition from the Amazonian to the extra-Amazonian patterns. The best known of these areas is in the Guianas (northern meeting of the *green, pink* and *purple-blue* forms), for many hundreds of 'varieties' have been exported from there by a professional dealer (Joicey & Kaye 1917); other areas are on the upper Rio Putumayo in southern Colombia (northern meeting of *red* and *blue*), the eastern Andean valleys in Ecuador (meeting of *blue* and *blue-triangle*), the lower Rio Huallaga in Peru (southern meeting of *red* and *blue*), the upper Rio Madeira (Rio Grande) in Bolivia (meeting of *blue, green* and *orange*), and the state of Maranhão in northeast Brazil (meeting of *green* and *orange*); there is also a region of polymorphism on the lower Rio Amazonas (the small pink areas).

PLATE II.I

Geographical distribution of the chief races of *H. melpomene* and *H. erato*, simplified by the omission of hybrid populations and zones of gradation. The subspecies of *erato* indicated by a yellow area is not illustrated, and has no parallel form in *melpomene*, which seems to be absent from that valley. Modified from Turner (1970), by permission of the editors of *Science Progress*. For a detailed map of the hybrid populations of Guiana, see Turner (1971c).

The names of the subspecies are believed to be as follows:

colour code	*melpomene*	*erato*
red (Central America)	*rosina*	*petiverana*
red (Colombia)	*bellula* (?)	*demophoon*
red (Rio Huallaga)	*amaryllis*	*favorinus*
pink (northern)	*melpomene*	*hydara*
pink (Peru)	*euryades*	*amphitrite*
grey (W. Ecuador)	*cythera*	*cyrbia*
grey (W. Colombia)	*vulcanus*	*venus*
yellow	—	*chestertonii*
blue-triangle	*plesseni*	*notabilis*
red-triangle	*xenoclea*	*microclea*
blue	*aglaope*	*estrella*
dark-blue (northern)	*vicina*	*reductimacula* (?)
dark-blue (Bolivia)	*penelope*	*venustus*
purple-blue	*meriana*	*amalfreda*
green	*thelxiope*	*erato*
orange (central)	*burchelli* ⎱	*phyllis*
orange (coastal)	*nanna* ⎰	

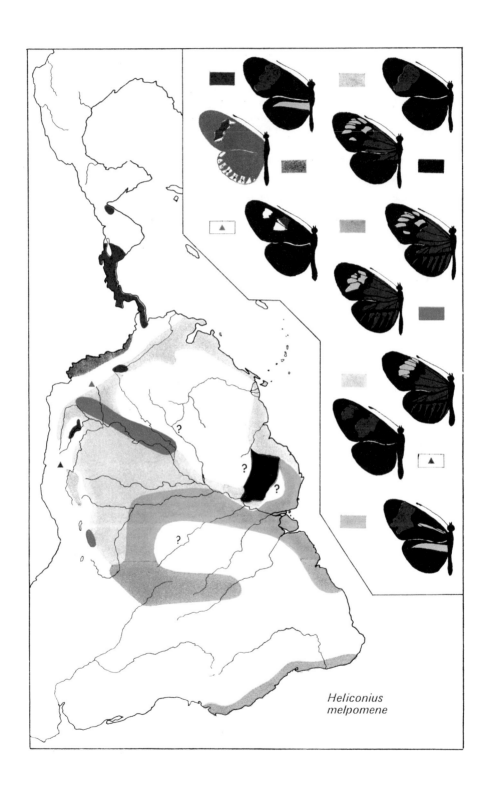

Heliconius melpomene

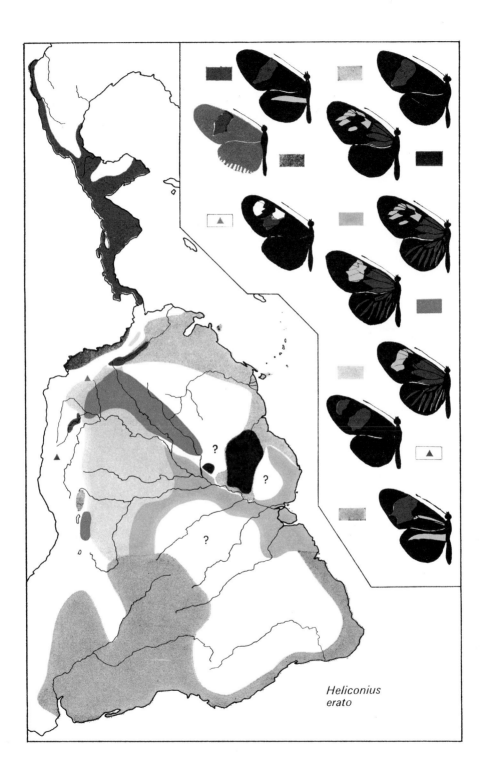

*Heliconius
erato*

The position of these polymorphic populations, at the meeting places of much more extensive monomorphic populations, implies that they are hybrids produced by the joining of previously isolated geographical races. The phenotypes of the butterflies in these areas give support to this idea. For example phenotypes found where the *blue-triangle* and *blue* forms meet in Ecuador include butterflies with the pattern of the *blue-triangle* forms, but with the forewing marks in various mixtures of red, yellow and white, butterflies with the two forewing marks of the *blue-triangle* form, with the red basal marks and hindwing rays of the Amazonian forms, and butterflies of this same pattern, but with the two outer marks on the forewing entirely yellow. These are among those we would expect to find among hybrids of the presumed parental subspecies. Genetic work on the Guianian forms produces further favourable evidence.

TABLE 11.4. Numbers of red and blue forms of *H. doris* in wild populations

Place	Date	Blue	Red	Total
Trinidad, West Indies	– ix 1962	15	1	16*
Moengo, Suriname	21 vii–3 viii 1962	12	3	15*
Moengo, Suriname	18–19 ix 1964	27♂ 5♀	6♂ 0♀	38†

* Sheppard 1963, captures and sightings.
† Original data, captures only.

GENETICS

Experimental crosses have been performed with *melpomene, erato, ethilla* and *doris*; the most extensive work has been on *melpomene* and *erato* in the Guianas. The *pink* subspecies from Trinidad have been crossed with polymorphic forms from the Guianas. The latest crosses in *melpomene* (Turner 1971a) used a stock derived from the interior of Surinam where the species is almost of the pure *purple-blue* form, with a small mixture of genes from the *green* and *pink* subspecies; these experiments have probably discovered most of the major genes differentiating these three subspecies.

The pattern of red rays and blotches of the *green* race is produced by an allele (D^R) dominant to the allele (D) producing the red marks of the *purple-blue* race, which in turn is dominant to the allele (d) producing no red marks in these areas, as in the *pink* race. The difference between the solid red band of the forewing in the *pink* race, and the broken yellow band of *green* and *purple-blue* races is produced by two loci *B* and *N*, one showing complete dominance, the other having intermediate heterozygotes; their interaction is shown in fig. 11.8 (Turner & Crane 1962, Sheppard 1963, Turner 1971a,c). In addition there is a locus (*F*) which controls the form of the yellow band on the forewing, so that *FF* and *Ff* individuals have the broken band of the *green* subspecies, *ff* individuals

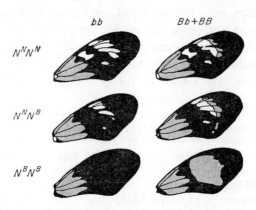

FIGURE 11.8. The interaction of the *B* and *N* loci in *H. melpomene* in determining the form of the band on the forewing.

have a fused band like the *dark-blue* race; the *green* and *purple-blue* races are *FF*, the *pink* race in Trinidad is *ff*, and because of the dominance relations the fused phenotypes appear only in F_2 broods. At first sight it is strange that the Trinidadian butterflies should be homozygous for a gene affecting the distribution of yellow marks which they do not possess; in this race the locus controls the distribution of white scales on the underside of the forewing, so that in *ff* genotypes the whole of the band which would otherwise be red is covered with white scales. This white reflects ultraviolet and probably appears white to the butterfly (Crane 1954) but as *erato*, which has similar white scales, does not respond to ultraviolet white when courting (Crane 1955), it seems likely that the colour has no function in intraspecific behaviour; possibly it simply makes the underside of the butterfly more cryptic, for *Heliconius* are usually much more drab underneath than above. The loci *B* and *D* are linked with an undetermined crossover rate, probably less than 16 per cent; *F* and *N* are in two further separate linkage groups, although there is some equivocal evidence that D and F show irregular assortment in back-crosses where the heterozygous parent is female (Turner & Crane 1962, Sheppard 1963, Turner 1971a).

There are a few complications and uncertainties. The allele N^N produces not only the yellow banding or spotting in the outer part of the forewing, but a yellow line running along the centre of the wing from its base, like that found in the *orange* race; its base can be seen as a yellow spot in fig. 11.8, although the rest of it is masked by the red markings. In addition, in some F_2 broods N^N controls one character that is found in none of the parental subspecies, traces of the yellow bar on the hindwing (found in the *orange* and *red* races), and several characters not normally found in *H. melpomene*, but found in related species, namely a row of white dots round the hind edge of the hindwing and a group of small white dots at the apex of the forewing (found in *ethilla* and *elevatus*) and a white spot at the outer angle of the hindwing (found in *ethilla*). In some broods the genotype bbN^BN^B has no forewing band, as in fig. 11.8 (Turner 1971a), in others it has a narrow red band like that of the genotype $B\text{-}N^NN^B$ (Turner & Crane 1962). The most likely explanation is that there is a so far undetected locus which enhances the amount of red in this genotype; this would also account for some wild-caught butterflies whose yellow marks show them to be N^NN^N, but which have a wide red band (Turner 1971c).

Thus the *green, purple-blue* and *pink* races are homozygous for the genotypes bFN^ND^R, bFN^ND and BfN^Bd. Hybridization of these three races would produce a great variety of patterns, and indeed almost all the varieties caught in the Guianas in the wild can be explained in terms of the phenotypes produced in the breeding experiments (with permutations of the basal markings with the various kinds of bands), although a few bizarre and rare forms of the forewing band remain to be explained. There is also a little circumstantial evidence that the frequencies of the forms in the polymorphic populations of the Guianas are roughly those expected from random mating of the parental subspecies (Turner 1971c).

Breeding experiments with *H. erato* have so far used stocks from Trinidad and a hybrid population in Surinam which is predominantly of the *pink* race; it is therefore possible that we do not know all the loci distinguishing the *green* and *pink* races. The red marks at the base of fore- and hindwings are produced by a dominant allele (D^R), and the shape of the forewing band by another locus, with the broken form (as in the *green* race) dominant to the solid form of the *pink* race; these loci are unlinked. A yellow band is recessive or hypostatic to a red one (number of factors unknown) (Beebe 1955, Turner & Crane 1962, Sheppard 1963). It seems likely that the polymorphism of *erato* in the Guianas is due to hybridization also.

Crosses between the *pink* races of *melpomene* and *erato* from Trinidad, and the *grey* and *blue-triangle* forms from Ecuador (Emsley 1964) are difficult to interpret, as the broods are small and it is not easy, if a character is hard to score, to distinguish continuous variation from segregation due to a single locus, and recombination from pleiotropism with variable expression. The results indicate that racial differences are inherited as follows:

Grey × *pink* races in *erato*: blue iridescence and shape of red band are polygenic, with little directional dominance; yellow bar on hindwing (underside, not shown in plate 11.1) is a single locus, the yellow bar being recessive but detectable in the heterozygote as a change in the shininess of the scales; the white marignal marks on the hindwing and a difference in the colour of spots on the head appear to be controlled by single loci with intermediate heterozygotes; the absence of a small red costal mark on the forewing (underside) (*grey* form) is very nearly recessive to its presence (*pink* form). The genes controlling the yellow bar and the white marks are linked, but show some recombination (or variability of expression).

Grey × *pink* races in *melpomene*: shape of band and blue iridescence are probably polygenic; the yellow bar, the white margin and the red costal spot are inherited as they are in *erato*; the white edging of the red forewing band is controlled by a single recessive gene; and in the F_2 a yellow forewing line, like that of the *orange* race, appears. All the single-locus characters, and the colour of the spots on the head, are affected by the same chromosome; the behaviour of the white dots and yellow bar in the *purple-blue* × *pink* cross makes one suspect that apparent recombination in the *grey* × *pink* cross may be due to variable penetrance, but if it is recombination then head colour and costal spot are loosely linked to the others, which form a tight cluster.

Blue-triangle × *pink* in *erato*: the white colour of the band is recessive or hypostatic to red colour (the number of factors is not known); the shape of the band can be scored as three characters—length, invasion of black into the middle, and roundness of the anterior margin; long bands and flat margins are recessive and probably unifactorial; it is not clear how the invasion of black is inherited.

In Trinidad, and in parts of South America (Brown 1972) *ethilla* (fig. 11.3) is polymorphic for the amount of yellow on the wings; the difference is due to a single locus, with the yellow form recessive. The patterns controlled by this locus are closely similar to those controlled by the *N* locus in *melpomene*, as the locus in *ethilla* influences not only the yellow forewing line and hindwing bar, but the size of the white dots on the hindwing (Turner 1968a).

H. doris (fig. 11.6) is not an easy species to breed; some results indicate that red is dominant to blue (Sheppard 1963), others, including 3:1 segregations, that red is recessive (Cook & Brower 1969). The green form appears to be recessive to blue, the colour change being produced by a change from yellow scales mixed with black in the green form to white scales mixed with black in the blue form, a change which is particularly easy in *Heliconius*, for example in the forewing band of *melpomene*, where the scales are white if underneath the red band, but otherwise are yellow. However, Cook & Brower find green segregating in numbers significantly lower than one in three, so that inheritance is not simple; they also find an overall excess of males over females in all their broods.

BEHAVIOUR, SENSORY PHYSIOLOGY AND DEMOGRAPHY

In a heliconid courtship, when both male and female are in flight at the beginning, the male initiates the courtship by chasing the female; if she is receptive she flies more slowly than normal, and the pair pursue each other in slow flight, with the male bumping into the female from above or below, or with the pair circling round each other. The female will then settle and the male hover over her, or in front of her, or settle beside her beating his wings. Once the female has closed her wings the male will settle by her, if he has not already done so, and copulate with her from a lateral position, moving from there into the normal "back to back" position. There are considerable variations between species, and some may omit some of the phases (Crane 1957).

In Trinidadian *H. erato* Crane (1955) has shown that red colour of the hue found on the butterfly's wing stimulates courtship behaviour in both sexes; the butterflies prefer orange flowers when feeding. The response to red is innate, being shown by butterflies which have never seen this colour. Swihart (1963, 1964, 1965, 1967a,b, 1968) has studied the way in which the sense-organs control this response to colour. At night the electroretinogram (ERG) of *erato* shows that the eye is responding to weak light, with a wave-form which does not alter with the wavelength of the light; the flicker-fusion frequency is about 90 cycles per second. During the day, under the stimulus of the butterfly's metabolic clock, the ERG changes to a different form, not so responsive to weak light, but showing a change in wave-form with wavelength (i.e. an ability to detect colour). The flicker-fusion frequency is about 165 cps, and in addition at about 20–26 cps the ERG changes from an 'on' response (the nerves are stimulated when the flash begins) to an 'off' response. The wingbeat of a courting male is 26 strokes per second (Swihart 1963). The retina itself contains at least two colour-receptors, one responding mainly to blue-green light and the other to red light, but deeper in the optic nervous system the ERG shows its strongest response to red light; Swihart (1964, 1965) suggests that the enhancement of the response to red is produced actively under the influence of nervous impulses coming from the central nervous system. On the other hand Bernhard *et al.* (1970) have found a layer of tracheoles round the retinal receptors of *H. erato* which reflect red light; they regard this as the cause of the extra sensitivity of the retina to red light, but have not attempted to explain the different responses discussed below between or within other species.

It seems therefore (Swihart 1963) that the metabolic clock of *H. erato* causes the ERG to change in the daytime to a form in which it will respond to colour and to the wingbeat of a courting butterfly, and that the red reflecting layer and selection of impulses within the optic tract cause this response to be particularly to red, the colour of the butterfly's own wings. The ERG of two colaenids shows, in the deeper layers of the optic tract, an enhanced response to their own colour (Swihart, 1967a). Thus the orange *A. vanillae* and the green *Philaethria dido* have peaks respectively at orange and green; the blue *Morpho peleides*, a member of another subfamily, peaks in the blue. *H. sara* and *H. ricini*, respectively black, blue and yellow (fig. 11.3) and black, red and yellow (fig. 11.7) have the same type of ERG, peaking in the red, but being much more responsive than that of *erato* to shorter wavelengths, the ERG is in effect responding to yellow. Both species also have an alternative ERG with a maximum in the green; in *sara* the yellow-red ERG develops while the butterfly is in the pupa, and the green one up to twelve hours after emergence (Swihart 1967b), but it is not known what causes the change from one to another form in the adult butterfly, although one *ricini* was observed in the act of change (Swihart 1967a).

A mark-release-recapture experiment on a dense population of *H. erato* in an overgrown coconut plantation in Trinidad revealed three very interesting and unusual things about these butterflies (Turner 1971b). First, they are exceptionally long-lived, with a daily survival rate for both sexes of around 99 per cent of the population, this giving a life-expectation of between 50 and 100 days, which is of the same order as that recorded in captivity. In contrast, temperate butterflies normally live longer in captivity than in the wild. Second, the butterflies have a restricted home-range. The experiment was carried out in three areas, the first two of which touched at their edges, and had centres only thirty yards apart; if a butterfly was captured in one of these areas it was highly probable that the next capture would be in the same area, only 34 out of 166 recaptures being of a butterfly caught previously in the other area. There was virtually no movement between these two areas and the third one, 200 yards away (one butterfly out of twenty had moved area). This lack of movement was not due to any weakness of flight by these butter-flies, and probably resulted from their maintaining a learned home-range of comparatively small extent (for evidence of learning-power, see works cited by Turner 1971b). As heliconids usually move no more than 30 or 40 yards when larvae (Alexander 1961a,b), an individual *H. erato* is likely to spend its whole life within an area of a few hundred square yards.

Home-range behaviour is obviously connected with the third interesting fact about wild *H. erato*: at night the butterflies return to a communal roost more or less within their home-range, up to several dozen *erato*, sometimes in company with *sara*, sleeping together on dried creeper hanging from the palms. Communal roosting has been noted also in *H. melpomene*, *sara* and *ricini* (Crane 1957), *H. ethilla* and *Dryadula phaetusa* (Turner, unpublished), *H. tales* and the *green* sub-species of *melpomene* and *erato* (Moss 1933), and in *H. charitonia*, which repeatedly 'homes' to the same roost (Beebe 1950; Jones 1930).

ALTRUISM

Being warningly coloured and nauseous is a form of altruism, for it is of no benefit to the individual, once he has been eaten, to have taught the predator to avoid other members of that species. As Hamilton (1964a,b) has shown, only if the beneficiaries are related to the martyr, so that they are likely to carry the same genes, will genes producing altruistic characters spread under natural selection. It is thus very interesting that *H. erato*, being warningly coloured and distasteful (and therefore altruistic), should have such a restricted home range, as when a predator learns by experience not to eat this species, the immediate beneficiaries (the *H. erato* within the predator's home range) will be relatives of the martyr.

Probably several other unusual features of *Heliconius*, unusual compared to those of many temperature butterflies at least, are related to altruism, warning colour and restricted home range. The extremely long adult life span, presumably assisted not only by the tropical climate but by relative freedom from predation, coupled with the very short-lived early stages (about three weeks from egg to adult in *melpomene* in Trinidad), must result in a butterfly flying in the same population as its parents and even grand-parents, which will again favour group-selection for the development of altruism. There must often be strong selection against a female laying all her eggs at once in the same place. A short-lived species will spread its eggs in space; a long-lived *Heliconius* is able to distribute its eggs in time, while remaining in a restricted area. Exceptionally, a few species lay their eggs in large clusters (see Beebe *et al.* 1960), and I have seen two wild female *H. sara* taking turns at laying their eggs in a shared batch. The significance of this unusual social behaviour is not known, but it may allow them to lay a large batch of eggs without having to mature all their eggs at once, such an ability being foreign to *Heliconius*. If the laying of mixed batches is a regular habit in the species it would have to be co-ordinated, as I have observed that eggs of *Dione juno* (a colaenid) layed by a female next to a previous batch of eggs do not survive, the first brood of larvae to hatch eating the second brood of eggs. The tendency of heliconids to carry out in a limited space the elaborate courtship flight described above may be connected with their restricted home-range (selection for choosing a vigorous mate will necessitate a long aerial courtship), and communal roosting may well be a protection to them, it being usually believed that it helps a distasteful species to be gregarious.

If the adult is less subject to dying than the disease and parasite-prone larvae and pupae, selection for rapid development in the vulnerable early stages may have delayed certain physiological developments until the long-lived adult stage, for as Labine (1968) points out, *Heliconius* are unable to lay eggs for several days after emergence, in contrast with a related North American fritillary (checkerspot) butterfly, which lays most of its eggs in the first few days of life.

If all the above features of heliconids are functionally related, then within the group there should be a correlation between distastefulness, restricted home range and range of courtship, communal roosting, longevity, and delayed sexual maturity. There is some evidence that this is so. To birds the most palatable of the species tested (Brower *et al.* 1963) were *D. iulia* and *A. vanillae*, then in increasing order of unpalatability come *H. doris*; *erato* and *sara*; and *melpomene* and *ethilla*. Of these only the last four are known to roost gregariously; the courtship flight of *vanillae* seems from a limited number of observations (unpublished) to involve a rapid chase in a straight line over some distance, rather than the circumscribed and prolonged fluttering of *Heliconius* and *D. phaetusa* (a colaenid which roosts communally); and the period between emergence and egg-laying is about 2–4 days in *A. vanillae*, *D. juno* and *D. phaetusa* (all colaenids), and about 7–11 days in *H. melpomene* and *ethilla*. The possibility of such functional correlations is worth investigating further.

DISCUSSION

As polymorphism is not an essential of Müllerian mimicry, one would not expect the study of polymorphic Müllerian mimics to reveal as much about the genetic architecture of Müllerian mimicry as is revealed by similar studies of Batesian mimicry (Sheppard 1965, 1969). The polymorphism of *Zygaena ephialtes* for spots, and for the two iso-alleles, and even the polymorphism of *H. ethilla* for brown and yellow forms will probably tell us little about the evolution of mimicry, although *H. doris* may be more profitable, as its polymorphism is like that of a Batesian mimic (two different mimetic forms, and a non-mimic), even though the species is unpalatable. One interesting thing about the brown-yellow polymorphism of *H. ethilla* is the apparently different viabilities of the morphs in males and females (Turner 1968a); it could be a coincidence, but its relative *H. nattereri* is sexually dimorphic, with brown females and yellow males (Brown 1970). More information about genetic architecture comes from studying differences between subspecies.

Most of the evolution of pattern in the heliconids comes from changes in the distribution of four basic colours: orange-brown (as in the colaenids or *ethilla*), red, yellow and white (blue is usually produced by iridescence, and *Philaethria* the only truly green genus is pigmented in the wing membrane, not the scales). From the species so far studied

(*D. iulia* and *H. erato* by Baust 1967, *H. sara* by Tokuyama *et al.* 1967) it seems that orange-brown and red are produced by similar pterin pigments, and yellow and white by a totally different pigment, 3-hydroxy-kynurenine, a cyclic amino acid probably produced by blocking a step in the synthesis of nicotinic acid from tryptophan (Brown 1967). (This pigment has now been found in forty-seven species—Brown & Domingues 1971.) Comparison of patterns between species show that red marks are frequently derived from brown (as in *ethilla* and *melpomene*) or yellow from white, but that one never sees a red mark which has become yellow. *Zygaena ephialtes* on the other hand directly substitutes yellow for red by a single mutation.

Crosses between subspecies of *H. melpomene* (and *H. erato*) reveal something of the genetic architecture behind this interplay of red and yellow patterns, the D and B loci of *melpomene* controlling chiefly red marks, the N and F loci chiefly yellow. In contrast with the Batesian mimic *Papilio dardanus* (Clarke & Sheppard 1960), differences between mimetic patterns are controlled not by supergenes but by a set of unlinked or loosely linked loci. Supergenes can form in two ways: by the reduction of linkage between two already polymorphic loci (Fisher 1930, Nei 1967, Turner 1967e), or by a situation in which two new mutations will establish themselves only if closely linked (Kojima & Schaffer 1964). As Batesian mimicry involves persistent polymorphism, and the Müllerian mimicry of these *Heliconius* involves the formation of geographical races, presumably by short-lived transient polymorphism (plate 11.1), it is tempting to conclude that the species differ in the time for which the polymorphism exists, this being too short in the Müllerian mimics for linkage to be substantially increased. This argues that supergenes causing mimicry arise by the first, rather than second method. However, this conclusion may not be valid, as we do not know that the patterns of our two subspecies have evolved directly one from the other; they may have gone through many intermediate stages, simply substituting genes at one locus at a time. Moreover, the locus controlling the red marks at the base of the wings in *melpomene* (*green* race) may be not a series of multiple alleles (D^R, D, d) but a supergene (DR, Dr, dr), one locus producing the red marks on the forewing, the other the rays on the hindwing; this is suggested by the polymorphism of *melpomene's* close relative *H. timareta*, which has patterns attributable to the combinations DR, dR and dr. It is possible that the N locus, which in *melpomene* (*purple-blue* race) controls yellow bars on both wings and white marginal and apical markings, and is probably responsible for similar bars and marginal marks in the *grey* race (in addition to controlling the polymorphism of *ethilla*, and perhaps the markings of the *orange* race and of *charitonia*) is itself a supergene, or it may simply increase the amount of 3-hydroxy-kynurenine produced, its distribution being controlled by other loci.

If the evolution of linkage has taken place in *Heliconius* species then it has probably occurred without changes in chromosome numbers, all but two out of seventeen species examined of the subgenus *Heliconius* having a haploid number of 21; *melpomene* and *erato* (four and five subspecies respectively) show no geographical variation in chromosome number (de Lesse 1967, Emmel 1969, Suomalainen, Cook & Turner 1971).

Z. ephialtes appears to have evolved its mimicry of *A. phegea* by means of two unlinked loci, controlling pattern and colour, and it is likely that the normally disadvantageous yellow mutant was unable to spread until the mimetic pattern was established. It is however possible that the *P* locus controlling pattern is a supergene, for Dryja (1959) reports some mutations of it that could in fact be crossovers.

It is difficult to say how much the mutual mimicry of heliconids involves retained ancestral patterns, as in *Zygaena*. Close examination of the patterns of *ethilla* and the *green* race of *melpomene* shows that, in addition to the yellow marks already discussed, these superficially very different butterflies are quite similar, requiring only a change of brown to red and a slight redistribution of markings to convert one to the other. The red ray marks of *melpomene* become, in some individuals back-crossed into the *pink* race, much wider, and not unlike the brown lunules on the hindwing of *ethilla*. The patterns of *erato* on the other hand, although superficially so similar to those of *melpomene*, prove on detailed examination to be very different, and not readily derived from them, as can be seen in plate 11.1 by comparing the red hindwing rays of the Amazonian races, or the forewing marks of the *blue-triangle* forms. Of these three species, *melpomene* and *ethilla* are more closely related to each other than they are to *erato*, and it is interesting that there are no species with the tiger pattern among the closest relatives of *erato*.

The extremely close parallel variation of *melpomene*, *erato* and the various other species presents us with a paradox: over most of the ranges of the butterflies the match is exact, but in other areas there is much less mimicry. The situation on the Rio Huallaga has already been mentioned; the *purple-blue* race of *erato* extends much farther west than that of *melpomene*; in Colombia *burneyi* has a pattern like that of the *purple-blue* races, but *erato* and *melpomene* have the *pink* pattern. Thus selection has been strong enough to cause the extreme convergence of *melpomene*, *erato* and the others, but is lax enough to allow anomalies of this sort.

The distribution of races in *melpomene* and *erato* is most readily explained as the result of subspeciation occurring as a result of the isolation of populations during climatic cycles accompanying glacial periods; Hester (1966) reviews the evidence for such cycles in the neotropics. Subsequent spread of the races would result in the polymorphic hybrid areas which we see today, the discovery of the monomorphic *purple-blue* race of *melpomene* having solved certain problems which originally made this interpretation difficult (Turner 1965). The hybrid area in the Guianas has persisted at least since the mid-eighteenth century, as is shown by some of the morphs being illustrated in works by Seba (1765) and Petiver (1767) (see also Turner 1967c) and is probably less than 50 kilometres wide (Turner 1971c). Likewise the similar hybrid zone of *Z. ephialtes* has existed at least since 1783, the year in which a yellow P-form was first described. These zones have thus persisted for at least 2000 generations in *Heliconius* (generation about 5 weeks) and 120 generations in *Z. ephialtes* (generation about $1\frac{1}{2}$ years), without speciation taking place, although one of the hybrid forms is probably selectively removed in *ephialtes*.

If speciation were to occur within *melpomene* and *erato*, and the ranges of the new species came to overlap, we would have a group of closely related species (all derived from *melpomene*), with diverse patterns, each mimicking a member of an equally diverse group of closely related species (from *erato*). It is likely that this is the way in which the pattern of diverse, related and mimicking species shown in table 11.3, has arisen within the genus *Heliconius*, as there would have been ample opportunities for this cycle of events to recur during the repeated glaciations of the Pleistocene. It seems likely that during the evolution of these patterns, selection for mimicry has been the driving force, but there may have been times when the properties of the optic nervous system, through sexual selection during courtship, have favoured the development of particular colours.

It would be interesting to know how much *Zygaena* species show home-range behaviour, and whether in these colony-forming insects there are any opportunities for the evolution of altruism, as described for *Heliconius*. The life-span of these moths seems to be no longer than

that of temperate, palatable butterflies, for the daily survival of
Z.filipendulae is estimated as 71 per cent (Manly & Parr 1968), which is
well within the range of 60 to 90 per cent for various British butterflies
(Cook, Frank & Brower 1971) and far below the estimate for *H. erato*.

Zygaena species release hydrocyanic acid from their tissues when
crushed, whether or not the larvae have fed on cyanogenic plants (Jones
et al. 1962). Some *Passiflora* are cyanogenic, *P. loncophora* for instance
is extremely so, and in view of the fact that no heliconid is known to feed
on any plant outside the Passifloraceae, it is likely that there is some
special chemical relation between the butterflies and their food plants
(Brower & Brower 1964, Ehrlich & Raven 1965), like that shown by
Rothschild, Reichstein, Brower and others between unpalatable danaid
butterflies and the Asclepiadaceae (Brower *et al.* 1967, Reichstein *et al.*
1968). If, as Dr Rothschild has suggested to me, the ability to feed on
poisonous Passifloraceae enabled the heliconids to enter a virtually
empty habitat, then this may have touched off the whole explosive
evolution of the group in the woodlands of tropical America.

ACKNOWLEDGEMENTS

My interest in the evolution of Lepidoptera owes a debt of inspiration to
E.B.Ford (from youthful reading of *Butterflies* and *Moths* through three
enjoyable years working in his laboratory) which cannot be repaid in
formal acknowledgements. The extensive research programme on heliconids has been directed by William Beebe and Jocelyn Crane-Griffin
of the New York Zoological Society, and this review owes much to a
period spent in Trinidad under Miss Crane's hospitality. I am grateful
to Dr A.H.D.Brown for criticizing sections of the draft, to Dr K.S.
Brown, jr. for much stimulating correspondence on *Heliconius*, and to
the staffs of the British Museum (Natural History) and the Hope
Department, Oxford, from whose collections came most of the information for the maps of *Heliconius* distributions.

REFERENCES

ALEXANDER A.J. (1961a) A study of the biology and behavior of caterpillars, pupae,
and emerging butterflies of the subfamily Heliconiinae in Trinidad, West
Indies. Part I. Some aspects of larval behavior. *Zoologica* (*N.Y.*) **46**, 1–24.

ALEXANDER A.J. (1961b) A study of the biology and behavior of the caterpillars, pupae and emerging butterflies of the subfamily Heliconiinae in Trinidad, West Indies. Part II. Molting, and the behavior of pupae and emerging adults. *Zoologica (N.Y.)* **46**, 105–123.

BAUST J.G. (1967) Preliminary studies on the isolation of pterins from the wings of Heliconid butterflies. *Zoologica (N. Y.)* **52**, 15–20.

BEEBE W. (1950) *High jungle*. Bodley Head, London; Duell, Sloan and Pearce Inc., New York.

BEEBE W. (1955) Polymorphism in reared broods of *Heliconius* butterflies from Surinam and Trinidad. *Zoologica (N. Y.)* **40**, 139–143.

BEEBE W., CRANE J. & FLEMING H. (1960) A comparison of eggs, larvae and pupae in fourteen species of Heliconiine butterflies from Trinidad, W.I. *Zoologica (N. Y.)* **45**, 111–154.

BEEBE W. & KENEDY R. (1957) Habits, palatability and mimicry in thirteen Ctenuchid moth species from Trinidad, B.W.I. *Zoologica (N. Y.)* **42**, 147–157.

BERNHARD C.G., BOËTHIUS J., GEMNE G. & STRUWE G. (1970) Eye ultrastructure, colour reception and behaviour. *Nature (London)* **226**, 865–866.

BIEZANKO C.M.DE (1938) Breves apontamentos sôbre alguns lepidópteros encontrados nos arredores de Posadas, em Missiones, na Argentina e de Villa Encarnación, no Paraguai, feitos durante excursões em 1931. *O Campo (Rio)* **9**, 1–7.

BIEZANKO C.M.DE (1949) Acraeidae, Heliconiidae e Nymphalidae de Pelotas e seus arredores. (Contribuição ao conhecimento da fisiografia do Rio Grande do Sul.) Biezanko, Pelotas, pp. 16.

BIEZANKO C.M.DE, RUFFINELLI A. & CARBONELL C.S. (1957) Lepidoptera de Uruguay: lista anotada de especies. *Revta Fac. Agron. Montivideo* **46**, 1–152.

BOVEY P. (1941) Contribution à l'étude génétique et biogéographique de *Zygaena ephialtes* L. (Lep. Zygaenidae). *Revue suisse Zool.* **48** (1), 1–90.

BOVEY P. (1942) Apparition de formes orangées dans un croisement interracial de *Zygaena ephialtes* L. *Arch. Julius Klaus-Stift. Vererb.Forsch.* **17**, 432–433.

BOVEY P. (1948) Determinisme génétique des formes orange chez *Zygaena ephialtes* (L.) *Arch. Julius Klaus-Stift. Vererb.Forsch.* **23**, 499–502.

BOVEY P. (1950) Deux formes nouvelles de *Zygaena ephialtes* (L.) obtenues par croisement. *Arch. Julius Klaus-Stift. Vererb.Forsch.* **25**, 35–38.

BOVEY P. (1966) Le problème des formes orange chez *Zygaena ephialtes* (L.). *Revue suisse Zool.* **73**, 16–218.

BROWER L.P. & BROWER J.VZ. (1964) Birds, butterflies, and plant poisons: a study in ecological chemistry. *Zoologica (N. Y.)* **49**, 137–159.

BROWER L.P., BROWER J.VZ. & COLLINS C.T. (1963) Experimental studies of mimicry. 7. Relative palatability and Müllerian mimicry among Neotropical butterflies of the subfamily Heliconiinae. *Zoologica (N. Y.)* **48**, 65–84.

BROWER L.P., BROWER J.VZ. & CORVINO J.M. (1967) Plant poisons in a terrestrial food chain. *Proc. natn. Acad. Sci. (Washington)* **57**, 893–898.

BROWN K.S.JR. (1967) Chemotaxonomy and chemomimicry: the case of 3-hydroxy-kynurenine. *Syst. Zool.* **16**, 213–216.

BROWN K.S.JR. (1970) Rediscovery of *Heliconius nattereri* in eastern Brazil. *Ent. News* 81, 129–140.

BROWN K.S.JR. (1972) The Heliconians of Brazil. Part IV. A rational proposal for the taxonomy of the Silvaniform Heliconians (in preparation).

BROWN K.S.JR. & DOMINGUES C.A.A. (1971) A distribuição do amino-ácido 3-hidroxi-L-quinurenina nos lepidópteros. *Anais Acad. bras. Ciênc.* suppl. 1970 (in press).

BROWN K.S. & MIELKE O.H.H. (1971) The Heliconians of Brazil (Lepidoptera: Nymphalidae). Part II. Introduction and general comments, with a supplementary revision of the tribe. *Zoologica (N.Y.)* (in press).

BULLINI L. & SBORDONI V. (1970) Evoluzione del mimetismo in *Zygaena ephialtes* (L.). *Atti Ass. Genet. Ital.* 15, 207–209.

BULLINI L., SBORDONI V. & RAGAZZINI P. (1969) Mimetismo mülleriano in popolazione italiane di *Zygaena ephialtes* (L.) (*Lepidoptera, Zygaenidae*). *Arch. zool. ital.* 44, 181–214.

BURTON M. (1962) Too hot to handle. *Daily Telegraph*, July 21st 1962, p. 11.

CLARKE C.A. & SHEPPARD P.M. (1960) Supergenes and mimicry. *Heredity* 14, 175–185.

CLARKE C.A., SHEPPARD P.M. & THORNTON I.W.B. (1968) The genetics of the mimetic butterfly *Papilio memnon* L. *Phil. Trans. R. Soc. London* B 254, 37–89.

COLLENETTE C.L. & TALBOT G. (1930) Observations on the bionomics of the Lepidoptera of Matto Grosso, Brazil. *Trans. ent. Soc. Lond.* 1929, 391–414.

COOK L.M. & BROWER L.P. (1969) Observations on polymorphism in two species of Heliconiine butterflies from Trinidad, West Indies. *Entomologist* 102, 125–128.

COOK L.M., FRANK K. & BROWER L.P. (1971) Experiments on the demography of tropical butterflies. I. Survival rate and density in two species of *Parides*. *Biotropica* (in press).

CRANE J. (1954) Spectral reflectance characteristics of butterflies (Lepidoptera) from Trinidad, B.W.I. *Zoologica (N.Y.)* 39, 85–113.

CRANE J. (1955) Imaginal behavior of a Trinidad butterfly, *Heliconius erato hydara* Hewitson, with special reference to the social use of color. *Zoologica (N.Y.)* 40, 167–196.

CRANE J. (1957) Imaginal behavior in butterflies of the family Heliconiidae: changing social patterns and irrelevant actions. *Zoologica (N.Y.)* 42, 135–145.

DRYJA A. (1959) Badania nad polimorfizmen Kraśnika Zmiennego (*Zygaena ephialtes* L.). (Genetical investigations on the polymorphism of *Zygaena ephialtes* L.). Panstwowe Wydawnictwo Naukowe, Warszawa (pp. 401 + folder of 12 plates and tables).

EBERT H. (1965) Uma coleção de borboletas (Lepid. Rhopal.) do Rio Amparí (Território do Amapá) com anotações taxonômicas sôbre Rhopalocera do Brasil. *Papéis Dep. Zool. S. Paulo* 18, 65–85.

EBERT H. (1970) On the frequency of butterflies in eastern Brazil, with a list of the butterfly fauna of Pocas de Caldas, Minas Gerais. *J. Lepid. Soc.* 23, suppl. 3, 1–48.

EHRLICH P.R. & RAVEN P.H. (1965) Butterflies and plants: a study in coevolution. *Evolution* 18, 586–608.

EMMEL T.C. (1969) Methods for studying the chromosomes of lepidoptera. *J. Res. Lep.* **7**, 23–28.

EMSLEY M. (1963) A morphological study of imagine Heliconiinae (Lep.: Nymphalidae) with a consideration of the evolutionary relationships within the group. *Zoologica* (*N.Y.*) **48**, 85–130.

EMSLEY M.G. (1964) The geographical distribution of the color-pattern components of *Heliconius erato* and *Heliconius melpomene* with genetical evidence for the systematic relationship between the two species. *Zoologica* (*N.Y.*) **49**, 245–286.

EMSLEY M.G. (1965) Speciation in *Heliconius* (Lep., Nymphalidae): morphology and geographic distribution. *Zoologica* (*N.Y.*) **50**, 191–254.

FISHER R.A. (1930) *The genetical theory of natural selection.* Clarendon Press, Oxford.

FORD E.B. (1937) Problems of heredity in the Lepidoptera. *Biol. Rev.* **12**, 461–503.

FORD E.B. (1946) *Butterflies.* Collins, London.

FORD E.B. (1953) The genetics of polymorphism in the Lepidoptera. *Adv. Genet.* **5**, 43–87.

FORD E.B. (1955) *Moths.* Collins, London.

FORD E.B. (1964) *Ecological genetics.* Methuen, London; Wiley, New York.

FRAZER J.F.D. & ROTHSCHILD M. (1960) Defence mechanisms in warningly-coloured moths and other insects. *XI. Int. Kongr. f. Entom. Wien.* 1960, Verh B, III, 249–256.

HAMILTON W.D. (1964a) The genetical evolution of social behaviour. I. *J. theoret. Biol.* **7**, 1–16.

HAMILTON W.D. (1964b) The genetical evolution of social behaviour. II. *J. theoret. Biol.* **7**, 17–52.

HESTER J.J. (1966) Late Pleistocene environments and early Man in South America. *Am. Nat.* **100**, 377–388.

JOICEY J.J. & KAYE W.J. (1917) On a collection of Heliconiine forms from French Guiana. *Trans. R. ent. Soc. Lond.* **1916**, 412–431.

JONES D.A., PARSONS J. & ROTHSCHILD M. (1962) Release of hydrocyanic acid from crushed tissues of all stages in the life-cycle of species of the Zygaeninae (Lepidoptera). *Nature* (*London*) **193**, 52–53.

JONES F.M. (1930) The sleeping Heliconias of Florida. *Nat. Hist.* **30**, 635–644.

KIRBY W.F. (1898) *European butterflies and moths.* Cassell, London.

KOJIMA K. & SCHAFFER H.E. (1964) Accumulation of epistatic gene complexes. *Evolution* **18**, 127–129.

LABINE P.A. (1968) The population biology of the butterfly, *Euphydryas editha.* VIII. Oviposition and its relation to patterns of oviposition in other butterflies. *Evolution* **22**, 799–805.

DE LESSE H. (1967) Les nombres de chromosomes chez les Lépidoptères Rhopalocères néotropicaux. *Annls Soc. ent. Fr.* (*n.s.*) **3**, 67–136.

LICHY R. (1960) Documentos para servir al estudio de los lepidópteros de Venezuela. IV. Una especie nueva del genero *Heliconius* Kluk (Rhopalocera, Nymphalidae): *Heliconius luciana* sp.n. *Rev. Fac. Agron.* (*Maracay*) **2** (3), 20–44.

MANLY B.F.J. & PARR M.J. (1968) A new method of estimating population size, survivorship, and birth rate from capture-recapture data. *Trans. Soc. Br. Ent.* **18**, 81–89.

Masters J.H. (1969) *Heliconius hecale* and *xanthocles* in Venezuela (Nymphalidae). *J. Lepid. Soc.* **23**, 104–105.

Michener C.D. (1942) A generic revision of the Heliconiinae (Lepidoptera, Nymphalidae). *Am. Mus. Novit.* **1197**, 1–8.

Moss A.M. (1933) The gregarious sleeping habits of certain Ithomiine and Heliconine butterflies in Brazil. *Proc. R. ent. Soc. Lond.* **7**, 66–67.

Moulton J. (1909) On some of the principal mimetic (Mullerian) combinations of tropical American butterflies. *Trans. R. ent. Soc. Lond.* **1908**, 585–606.

Nei M. (1967) Modification of linkage intensity by natural selection. *Genetics* **57**, 625–641.

Petiver J. (1767) *Jacobi Petiver Opera*. John Millan, London.

Povolný D. & Gregor F., Jr. (1946) Vřetenušky (*Zygaena* Fab.) v zemi Moravsko-slezské. *Entomologické Přírucky entomologických Listů* (*Brno*), No. 12, 1–100.

Povolný D. & Pijáček J. (1949) Příspěvek k otázce polymorphismu *Zygaena ephialtes* L. *Přírodovědeckého Sborníku Ostravského Kraje* **10** (4), 1–11.

Reichl E.R. (1958) *Zygaena ephialtes* L.: I. Formenverteilung und Rassengrenzen im niederöstereichischen Raum. *Zeit. wien. ent. Ges.* **43**, 250–265.

Reichl E.R. (1959) *Zygaena ephialtes* L.: II. Versuch einer Deutung der Rassen und Formenverteilung auf populationsgenetischer Basis. *Zeit. wien. ent. Ges.* **44**, 50–64.

Reichstein T., von Euw J., Parsons J.A. & Rothschild M. (1968) Heart poisons in the monarch butterfly. *Science* **161**, 861–866.

Seba A. (1765) *Locupletissimi rerum naturalem Thesauri accurata descriptio et inconibus artificiosissimis Expressio per Universam Physices Historiam.* H.C. Arksteum, H. Merkum and P. Schouten, Amstelaedamum.

Sheppard P.M. (1963) Some genetic studies of Müllerian mimics in butterflies of the genus *Heliconius. Zoologica* (*N.Y.*) **48**, 145–154.

Sheppard P.M. (1965) Mimicry and its ecological aspects. *Genetics Today* (*Proc. XI Int. Cong. Genet. Hague*) **3**, 553–560.

Sheppard P.M. (1969) Evolutionary genetics of animal populations: The study of natural populations. *Proc. XII Intern. Congr. Genet.* (*Tokyo*) **3**, 261–279.

Suomalainen E., Cook L.M. & Turner J.R.G. (1971) Chromosome numbers of heliconiine butterflies from Trinidad, West Indies (Lepidoptera, Nymphalidae). *Zoologica* (*N.Y.*) (in press).

Swihart S.L. (1963) The electroretinogram of *Heliconius erato* (Lepidoptera) and its possible relation to established behavior patterns. *Zoologica* (*N.Y.*) **48**, 155–165.

Swihart S.L. (1964) The nature of the electroretinogram of a tropical butterfly. *J. Insect Phys.* **10**, 547–562.

Swihart S.L. (1965) Evoked potentials in the visual pathway of *Heliconius erato* (Lepidoptera). *Zoologica* (*N.Y.*) **50**, 55–62.

Swihart S. (1967a) Neural adaptations in the visual pathway of certain Heliconiine butterflies, and related forms, to variations in wing coloration. *Zoologica* (*N.Y.*) **52**, 1–14.

Swihart S.L. (1967b) Maturation of the visual mechanisms in the neotropical butterfly, *Heliconius sarae. J. Insect Phys.* **13**, 1679–1688.

SWIHART S.L. (1968) Single unit activity in the visual pathway of the butterfly *Heliconius erato. J. Insect Phys.* **14**, 1589–1601.

TOKUYAMA T., SENOH S., SAKAN T., BROWN K.S. JR. & WITKOP B. (1967) The photoreduction of kynurenic acid to kynurenine yellow and occurrence of 3-hydroxy-L-kynurenine in butterflies. *J. Am. chem. Soc.* **89**, 1017–1021.

TURNER J.R.G. (1965) Evolution of complex polymorphism and mimicry in distasteful South American butterflies. *Proc. XII int. Congr. Ent. London* 1964, 267.

TURNER J.R.G. (1966) A rare mimetic *Heliconius* (Lepidoptera: Nymphalidae) *Proc. R. ent. Soc. Lond.* (B) **35**, 128–132.

TURNER J.R.G. (1967a) The generic name of *Papilio iulia* Fabricius, sometimes called the Flambeau (Lepidoptera, Nymphalidae). *Entomologist* **100**, 8.

TURNER J.R.G. (1967b) A little-recognised species of *Heliconius* butterfly (Nymphalidae). *J. Res. Lepid.* **5**, 97–112.

TURNER J.R.G. (1967c) Some early works on heliconiine butterflies and their biology (Lepidoptera, Nymphalidae). *J. Linn. Soc., Lond.* (*Zool.*) **46**, 255–266.

TURNER J.R.G. (1967d) Goddess changes sex, or the gender game. *Syst. Zool.* **16**, 349–350.

TURNER J.R.G. (1967e) On supergenes. I. The evolution of supergenes. *Am. Nat.* **101**, 195–221.

TURNER J.R.G. (1968a) Natural selection for and against a polymorphism which interacts with sex. *Evolution* **22**, 481–495.

TURNER J.R.G. (1968b) Some new *Heliconius* pupae: their taxonomic and evolutionary significance in relation to mimicry (Lepidoptera, Nymphalidae). *J. Zool.* (*London*) **155**, 311–325.

TURNER J.R.G. (1970) Mimicry: a study in behaviour, genetics, ecology, and biochemistry. *Sci. Prog.* (*Oxford*) **58**, 219–235.

TURNER J.R.G. (1971a) The genetics of some polymorphic forms of the butterflies *Heliconius melpomene* (Linnaeus) and *H. erato* (Linnaeus). II. The hybridisation of subspecies of *H. melpomene* form Suriname and Trinidad. *Zoologica* (in press).

TURNER J.R.G. (1971b) Experiments on the demography of tropical butterflies. II. Longevity and home-range behaviour in *Heliconius erato. Biotropica* **3**, 21–31.

TURNER J.R.G. (1971c) Two thousand generations of hybridisation in a *Heliconius* butterfly. *Evolution* (in press).

TURNER J.R.G. & CRANE J. (1962) The genetics of some polymorphic forms of the butterflies *Heliconius melpomene* Linnaeus and *H. erato* Linnaeus. I. Major genes. *Zoologica* (*N.Y.*) **47**, 141–152.

WICKLER W. (1968) *Mimicry in plants and animals* (English translation by R.D. Martin). Weidenfeld & Nicholson, London.

12 ❀ Avian Feeding Behaviour and the Selective Advantage of Incipient Mimicry

L. P. BROWER, J. ALCOCK
AND JANE V. Z. BROWER

INTRODUCTION

Studies of mimicry have contributed significantly to our understanding of the process of natural selection and are well reviewed by Rettenmeyer (1970). The recent genetical investigation by Clarke, Sheppard & Thornton (1968) on the Malaysian butterfly *Papilio memnon* L. is of special importance. It confirms for a wholly separate and very different geographical area the conclusions arrived at in earlier genetic analyses of the African mimic, *P. dardanus* Brown, so admirably summarized by Ford (1964). These genetical studies make it clear that the evolution of mimicry results from the accumulation of numerous changes each of which increases to a small extent the resemblance of the mimic to the model. Consequently the older idea of Punnett (1915) and Goldschmidt (1945) that mimicry evolves by means of saltatory macromutations is no longer tenable.

There remains, nevertheless, a substantially unresolved problem: what process provides a selective advantage to divergent phenotypes in the incipient period of departure of a species from its primordial appearance towards that of the model? In other words, while granting that a fully developed mimic does enjoy a selective advantage over a nonmimetic form, it is difficult to imagine how a micromutational change in the original nonmimetic form affecting only a small portion of the total colour pattern can provide any mimetic advantage. The following quotation from Ford (1964, p. 245), in discussing monomorphic mimicry, points out this problem lucidly:

'... small changes in the original pattern ... are almost certain to be harmful. Only a considerable step, producing something near enough in appearance to a protected form to give an advantage, is likely to become established. This could then be perfected by selection acting on the gene-complex.'

The question of how considerable the initial step must be is of great theoretical interest to understanding the evolution of both mono-morphic and polymorphic mimicry. Obviously, earlier workers have assumed that predators could not be easily deceived. At the other extreme is the idea that predators, even though they may possess keen discrimina-tory abilities, will under certain conditions generalize their learned avoidance of an unpalatable species to include even the remotely similar pattern of an incipient mimic. It is our contention that the question raised by the earlier workers and to a lesser degree by Fisher (1930), Sheppard (1961), and Ford (*l.c.*) can be resolved by a thorough study of the behaviour of predators likely to be involved as the selective agents in the evolution of mimicry.

Well before Goldschmidt's paper several rather informal experimental studies by Morgan (1900), Muhlmann (1934), Mostler (1935), and Windecker (1939) had shown that birds conditioned to avoid a particular model might also avoid a mimic the resemblance of which was not highly detailed. Working with chickens as predators in artificially contrived model-mimic experiments, Schmidt (1958, 1960) and Duncan & Sheppard (1963, 1965) extended the earlier findings that vague resemblances could be advantageous. The latter authors more fully explored the conditions under which this generalization takes place and found that the greater the penalty for making a mistake, the less the mimic needed to look like the model to be protected. The study of J.Brower (1958) was the first to show with natural butterfly prey under carefully controlled experimental conditions that wild-caught birds, after learning to reject a model butter-fly, not only rejected its bright orange mimic but also a darker mahogany coloured geographical race of the mimic. Another study by Brower, Brower & Collins (1963) with 62 silverbeak tanagers and several species of Heliconiine butterflies suggested that the visual cues to which the avoidance responses of the birds became conditioned were subject to extreme generalization. After learning to avoid unpalatable butterflies of one set of colour patterns, the birds went on to avoid others of the same size and shape but of completely different colour patterns. However, the recent work of Coppinger (1969, 1970) has shown that avoidance of butterfly prey never before seen may result from a conservative tendency of birds to be wary of novel food, and it is possible that Brower,

Brower & Collins' (*l.c.*) findings represented novel stimulus rejection rather than extreme generalization. Their experimental design was in fact such as not to preclude this possible alternative explanation. Consequently, it can be fairly stated that no one has yet conclusively demonstrated with a large sample of birds using a plausible incipient butterfly mimic that this mimic could derive an advantage from its colour pattern resemblance, albeit imperfect, to that of the model species.

To fill this gap, the experiment in this paper tested the response of a group of wild-captured adult insectivorous birds to two potential mimetic species after the birds had been presented with a short series of a model butterfly. The mimics were of an altogether different shape than the model and shared some, but not all, elements of the colour pattern in common. If the predators tended to avoid these imperfect mimics, this would indicate that an incipient mimetic butterfly could under certain conditions enjoy a selective advantage through a reduction in the pressure of predation.

MATERIALS AND METHODS

The experiments were conducted during the summers of 1965 and 1966 at Simla, the William Beebe Tropical Research Station in Trinidad, West Indies. The experimental predators were fork-tailed flycatchers, *Muscivora tyrannus monachus* Hartlaub, an abundant insectivorous migrant species from South America found in large numbers along dirt roads through the cane fields near Port-of-Spain, and silverbeak tanagers, *Ramphocelus carbo magnirostris* Lafresnaye, a common bird of forest edges and second growth, omnivorous in feeding habits. The birds were caught in mist nets, placed in cloth bags, and driven to Simla where they were put into separate experimental cages. The flycatchers were first fed live mealworms and live small brown moths, the latter captured at a light trap. When they had eaten these, they were fed dead small brown moths, which had been taken at the same source and frozen prior to use, and more live mealworms. After a day or two the birds adjusted to the cage environment and thereafter were fed regularly three times a day. The tanagers were given bananas, small brown moths, and mealworms. Their rate of acclimatization to the cages was slower than the flycatchers, and several died.

The experimental cages were made of $\frac{1}{2}$-in square wire mesh and were

30 in × 30 in × 30 in. They were located in a large, covered, screened-in enclosure on a ridge in the forest. Each cage had two horizontal perches at different heights, a sand-covered floor, a sheet metal front with an 8 in × 8 in pane of one-way glass and two petri dishes for food and water. The cages were separated from one another by sheet metal partitions and each was illuminated by a sixty-watt bulb which supplemented the low intensity natural back-lighting.

The butterflies (plate 12.1) offered to the birds were captured at locations throughout the island, stored alive in glassine envelopes, and then placed in a freezer at Simla. They were removed after freezing, their wings spread, and then re-frozen. *Cystineura cana* Erichs. and *Euptychia renata* (Cram.), two edible butterflies, males of *Heliconius erato hydara* Hewitson, the distasteful model, males of *Anartia amalthea* (L.) and *Biblis hyperia* (Cram.), two potential but imperfect mimics of the model, and males of *A. jatrophae* Johannson, a control butterfly unlike *H. erato*, were used in the experiment. *Cystineura* is small and white-brown; *Euptychia* is small and chocolate brown; *H. erato*, *Biblis*, and *A. amalthea* all have red and black patterned wings; *A. jatrophae* is the same shape and the same size or slightly larger than *A. amalthea* but is whitish-grey with brown markings.

The butterflies were offered to the birds singly by placing them dorsal side up in the food petri dish with twelve-inch forceps after opening a sliding door at the bottom-centre of the cage front. An experiment was begun after a bird regularly attacked small brown moths and *Cystineura* in less than two minutes. The birds were fed the night before and any remaining food was removed the following morning at 7.30 a.m. Each bird was offered its complete series of butterflies in a single morning from 8.30 to 10.30 a.m. The experimental birds were given the five *H. erato*

PLATE 12.1. An incipient mimetic advantage can be gained by stimulus generalization. Birds were offered the distasteful model *H. erato*, a black butterfly with red patches, which they rapidly learned to reject. Most birds then also rejected (1) normal *B. hyperia* which is black with red hindwing borders, (2) *Biblis* made uniformly black by painting out the red, and (3) *A. amalthea* which has red bands but otherwise differs considerably from the model. When *A. amalthea* was turned upside down so the birds could not see the red bands, they treated it as they did the white and brown *A. jatrophae*: both were eaten. Thus under certain conditions even a slight resemblance to the bicoloured model can lead to substantial mimetic advantage. (Approximately × 0.64.)

RED

Heliconius erato

Biblis hyperia
(red margins normal)

RED

Biblis hyperia
(red margins darkened)

Anartia amalthea

RED

Anartia amalthea
(Upside down)

Anartia jatrophae

PLATE 12.1

[to face p 264

models, then the two mimetic species, and finally the control *A. jatrophae* (table 12.1). The first member of each of the first five pairs of butterflies was determined by a random numbers table to ensure a random series of presentations. However, the order from 6a to 8b was fixed as in table 12.1.

TABLE 12.1. Experimental routine: order of presentations.

	Experimental group	Control group
1 a	*Cystineura* or *Heliconius*	*Cystineura* or *Euptychia*
b	*Heliconius* or *Cystineura*	*Euptychia* or *Cystineura*
	.	.
	.	.
	.	.
	.	.
	.	.
	.	.
5 a	.	.
b	.	.
6 a	*Cystineura*	*Cystineura*
b	*Biblis hyperia*	*Biblis hyperia*
7 a	*Cystineura*	*Cystineura*
b	*Anartia amalthea*	*Anartia amalthea*
8 a	*Anartia jatrophae*	*Anartia jatrophae*
b	*Cystineura*	*Cystineura*

The experimental procedure was designed to test the hypothesis that birds which avoid *Heliconius erato* would also avoid butterflies whose resemblance to this model was far from perfect. In other words, would the Experimentals avoid *Biblis* and *A. amalthea*, but accept *A. jatrophae*, after having been presented with five *H. erato* butterflies?

The Control birds were given an innocuous edible insect, *E. renata*, in the place of the model, *H. erato* (table 12.1). Their response to the mimics would indicate the normal baseline of avoidance of *Biblis* and *A. amalthea*. This response could be the result of prior experience with the unpalatable *H. erato* or similar appearing Heliconiines (or even with *Biblis* which proved to be distasteful to some of the birds, table 12.4), or because of prior experience with other distasteful aposematic insects, or because the mimics were novel prey and therefore potentially avoidance-inducing (see Thorpe 1963, Coppinger *l.c.*).

A second group of Experimentals (Experimental Group 2) was tested in the summer of 1966 in an effort to determine whether the colour red, present on the wings of *H.erato*, *Biblis*, and *A.amalthea* and strikingly noticeable on all three, was stimulating avoidance of the mimics. These birds were offered the same series of presentations given to Experimental Group 1 except that (a) the *Biblis* they received had its red border on the hindwing blacked out with permanent magic marker ink and (b) *A.amalthea* was presented upside down, exposing its nonmimetic underside of brown-orange, white, and pale brown-red. If red were the only sign-stimulus eliciting avoidance, then the birds of Experimental Group 2 should avoid neither *Biblis* nor *A.amalthea* despite their training to avoid *Heliconius*.

It should be noted that the experiment was designed in such a way that the Control and Experimental birds received equal numbers of insect presentations. However, because the experimental birds found the models unpalatable or rejected them from the start this meant that they ate about one-half as many insects prior to receiving the three test insects. It might be argued that the greater acceptance of the mimetic insects by the control birds was a result of this difference in experimental conditioning. If this were the case then their acceptance of the final test butterfly (the Control *A.jatrophae*) should have been greater than that by the Experimental Group, which it clearly was not (table 12.3c).

RESULTS

A. EXPERIMENTAL GROUP I AND CONTROL GROUP I

Table 12.2 compares the responses of the two predator species in all three groups (Control, Experimental 1, Experimental 2) to *Biblis*, *A.amalthea*, and *A.jatrophae*. It is clear that the differences in the responses of the flycatchers and tanagers are not significant so that subsequent analyses will lump the data for the two bird species.

Table 12.3 compares the Control and Experimental Group 1 birds. Although the Controls did tend to hesitate before attacking the mimics, with only a few exceptions they eventually did attack them. In contrast, the Experimentals which had had experience with *H.erato* generally refused to touch the mimics in the two-minute presentation period and some of those that did attack did so with great hesitation. The refusal of

TABLE 12.2. Summary of the number of attacks of the 60 birds on their three successive test butterflies. Each bird had 2 minutes to attack each butterfly.

	Number of birds attacking		
Bird categories	First mimic (*Biblis hyperia*)	Second mimic (*Anartia amalthea*)	Control butterfly (*Anartia jatrophae*)
Controls			
10 Fork-tailed flycatchers	10	8	10
10 Silverbeak tanagers	7	9	8
Experimental Group 1			
10 Fork-tailed flycatchers	3	4	10
10 Silverbeak tanagers	3	5	10
Experimental Group 2			
10 Fork-tailed flycatchers	3*	9†	10
10 Silverbeak tanagers	4*	9†	8

* With red border blacked out.
† Presented upside down.

the majority of Experimental Group 1 birds to peck *Biblis* and *A. amalthea* cannot then be due to a tendency to avoid novel stimuli, aposematic insects in general, or because of prior experience with the model in nature, *per se*; rather, the birds, experimentally conditioned to avoid the model, *H. erato*, either confused the mimics with it, or were unwilling to risk an attack because of the resemblance. However, one-half of the Experimental Group 1 and 2 birds (10 of each species) did not attack a single one of their 5 *H. erato* butterflies. This was also found in an earlier study (Brower, Brower and Collins, *l.c.*) in which 9 out of 19 silverbeak tanagers rejected their first *H. erato* (and *H. melpomene*, a nearly identical Müllerian mimic). An interpretation which is most consistent with the experimental data is that the birds based their rejection of the models on prior experience in the wild. The fact that they saw but did not attack the models and then went on to reject the mimics suggests that simply seeing a butterfly species with which they have had previous noxious experience is a sufficient stimulus to reinforce their rejection response to the models, and consequently to the mimics. Otherwise, a comparable

10

number of the control birds should also have rejected the mimics, which they did not do.

POSSIBLE CRITICISM OF THE EXPERIMENT

One objection to the experiment ought now to be examined critically. It may be argued that it would have been preferable if the Controls had

TABLE 12.3

(a) The reactions of Experimental Group I and Control Group I birds to the first mimic, *Biblis hyperia*.

	Experimental birds	Control birds	Totals
Attacked	6 (30%)	17 (85%)	23
Not attacked	14	3	17
Totals	20	20	40

$$\chi^2_{(1)} = 10 \cdot 23^*, P < 0 \cdot 005$$

(b) The reactions of the same predators to the second mimic, *Anartia amalthea*.

	Experimental birds	Control birds	Totals
Attacked	9 (45%)	17 (85%)	26
Not attacked	11	3	14
Totals	20	20	40

$$\chi^2_{(1)} = 5 \cdot 39^*, P < 0 \cdot 025$$

(c) The reactions of the same predators to the third test butterfly, a non-mimetic control, *Anartia jatrophae*.

	Experimental birds	Control birds	Totals
Attacked	20 (100%)	18 (90%)	38
Not attacked	0	2	2
Totals	20	20	40

$$\chi^2_{(1)} = \text{insignificant by inspection}$$

* With Yates' correction.

TABLE 12.4. Comparative palatability of *Biblis hyperia* and *Anartia amalthea* to Control Group 1 birds.

	Biblis hyperia	*Anartia amalthea*	Total
Attacked and eaten	8 (47%)	15 (88%)	23
Attacked but not eaten	9	2	11
Totals	17	17	34

$$\chi^2_{(1)} = 48.4^*, P < 0.05$$

* With Yates' correction.

received (instead of the edible *Euptychia*) an unpalatable model, but with a colour pattern distinct from that of the two mimic butterflies. This argument could have some validity inasmuch as conditioning a bird to avoid *any* model might cause it to tend to avoid all other butterflies except those in nature which it had found palatable.

To control against this possibility, all birds were given a nonmimetic butterfly, *A. jatrophae*, after presentation of the mimics. This is a large, primarily white insect with scattered dark spots; its colour pattern, though somewhat like *Cystineura*, was totally unlike either mimic. Importantly, it was the same general size and shape as *A. amalthea* (see plate 12.1). If rejection of the mimics had been based on the birds' refusal to attack all butterflies other than the edible *Cystineura* with which they were familiar, then they should have avoided *A. jatrophae* which was about three times as large and of a quite different pattern. Table 12.3c shows that, on the contrary, all the Experimentals attacked *A. jatrophae*. Moreover, the birds did not hesitate to attack *A. jatrophae*, the average being 18 seconds compared to 25 seconds to attack *Cystineura*. Therefore, the rejection of the two mimics by birds which had experienced the model must have been the result of the similarities between the three species, and neither the result of generalization from prior experience nor rejection of novel prey species.

B. EXPERIMENTAL GROUP 2

We predicted that *Biblis* would prove to be more effective in eliciting avoidance in the Experimental birds than *A. amalthea* because *Biblis* is bicoloured and simply patterned like *H. erato*, whereas *A. amalthea* not only is more complexly patterned, but has white spotting as an additional colour. In fact *Biblis* was rejected 15 per cent more often than *A. amalthea* (compare Tables 12.3a and b). However, the experiment was carried out on too small a scale to show a significant difference in the effectiveness of mimicry of the two species $\chi^2_{(1)} = 0.43$, $P > 0.50$). Alternatively, it is possible that the birds were reacting only to the presence of red patches on the wings of the two mimicking species, a striking element of the patterns of both mimics and model.

In order to test the hypothesis that the red portions of the patterns of the mimics were the key component responsible for the predators' reluctance to attack the mimics, 20 birds were given *Biblis* with the red border of the hindwing blacked out, followed by *A. amalthea* presented

TABLE 12.5

(a) The reactions of predators of Experimental Groups 1 and 2 to *Biblis hyperia* butterflies with the red band intact *versus* those with the red band blacked out with 'magic marker' ink.

	Biblis hyperia + red (normal)	*Biblis hyperia* − red (blackened)	Totals
Attacked	6 (30%)	7 (35%)	13
Not attacked	14	13	27
Totals	20	20	40

$\chi^2_{(1)}$ = insignificant by inspection

(b) The reactions of the same predators to *Anartia amalthea* butterflies in a dorsal position showing the red band *versus* those in a ventral position in which the red band is indistinct.

	A. amalthea + red (dorsal side up)	*A. amalthea* − red (ventral side up)	Totals
Attacked	9 (45%)	18 (90%)	27
Not attacked	11	2	13
Totals	20	20	40

$\chi^2_{(1)} = 7.29^*$, $P < 0.01$

* With Yates' correction.

upside down. The birds, as in Experiment 1, were previously conditioned to avoid *H. erato* and were given a final *A. jatrophae* control. Table 12.5 shows the results of this experiment. *Biblis*, despite the obliteration of red, was avoided to the same extent as it had been with the red present (table 12.5a). On the other hand, the upside down *A. amalthea* were now taken as freely as the *A. jatrophae* controls (compare tables 12.5b and 12.3c).

Thus we see that red was not necessary for *Biblis* to be an effective mimic of *Heliconius*. The black component of the simple colour pattern of *Biblis*, which was similar in tone and texture to that of the model, was sufficient alone to induce avoidance in the experimental birds. This recalls Schmidt's (1960) finding that black was the critical colour in eliciting avoidance in the chickens he tested with multicoloured red, white, and black models.

The underside of *A. amalthea* is brown-orange in over-all appearance and can be considered intermediate in colour between the upper surfaces of the two *Anartia* species. In the absence of a distinct red band set off against a black background the mimetic advantage is lost. It seems probable that the mimetic advantage of the upper surface of this species derives from the red bands but further work has to be done to establish this point.

DISCUSSION

Whether or not a predator, the selective agent in the evolution of mimicry, will attack a Batesian mimic depends on the interaction of a number of factors. The critical points are listed in table 12.6. A major distinction here is between the *capability* of the predator to make a discrimination between similar prey and its *likelihood* to do so. The tanagers and fly-catchers tested in the experiment were almost certainly able to distinguish between *H. erato*, *B. hyperia*, and *A. amalthea*. In other mimetic complexes the similarity between the colour patterns of the models and mimics is so great that some predators might well be unable to detect any difference.

However, morphological cues are not the only ones available to a potential predator. Behaviour is probably at least as relevant. Beal (1918) found that the stomachs of swallows and kingbirds contained only drone bees. Despite the extremely close resemblance between drones and workers, certain birds were able and willing to discriminate between the

TABLE 12.6. Factors influencing a predator's response to a Batesian mimic.

I. Neurophysiological capacity to discriminate between sets of similar stimuli and to retain the formed discrimination
   A. The effectiveness of the model's pattern and behaviour in promoting conditioned avoidance
   B. The degree of morphological and behavioural similarity of the mimic to the model

II. Likelihood of attack once having made a discrimination
   A. The hunger of the predator or of its dependent progeny
   B. The abundance and relative palatability of alternative food
   C. The expectation of successful capture
   D. Previous experience with the model or aposematic prey similar to the model and with the mimic
      1. The expectation of a remembered penalty for an error in discrimination based on the sensitivity of the predator to the model's chemical or other defences
      2. The relative abundance of the mimic to the model

two. In our experiment we were unable to demonstrate that *Biblis* was at a significantly greater advantage than the other mimic, *A. amalthea*, despite its greater resemblance to *Heliconius*. However, under natural conditions it is highly probable that *Biblis* is the more effective mimic. When flying, it closely resembles *H. erato* because its slow and fluttering flight enhances the mimetic effect of the red borders on the hindwing, whereas *A. amalthea* behaves very differently.

CONCLUSION

Fork-tailed flycatchers and silverbeak tanagers are probably capable of discriminating between *H. erato* and three imperfect mimics, *Biblis* with and without red, and *A. amalthea*, but were effectively unwilling to do so. In this experiment, at least two of the mimetic colour patterns were so imperfect that either could be the product of a single minor genetic event in many species of present day butterflies.

Clearly, the initial step in the evolution of mimicry need not be a large change in the direction of the colour-pattern of the model. We can therefore conclude that the nature of avian feeding behaviour provides a selective *milieu* which is entirely consistent with the origin of mimicry by micromutation, and the old hypothesis demanding macromutation is obviated. However, the gradual perfection of mimicry through a series of small steps depends upon finer discrimination and greater willingness to attack than was demonstrated by the predators in this experiment, and therefore requires further study.

SUMMARY

1. The feeding behaviour of birds was investigated as a possible selective basis for the origin of mimicry as demanded by modern genetical theory. Experiments were carried out on 60 individual birds belonging to two species in widely separated families.

2. Birds trained to avoid an unpalatable model butterfly also avoided three butterflies of varying degrees of resemblance to the model, but accepted one of these when a major colour in common was obliterated by offering the butterfly in an upside down position. Nearly all of these birds accepted a final test insect which differed from the model and mimics, and nearly all birds in a control series which were not conditioned to avoid the model ate the three mimics.

3. It was concluded that the feeding behaviour of birds is such that the initial step in the evolution of mimicry need not involve a great change in the incipient mimic's appearance and can clearly be the result of micromutation.

ACKNOWLEDGMENTS

We are grateful to Jocelyn Crane Griffin and Professor Donald R.Griffin, for making the superb facilities of Simla, the William Beebe Tropical Research Station in Trinidad, West Indies, available to us during the summers of 1965 and 1966. Professor E.B.Ford F.R.S., Professor E.Mayr and Dr Raymond P. Coppinger offered helpful comments on the manuscript. We also wish to thank Kenneth Frank, John Hernandez, Asa Wright, Lorna Coppinger, Susan Glazier and Helen Sullivan for help in various aspects of the study. The work was supported by a grant from the Harvard University Committee on Evolutionary Biology and by U.S. National Science Foundation Grants 8707, GB4924, and GB7637 with Professor L.P.Brower as principal investigator.

REFERENCES

BEAL F.E.L. (1918) Food habits of the swallows, a family of valuable native birds. *U.S. Dept. Agr. Bull.* **619**, 1–28.

BROWER J.VZ. (1958) Experimental studies of mimicry in some North American butterflies. 1. The Monarch, *Danaus plexippus*, and Viceroy, *Limenitis archippus archippus. Evolution* **12**, 32–47.

BROWER L.P., BROWER J.VZ. & COLLINS C.T. (1963) Experimental studies of mimicry. 7. Relative palatability and Müllerian mimicry among neotropical butterflies of the subfamily Heliconiinae. *Zoologica* **48**, 65–85.

CLARKE C.A., SHEPPARD P.M. & THORNTON I.W.B. (1968) The genetics of the mimetic butterfly *Papilio memnon* L. *Phil. Trans. R. Soc. London* B, **254**, 37–89.

COPPINGER R.P. (1969) The effect of experience and novelty on avian feeding behavior with reference to the evolution of warning coloration in butterflies. I. Reactions of wild-caught adult blue jays to novel insects. *Behaviour* **35**, 4–60.

COPPINGER R.P. (1970) The effect of experience and novelty on avian feeding behavior with reference to the evolution of warning coloration in butterflies. II. Reactions of naive birds to novel insects. *Am. Nat.* **104**, 323–335.

DUNCAN C.J. & SHEPPARD P.M. (1963) Continuous and quantal theories of sensory discrimination. *Proc. R. Soc. London* B. **158**, 343–363.

DUNCAN C.J. & SHEPPARD P.M. (1965) Sensory discrimination and its role in the evolution of Batesian mimicry. *Behaviour* **24**, 269–282.

FISHER R.A. (1930) *The genetical theory of natural selection.* Clarendon Press, Oxford.

FORD E.B. (1964) *Ecological genetics.* Methuen & Co., London.

GOLDSCHMIDT R.B. (1945) Mimetic polymorphism, a controversial chapter of Darwinism. *Q. Rev. Biol.* **20**, 147–164, 205–230.

MORGAN L.P. (1900) *Animal behaviour.* London: Edward Arnold.

MOSTLER G. (1935) Beobachtungen zur Frage der Wespenmimikry. *Z. Morph. Öekol. Tiere* **29**, 381–454.

MUHLMANN M. (1934) Im Modellversuch küntslich erzeugte Mimikry und ihre Bedeutung für den 'Nachamer'. *Z. Morph. Öekol. Tiere* **28**, 259–296.

PUNNETT R.C. (1915) *Mimicry in butterflies.* Cambridge University Press.

RETTENMEYER C.W. (1970) Insect mimicry. *A. Rev. Ent.* **15**, 43–74.

SCHMIDT R.S. (1958) Behavioral evidence on the evolution of Batesian mimicry. *Anim. Behav.* **16**, 129–138.

SCHMIDT R.S. (1960) Predator behavior and the perfection of incipient mimetic resemblances. *Behaviour* **15**, 244–252.

SHEPPARD P.M. (1961) Recent genetical work on polymorphic mimetic *Papilios. Symposia Royal Entomological Society* **1**, 23–30.

THORPE W.H. (1963) *Learning and Instinct in Animals.* Methuen & Co., London.

WINDECKER W. (1939) *Euchelia jacobeae* L. und das Schutzrachtenproblem. *Z. Morph. Öekol. Tiere* **35**, 84–139.

13 ❋ An Analysis of Spot Placing in the Meadow Brown Butterfly *Maniola jurtina*

K. G. MCWHIRTER AND E. R. CREED

INTRODUCTION

The first study of the variation of spot-number in the Meadow Brown Butterfly, *Maniola jurtina*, was reported from the uninhabited island of Tean, Isles of Scilly, in 1946 (Dowdeswell, Fisher & Ford 1949). The work was extended to other islands of the Scilly archipelago and to sites in mainland Britain from 1950 onwards (Dowdeswell & Ford 1953). From that year the area of interest in Britain has steadily widened; in addition collections of imagines from virtually the whole geographic range and dating from between 1890 and 1935 have been analysed for spot-number distribution by Dowdeswell & McWhirter (1967). A review of our present knowledge in respect of the distribution of spot-numbers may be found in Ford (1971). Heritability of spot-number at 15°C has been estimated for females by two different methods at 63 and 83 per cent; corresponding male progeny did not show significant heritability (McWhirter 1969). At 22°C females show approximately 78 per cent heritability and males about 48 per cent (McWhirter, in preparation).

Spots can only occur at defined sites on the hind wing. Therefore variability resides both in spot-number and in spot-placing. Since 1958 most samples have been scored for both these variants, though previous publications have been restricted to spot-number only. In this paper an analysis is made of the variability in spot-placing; this too shows the existence of regular geographic patterns or trends, as well as interactions with spot-number and sex.

SPOT-PLACING

In adult *M. jurtina* of both sexes the outer third of the underside of the hindwings is lighter in colour than the more proximal part. This border

is divided by the veins of the wing, between each of which is a distinct
fold; the spots are always found on these folds with never more than one
per fold. Although this gives a maximum of seven possible positions, only
five of them are at all commonly occupied (see fig. 13.1), thus the normal

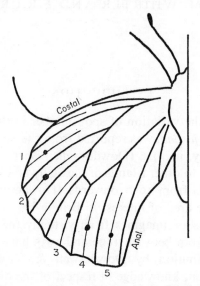

FIGURE 13.1. Diagram of the underside of the hindwing of *M. jurtina*
showing the five positions most commonly occupied by spots.

range of spot-numbers is from nought to five. In addition, the spots are
only found in certain combinations, so that of the ten possible ways in
which two spots might occur, only three are usually encountered. The
most frequent arrangements are shown in diagrammatic form in table
13.1, from which it may be seen that they can be classified according to
whether there is a preponderance of spots towards the costal or the anal
border of the wing, or an equal division as in the 'splay' and 'median'
arrangements. For convenience this bias, or the lack of it, is reflected in
the names adopted. In the analysis that follows, only butterflies having
these common arrangements of their spots are included; those rare
individuals with other arrangements in the five normal positions or in
which the position between the second and third is occupied are ignored.

    There is a very large number of comparisons that could be undertaken
on the basis of spot-placing. The relative proportions of each spot-
placing type within each number class is perhaps the simplest and

reveals several clear trends. However, it soon becomes obvious that many of these trends are common to all spot-number classes; a preponderance, or a lack, of the more costal spots is usually seen as a feature of the sample as a whole, extending, though in different degrees, to both sexes. We have therefore adopted a costality index as the most easily computed general parameter with regard to spot-placing. This is the proportion,

TABLE 13.1. The spot placing arrangements most commonly found in *M. jurtina* together with the nomenclature adopted. The five possible positions of spots are shown in fig. 13.1.

		◄— Costal			Anal —►	
		1	2	3	4	5
	0	—	—	—·	—	—
Costal	1	—	•	—	—	—
Anal	1	—	—	—	•	—
Costal	2	•	•	—	—	—
Splay	2	—	•	—	•	—
Anal	2	—	—	—	•	•
Costal	3	•	•	—	•	—
Median	3	—	•	•	•	—
Anal	3	—	•	—	•	•
Costal	4	•	•	•	•	—
Splay	4	•	•	—	•	•
Anal	4	—	•	•	•	•
	5	•	•	•	•	•

expressed as a percentage, that the total number of costal spots (positions 1 and 2) in the individuals of one sex represent of all costally and anally biased spots (positions 1, 2, 4, and 5); the almost centrally placed third spot is not included. Thus a Median 3 is credited with one costal and one anal spot, while a Costal 4 has two costal and one anal. A disadvantage of this index is that its sensitivity is reduced by increasing numbers of individuals with the neutral splay and median arrangements, though this in itself may be an important element of variation. Thus in males where a large number of insects may be splay 2, the index provides less discrimination between areas than in the females.

TABLE 13.2. The distribution of male *M. jurtina* by spot placing types and the calculated value of the costality index for fifteen areas. Individuals with nought, five or any unusual arrangement of spots are excluded.

	Cos 1	An 1	Cos 2	Spl 2	An 2	Cos 3	Med 3	An 3	Cos 4	Spl 4	An 4	Costality Index	S.E.
Isles of Scilly	201	60	25	2109	5	936	212	212	181	447	47	54·8	0·5
Channel Islands	2	—	—	35	—	12	—	4	1	5	1	53·4	4·1
West Cornwall	4	8	—	244	—	52	17	57	5	27	8	49·4	1·6
Lundy	—	2	—	77	—	8	—	8	—	6	1	49·4	3·3
Boundary: Cornish	13	139	5	2735	7	334	78	356	22	121	30	49·0	0·5
Boundary: Intermediate	7	74	5	1341	8	150	36	172	11	69	11	48·9	0·8
Boundary: English	24	163	5	2833	7	318	77	362	16	132	29	48·9	0·5
Southern England	27	85	—	987	11	100	25	171	6	68	14	47·6	0·9
Scotland	—	27	—	75	17	2	2	59	—	5	8	35·7	2·3
Ireland	3	8	—	36	—	3	1	2	—	1	—	48·1	4·9
General European: N.C.	5	5	—	128	2	9	2	3	1	6	—	50·4	2·7
General European: S.W.	3	13	1	124	—	28	7	20	1	17	2	49·9	2·2
Spain	3	10	—	70	3	25	5	36	—	32	2	47·3	2·3
Rhodes	—	—	—	49	—	13	16	39	16	42	23	47·1	2·1
North Africa	—	11	—	99	5	59	14	57	9	102	7	49·2	1·5

TABLE 13.3. The distribution of female *M. jurtina* by spot placing types and the calculated value of the costality index for fifteen areas. Individuals with nought, five or any unusual arrangements of spots are excluded.

	Cos 1	An 1	Cos 2	Spl 2	An 2	Cos 3	Med 3	An 3	Cos 4	Spl 4	Costality Index	S.E.
Isles of Scilly	2327	64	2603	353	3	737	31	8	144	88	86·6	0·3
Channel Islands	36	1	38	11	—	22	1	1	—	—	82·0	2·7
West Cornwall	181	22	110	158	—	113	3	7	13	8	71·2	1·3
Lundy	3	1	2	3	—	2	—	—	—	—	70·0	10·2
Boundary: Cornish	1058	370	686	1527	13	641	41	31	32	46	66·2	0·5
Boundary: Intermediate	693	245	291	624	5	272	17	16	7	17	67·0	0·8
Boundary: English	1564	426	479	907	8	521	27	39	24	37	69·2	0·6
Southern England	550	97	183	233	2	112	3	4	6	11	74·5	1·0
Scotland	6	19	6	13	—	2	—	3	—	—	48·7	5·7
Ireland	3	5	—	1	—	1	—	1	—	—	43·8	12·4
General European: N.C.	55	5	15	7	—	5	—	—	—	—	85·7	3·2
General European: S.W.	66	5	50	27	—	13	1	11	1	5	75·5	2·4
Spain	27	7	33	8	—	9	—	1	—	—	82·2	3·2
Rhodes	27	2	13	11	1	18	—	5	—	11	66·1	3·4
North Africa	26	7	56	7	—	23	2	—	2	—	82·8	2·4

Some samples caught in Britain in 1958, and virtually all since 1959, have been scored for spot-placing. For the purposes of this paper the samples have been grouped in broad geographical areas and the results within each area are summed and listed in tables 13.2 and 13.3 for the males and females respectively. The data for Britain are restricted to collections made during the main period of emergence; that is late July and August. These data are not always homogeneous when compared year by year, and the most extensive samples, those from the Isles of Scilly and the Boundary region in Devon and east Cornwall, suggest a decrease in costality in the females (see below); however, since in general the same relationships are observed between the areas in each year, we feel justified in taking the totals recorded in tables 13.2 and 13.3 as giving the best indication that we have of major characteristics. A further assumption that we have had to make is that the data from Europe and North Africa relating to insects caught between 1890 and 1935 are directly comparable with the British and other data subsequent to 1958.

Samples from Devon and east Cornwall have been allocated to the categories Boundary: Cornish, Boundary: Intermediate, and Boundary: English on the basis of the distribution of their spot numbers, and not according to their geographic position. Thus one locality may fall in different categories in successive years. The Scots samples are the same as those published by Forman, Ford & McWhirter (1959) and Creed, Dowdeswell, Ford & McWhirter (1962) together with an earlier museum collection caught in 1929. The European and North African samples are the same as those recorded by Dowdeswell & McWhirter (1967), whose classification into General European stabilization areas is followed here, and also include some unpublished material from Rhodes kindly provided by Dr The Hon. Miriam Rothschild.

Following the accumulated results for each area in tables 13.2 and 13.3 is the costality index, calculated from the combined sample. The relative values of the index for each area, and in each sex, are shown diagrammatically in fig. 13.2 in which may be seen the marked degree of correlation between the two sexes in most samples. For ease of comparison scales adopted in each sex have been adjusted so that the standard deviation of index values is the same in each. The correlation between the sexes is given by $r = 0.81$, with 11 degrees of freedom, for which $P < 0.001$ (the small samples from Lundy and Ireland are not included).

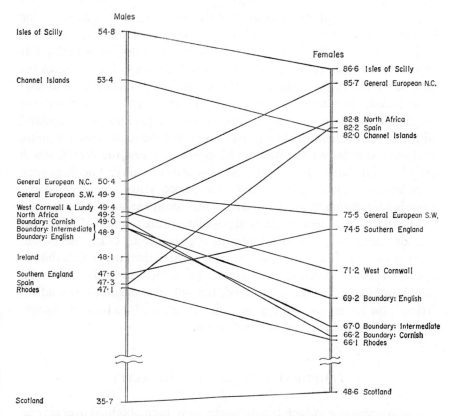

FIGURE 13.2. The relative values of the costality index in male and female samples from the fifteen areas investigated; female values for Lundy and Ireland are not included because the sample sizes are so small. The scales have been adjusted to give the same standard deviation of observed values in each sex.

This situation is directly comparable with the spot number distributions. There the correlation between the average spot number in each sex has been estimated at 0·71 ($P < 0·001$) and while the mean of male spot number is consistently higher than in the female, its variance is much less (McWhirter 1957).

Ranking of costal and anal 'features' of spotting, by comparing spot-types within given spot-numbers, produces a strikingly coherent pattern, at least over the northerly parts inhabited by *M. jurtina*. The General European (G.E.) area is stable and few populations show any marked bias; the north-central part of the General European area

(G.E.N.C.) has males which are slightly more costal than those in the southwestern part (G.E.S.W). The extension of G.E. into southern England shows a slightly more anal bias in the males; historically this 'Southern English' region has been less stable than G.E., but since the general shift of high-spotting in 1954–58, and subsequent return to near the old level, only one abnormal sample has been found in it. Towards the southwest, the main southern English region merges into the 'boundary' with East Cornish and the females, though not the males, become more anal, at least as far west as Bodmin Moor; farther west, the West Cornish samples shift markedly in both sexes towards costality. The Scillonian populations, isolated from Britain for 70,000 years (i.e. over 60,000 years before Britain was separated from the Continent) are the most costal so far studied. The next most costal populations are found in the Channel Islands; these imagines are similar in appearance to G.E. insects, but most of the islands are much more highly spotted than the G.E. stabilization. Northwards from south England, the transition to the most anal populations of all, those in Scotland, has only just begun to be studied. Ireland, the lowest spotted region so far found, tends to be anal, though sites examined are few and numbers small.

TEMPORAL CHANGES IN COSTALITY

The only areas from which large samples have been obtained over several years are the Isles of Scilly, collected every year except 1960 (when only an early collection was made), and the Devon–Cornwall 'Boundary' area. Costality indices for each year are given in table 13.4 together with the χ^2 value obtained from a $2 \times n$ comparison of the numbers of costal and anal spots in each year's sample. As we have already pointed out the index is more sensitive in females than in males, but this is not sufficient to account for the very much higher χ^2 value for the females from each area than for the males. As with spot number distribution, male spot placing is more highly stabilized than the female.

Much of the heterogeneity in the females is seen to be due to a general decrease in costality during the period of study; this trend is hardly apparent in the males and only in Scilly do they show any significant correlation with the females ($r = 0.68$; $0.02 > P > 0.01$). This is to be contrasted with the normally high correlation between male and female indices. Female variability is shown by the greater spread of their values

TABLE 13.4. Annual costality index values for males and females in southwest England. The χ^2 values are based on a $2 \times n$ comparison of the number of costal and anal spots each year.

| Year | Isles of Scilly | | Boundary | | | | | |
| | | | Cornish | | Intermediate | | English | |
	♂	♀	♂	♀	♂	♀	♂	♀
1958	56·2	88·2	—	—	—	—	—	—
1959	59·1	91·1	48·6	67·9	50·4	66·5	48·1	70·6
1960	—	—	49·8	72·1	48·8	74·1	49·0	73·7
1961	55·0	90·8	50·5	65·9	49·1	72·1	48·2	71·4
1962	53·3	86·6	47·9	64·4	48·2	67·8	47·6	68·4
1963	54·8	86·6	48·5	67·5	47·6	64·5	49·7	67·8
1964	55·3	88·4	50·4	71·8	50·0	67·2	49·2	76·4
1965	54·8	82·8	50·2	70·3	51·0	69·1	50·0	70·3
1966	56·0	87·8	49·7	65·2	50·0	68·7	48·3	62·6
1967	56·4	88·5	49·8	64·2	50·0	61·2	49·2	65·9
1968	54·6	85·1	48·6	60·7	47·1	65·6	48·6	64·7
1969	53·8	82·9	51·9	63·2	50·5	65·8	49·6	66·3
1970	53·9	85·5	47·4	62·2	48·0	61·4	47·6	61·8
χ^2	4·87	54·24	4·08	59·46	2·69	22·82	2·01	52·45

in fig. 13.2 (regardless of the scales chosen) and the much greater χ^2 values in table 13.4. We conclude that the decrease in costality in southwest England is part of a temporary fluctuation and that the mean values best characterize these populations.

DISCUSSION

The Devon–Cornwall boundary shows a deficiency of females with one spot on the Cornish side compared with the English, and a deficiency of females with two spots on the English side compared with the Cornish. If spot-placing types are studied in the females, significant differences are found in the frequencies of costal 1 relative to anal 1 and of costal 2 relative to splay 2 (table 13.3). These differences can be analysed by Woolf's method (1955) which allows convenient estimation of deficiencies or concentrations in a cell of a 2×2 table and their fiducial limits. Moving westwards from clearly S.E. to clearly E.C. populations in any year (irrespective of where the change-over occurs geographically), we find that as part of the decline in females with one spot there is a significantly greater deficiency of costal 1 specimens, while anal 1 frequencies are not so much affected. The relative deficiency of costal 1 is 22·1 per cent (95 per cent fiducial limits at 33·6 and 8·6 per cent) and the significance is measured by $\chi^2_{(1)} = 9·4; P < 0·01$. Again, if we move from the region of clearly E.C. populations to the Southern English (S.E.) area, whatever the geographical location at the time, we find that two-spotted females decline and the deficiency at 2 is disproportionately due to an absence of splay 2 females. Thus, within the category of two-spotted females, the relative deficiency of splay 2 is 14·9 per cent (95 per cent fiducial limits at 26·2 and 1·9 per cent), while the significance is measured by $\chi^2_{(1)} = 4·9; 0·05 > P > 0·02$. However, these changes in the frequency of spot-placings do not by themselves account for the whole of the change in spot-number observed at the boundary; we have here two interacting systems with a degree of interdependence.

Table 13.2 shows that there is close uniformity across the boundary region in the proportions of costal to anal 1 males and costal to anal 3 males. The homogeneity of spot-placings in this sex is indeed remarkable and, in the geographical context, the costal–anal balance fits neatly between the high costality of Scilly and West Cornwall and the relative anality of the main body of south England.

The boundary region can thus be seen as a considerable area, a rectangle with sides of some 80 km (W.S.W.–E.N.E.) by 60 km (N.N.W.–S.S.E) surrounding the unoccupied heights of Dartmoor, within which male *jurtina* populations show minimal variations, both as to spot-number and spot-placings, while female populations depart from the regular west–east costal-anal cline and exhibit orderly gradations of spot-type within at least two classes, the one-spots and the two-spots. At first sight it would seem remarkable that these gradations seem practically independent of the exact geographical location since the area of transition in spot numbers from S.E. to E.C. swings back and forth over Devon (Creed *et al.* 1970). However, if one visualizes the whole area as being occupied by a stabilized population (comparable with stabilizations such as the General European one) selective elimination can throw up each year a pattern of female survivors which varies according to some unknown element of the habitat, which itself differs in most years according to a simple geographical pattern. This element has been assumed to act selectively upon some stage of the insect; now that characteristic disturbances can be demonstrated in two genetically influenced features, spot-number and spot-placing, a selective hypothesis becomes even more clearly indicated.

The situation at the boundary is to be contrasted with the transition between mainland English populations and those few that we have studied from Scotland. The Scots samples differ little in distribution of spot-number from the English and indeed from the G.E. stabilizations, though the mode at o in the female may be slightly greater. However, the difference in spot-placing is extreme. The intermediate anal-costal system characteristic of mainland England and G.E. is converted to the most anal system so far found. Thus a very sharp reversal of spot-placing is not reflected in any marked change in the distribution of spot-numbers, indicating again the degree of independence between these two systems. The variability of populations peripheral to G.E. is well illustrated in the north and west of Britain, but is also exemplified in the Mediterranean, where our limited knowledge indicates stabilization areas in Iberia, northwest Africa and west Asia. Numbers in Iberia (excepting South Portugal) and in northwest Africa are adequate for comment; data from west Asia are too few. Previously unpublished figures from Rhodes give a reasonably clear view of the spot-placing situation. Spanish populations tend to be more anal than the G.E. norm, but among the females costal 2 is disproportionately frequent. Samples from Rhodes also tend in the anal direction, but, in this case, females at costal 1 are unexpectedly high. Northwest African populations have a high costality index, but among them males at anal 1 appear to be common. Insufficient data are available from Italy, Sardinia, Sicily, Malta and Greece to reveal any coherent picture of spot-placing. Probably all these peripheral areas have, as Dowdeswell & McWhirter (1967) suggested on the basis of spot numbers, developed different interactions and unusual features of the gene-pool as compared with the G.E. area.

In an animal with no fossil relics, evolutionary speculation must be hazardous. It is not known whether the trend to anality in the north is parallel with similar events in Scandinavia, though a small sample from Finland is non-significantly rather more anal than the bulk of G.E.N.C. Ireland may have been colonized from an already analized north British population, while the high costal areas of Scilly, Channel Isles, and also a small sample from the Canary Isles, could represent isolated and specialized detachments from an old stabilized core. However, similarities in spot-placing, or in the distribution of spot numbers, may reflect uniformity of the selective agents operating on these systems in recent times.

Traditional concepts of gene-flow and selection, which are based largely on one-locus models, would lead us to expect that, even in a somewhat immobile species like *M. jurtina*, a boundary effect would gradually be smoothed out into a uniform cline, maintained by a flow of genes into the populations at the ends of the transition zone. But there are here three genetic systems, determining sex, spot-number and spot-placing; these must be palaeogenic systems of great antiquity (long pre-dating the evolution of the species under study and carried over from the general Satyrine stock) so that in the course of time many interactions with the remainder of the genotype and with features of the environment will have become established and they will have become resistant to small changes in the direction of selection. They are 'viscous'. Within the genetic environment of the male sex, natural selection seems indeed to have procured transitions from east to west of slowly increasing costality and spot-number, which in the British area, at least, seem generally to go together. Some local abnormality in the way the gene-pool operates, however, seems to have imposed an unusual degree of high anality upon the females in the region of the boundary; and this further requires most populations to take up sharply differing options (or quantum-positions) as to spot-number.

The relative impotence of small or irregular changes in selective forces to secure directional changes in a viscous genetic structure is quite compatible with the existence of strong endocyclic selection in each generation, acting in various directions at successive stages of development. Such endocyclic selectional systems have been described in snails (Goodhart 1962 and Dowdeswell, chapter 19), postulated in Man (Kirby *et al.* 1967) and are now being recognized in various apparently stable populations of those organisms which have been adequately studied by sampling developmental series. Long-lasting distortions of clinal patterns such as those in the *M. jurtina* populations of the boundary region suggest that a rather complex pattern of endocyclic selection underlies them and offer an opportunity to study, at the population level, the fine detail of the mode of transition from one quantum of gene-frequencies to another.

Collaboration between E.B.Ford and the late Sir Ronald Fisher inaugurated the era of the mathematical approach to evolution of natural populations in the field. The early single-locus approaches to natural selection lead on to the understanding of more complex systems. The study of interacting gene systems at the population level, both in the wild and by breeding and experiment in the laboratory, and the elucidation, in appropriate organisms, of the ways in which selective elimination achieves stabilized patterns are most necessary concomitants of that vigorous development of multiloci mathematical and computer models of evolution which is now under way. Although such models are proliferated with much mathematical elegance, correct biological parameters can only be obtained through studies, which must often be protracted, of natural populations in the field.

SUMMARY

1. The character previously studied by Ford and his colleagues in the butterfly *M. jurtina* has been the number of spots occurring on the underside of the hindwing. This paper considers the relative positions of these spots.

2. Only a limited number of arrangements of the spots is commonly found. These may be classified according to whether there is a preponderance of spots towards the costal border of the wing, towards the anal border, or no bias at all. By grouping all individuals from one area an index of costality versus anality may be calculated for each sex; this index usually gives similar results in the two sexes.

3. A general decrease in costality has occurred in southwest England since 1958. Data are not sufficient to show whether such changes have occurred elsewhere.

4. The costality index, like the distribution of spot numbers, shows a clear pattern of geographic variation. Though the major changes in the two systems do not occur at the same places, there is clear evidence of interaction between them and with sex.

ACKNOWLEDGEMENTS

We would particularly like to thank Professor E.B.Ford and W.H.Dowdeswell and also all those who have provided samples or helped to collect them. D.R.Lees has kindly commented on a draft of the manuscript.

REFERENCES

CREED E.R., DOWDESWELL W.H., FORD E.B. & MCWHIRTER K.G. (1962) Evolutionary studies on *Maniola jurtina*: the English mainland, 1958–60. *Heredity* **17**, 237–265.

CREED E.R., DOWDESWELL W.H., FORD E.B. & MCWHIRTER K.G. (1970) Evolutionary studies on *Maniola jurtina* (Lepidoptera, Satyridae): the 'boundary phenomenon' in southern England 1961 to 1968. In *Essays in Evolution and Genetics in Honor of Theodosius Dobzhansky* (Hecht & Steere, eds.). Appleton-Century-Croft.

DOWDESWELL W.H. & FORD E.B. (1953) The influence of isolation on variability in the butterfly *Maniola jurtina* L. *Symp. Soc. exp. Biol.* **7**, 254–273.

DOWDESWELL W.H. & MCWHIRTER K.G. (1967) Stability of spot distribution in *Maniola jurtina* throughout its range. *Heredity* **22**, 187–210.

DOWDESWELL W.H., FISHER R.A. & FORD E.B. (1949) The quantitative study of populations in the Lepidoptera. 2. *Maniola jurtina* L. *Heredity* **3**, 67–84.

DOWDESWELL W.H., FORD E.B. & MCWHIRTER K.G. (1960) Further studies on the evolution of *Maniola jurtina* in the Isles of Scilly. *Heredity* **14**, 333–364.

FORD E.B. (1971) *Ecological Genetics* (3rd ed.). Methuen, London.

FORMAN B., FORD E.B. & MCWHIRTER K.G. (1959) An evolutionary study of the butterfly *Maniola jurtina* in the north of Scotland. *Heredity* **13**, 353–361.

GOODHART C.B. (1962) Variation in a colony of the snail *Cepaea nemoralis* L. *J. anim. Ecol.* **31**, 207–237.

KIRBY D.R.S. MCWHIRTER K.G. TEITELBAUM M.S. & DARLINGTON C.D. (1967) A possible immunological influence on sex ratio. *Lancet* **2**, 139–140.

MCWHIRTER K.G. (1957) A further analysis of variability in *Maniola jurtina* L. *Heredity* **11**, 359–371.

MCWHIRTER K.G. (1969) Heritability of spot-number in Scillonian strains of the Meadow Brown butterfly (*Maniola jurtina*). *Heredity* **24**, 314–318.

WOOLF B. (1954) On estimating the relation between blood groups and disease. *Ann. hum. Genet.* **19**, 251–253.

14 ❀ An Esterase Polymorphism in the Bleak, *Alburnus alburnus* Pisces

P.T.HANDFORD

INTRODUCTION

Polymorphic protein systems have received much attention since the development of gel electrophoresis and the 'zymogram' techniques (Smithies 1955, Hunter & Markert 1957), but it was not until quite recently that their study developed much beyond the descriptive level, and use was made of their potential as valuable tools in the study of ecological and evolutionary genetics (Fujino & Kang 1968a,b, Hubby & Throckmorton 1965, Koehn & Rasmussen 1967, Prakash, Lewontin & Hubby 1969, Semeonoff & Robertson 1968, Tamarin & Krebs 1969).

This paper reports an esterase polymorphism in a small cyprinid fish, the bleak, *Alburnus alburnus*, and describes the initial results of an attempt to use such a polymorphic system in estimating the nature of the effects of the selection to which the fish are subject at different stages of the life history. Here, attention is centred on fish aged between one and three years although some data on older fish are included.

MATERIALS AND METHODS

All fish were taken by seine net from a 200-yard stretch of the Seacourt Stream, a tributary of the River Thames, approximately 3 miles from Oxford, England. Fish were transported to the laboratory where they were either kept alive in tanks or deep frozen until they could be analysed.

Enzyme extracts were made by homogenizing tail muscle with an equal volume of distilled water in a Gallenkamp electric homogenizer. Distilled water was used as it provided better extraction than any of the buffers tried (compare Clements 1967). The homogenates were stored at −25°C until they were assayed.

In fish of greater length than 90 mm, the enzymes are at a much lower concentration in the muscle and therefore the extracts from these fish were concentrated by freeze drying.

Tests have shown that the esterases studied here are quite stable and patterns can be reproduced after up to ten months storage and six thawings and refreezings, which is of considerable use when comparison of fractions from individuals derived from different catches is necessary.

ELECTROPHORESIS AND STAINING

The technique used throughout this study was vertical DISC-electrophoresis on acrylamide gel columns using Quickfit and Quartz equipment. Electrophoresis conditions and procedure were modified after those of Davis 1964: compartmental buffer: Glycine/HCl pH 8·2; gel buffer: TRIS/HCl pH 8·9; separating gel concentration, 10 per cent Acrylamide. A separating gel only was used, tests having showed that there was no loss of resolution on omission of a large-pore spacer gel.

Samples were applied to the gels absorbed on to filter paper discs and sixteen tubes, cooled by iced water, were run together with a combined current of 75 mA. In order that the proteins should be run over the same distance in each tube, a running marker is used which migrates well ahead of the sample and electrophoresis is stopped when this marker coincides with a line set at a prescribed distance from the gel origin. Davis uses a fixed mark on the tube wall, and pipettes an accurate volume of gel solution into the tube, thus building gels of equal length; in the present technique, this time consuming pipette operation is obviated by using a movable rubber ring which is fixed at 35 mm from the gel origin once the gel has solidified. Bromophenol Blue was used as the running marker.

Gels were incubated in the following solution for visualization of esterase bands:

> 70 mg Fast Blue B
> 1 ml 2 % α-naphthylacetate in acetone
> 1 ml 2 % β-naphthylacetate in acetone
> 100 ml Tris/HCl buffer pH 7·1

Staining took 30 minutes at 25°C. Although most gels were readily scored for their phenotype, composites of extracts of known and unknown constitution were run in the same tube to establish the identity of doubtful bands.

RESULTS

The extent of variation observed is summarized in fig. 14.1. No variation has been found in the highly active fraction migrating in front of the polymorphic region.

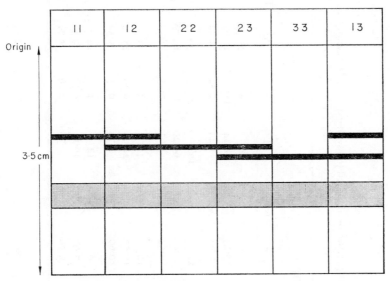

FIGURE 14.1 The relative mobilities of the three esterases and their arrangement in the six observed phenotypes.

As the fish is oviparous, no direct progeny data are as yet available which will demonstrate the reality of the suggested genetic basis of the system. Further, all groups, whether individually or summed show significant distortion from the Hardy–Weinberg expectations calculated on the model suggested. However the nature of the phenotypes observed, together with the total absence of individuals with more than two bands makes it likely that the genetic model most compatible with observations is that of three codominant alleles at a single locus.

Table 14.1 shows the summed length class data together with Hardy–Weinberg expectations and allele frequencies. The relationship between

TABLE 14.1. Allele frequencies and comparison of observed (O) with expected (E) genotype numbers in the esterase system of the bleak, *Alburnus alburnus*.

Length class with mean lengths in mm		Genotypes							Alleles		
		11	12	22	23	33	13	Total	1	2	3
O 38.89	O	2	9	2	26	8	25	72	0.264	0.271	0.465
	E	5.01	10.29	5.28	18.14	15.58	17.67				
	O/E	0.40	0.88	0.38	1.43	0.51	1.42				
I 60.33	O	9	21	6	108	20	99	263	0.262	0.268	0.470
	E	18.09	36.98	18.88	66.20	57.97	64.78				
	O/E	0.50	0.57	0.32	1.63	0.35	1.53				
II 85.70	O	6	10	1	41	3	43	104	0.313	0.255	0.433
	E	10.16	16.57	6.75	22.93	19.46	28.12				
	O/E	0.59	0.60	0.15	1.79	0.15	1.53				
III+ 10.062	O	1	6	3	14	1	16	41	0.293	0.317	0.390
	E	3.51	7.61	4.12	10.14	6.24	9.36				
	O/E	0.29	0.79	0.73	1.38	0.16	1.71				

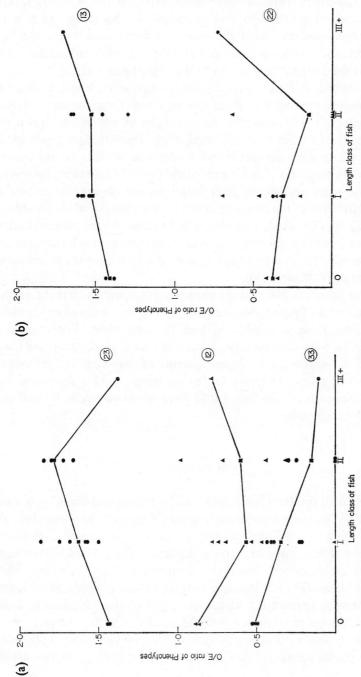

FIGURE 14.2. The relation of the *O/E* ratio of five phenotypes to the length of fish. (a) Phenotypes 23, 12 and 33. (b) Phenotypes 13 and 22. *O/E* points for individual catches within a length class are represented by discs or triangles; summed group points by squares.

length of fish and the ratio of observed and expected numbers (O/E ratio) is represented graphically for five of the phenotypes in figs. 14.2a and 14.2b. Varying numbers of individual nettings were involved in producing the total numbers shown in table 14.1 and the O/E ratios calculated from such individual catches are represented in figs. 14.2a and 1.42b.

Most catches fall into fairly discrete groups when length is plotted against abundance and it is these groups which form the basis of the length classes O, I, II etc., the mean lengths of which are shown in table 14.1. The Petersen method of ageing fish (Tesch 1968) depends upon fish falling into these distinct length modes and one can be quite confident that the majority of fish in a given mode are the same age; however, fish at the extreme ends of each length group cannot be definitely allocated to either of the two neighbouring year classes and so, considering the numbers involved, it is thought best to describe the present results in terms of length groups only, although these groups can be thought of as being equivalent to age groups as there is a high correlation between length and age (Macan 1970).

Though there are fish ageing techniques described which make use of annuli on scales (Tesch 1968) there is evidence that without exhaustive investigation, these methods can lead to some error. For example, Geen et al. (1966) working on two species of Catostomid, concluded that scales did not supply a reliable age measure of spawning fish, estimates generally being low. Hogman (1968) studying annulus formation in captive Coregonus species found that there were two marks formed on most scales each year.

DISCUSSION

Compared with similar work carried out by Fujino and Kang (1968a,b) on variation in serum transferrin and serum esterase in a number of species of tuna fish, the present system appears to be one of some complexity: no simple approach to or departure from Hardy–Weinberg ratios can be described; however all phenotypes do show an initial divergence from $O/E = 1$ between Groups O and I, implying a closer agreement with expectation at earlier stages of the life history. This apparent divergence contrasts with the situation in the skipjack tuna transferrin system where the genotype distributions steadily approach expectancy with increasing size of fish such that $O/E = 1$ at the usual maximum length of the species. In the serum esterase system of the same species, the genotype frequencies are almost exactly according to expectancy and this situation appears to be quite independent of age or sex. In the bleak, most work has been carried out either on immature fish or on older fish out of breeding condition, and as such their sex is not usually determinable with any certainty. However, of fish whose sex is known, no obvious non-uniformity of genotype distribution between the sexes has been found. Linear regressions of O/E on length show that only in genotype 33 is the relationship clearly significant ($t_{(11)} = -4.18$, $0.01 > P > 0.001$), although others are bordering on significance, and in comparisons between observed genotype numbers of successive length classes, only one made between Group O and Group I approaches significance.

Thus, there is precedent for both age dependency and age independency of genotype distribution within polymorphic protein systems, and the situation in the bleak cannot be categorically described as either. However, the appearance of the graphs and the fact that certain comparisons are approaching significance, gives credibility to the suggestion that there is differential selection of genotypes over the life-history. The large departure from Hardy–Weinberg equilibrium observed in even the smallest fish makes it probable that selection is operating at some stage, if only in the form of non-random mating. The fact that two of the heterozygotes show consistent excess suggests that heterosis may be operating, and Koehn's work (1969) on the esterase system in *Catostomus clarki* has clearly shown that selection for heterozygotes can be an important feature in the maintenance of such polymorphisms. The heterozygote excess could also be due to differences in gene frequency between the sexes, but as mentioned above, no indication of such sex differences have been observed. The changes in genotype abundance could be related to any contamination of the study area by neighbouring populations with different gene frequencies, although this is not considered likely as there has not been significant change in gene frequency over the period of study. The genotype abundance changes could also be manifestations of a directional evolutionary change and not of an endocyclic process; samples of fish derived from this year's spawning will help in deciding which situation obtains. That the polymorphism is stable is suggested by its presence in small samples of bleak from the river Welland in Lincolnshire and the river Stour in Dorset.

SUMMARY

An esterase polymorphism in the bleak, *Alburnus alburnus* is described and a genetic model suggested. All length groups show significant distortion from Hardy–Weinberg expectancies and an excess of two of the heterozygotes, and it is suggested that the data support the proposition that the genotype frequencies are age dependent and that heterosis may be involved in the maintenance of the polymorphism.

ACKNOWLEDGEMENTS

This work was carried out in the Genetics Laboratory of the Zoology Department, Oxford, whilst in receipt of an N.E.R.C. Research grant. Thanks are due to Professor E.B.Ford for the provision of research facilities, to Mr A.J.French and Miss C.J.Farley for technical assistance, and to Mr T.Reimchen for useful discussion. I am also grateful to Mr D.R.Lees for reading the manuscript.

REFERENCES

CLEMENTS A.W. (1967) A study of soluble esterases in *Pieris brassicae*. *J. Insect Physiol.* **13**, 1021–1030.

DAVIS B.J. (1964) Disc Electrophoresis II. Method and application to human serum proteins. *Ann. N. Y. Acad. Sci.* **121**, 404–427.

FUJINO K. & KANG T. (1968a) Serum esterase groups in Pacific and Atlantic Tuna. *Copeia* **1**, 56–63.

FUJINO K. & KANG T. (1968b) Transferin groups in tunas. *Genetics* **59**, 79–91.

GEEN G.H., NORTHCOTE T.G., HARTMAN G.F. & LINDSEY C.C. (1966) Life history of two species of Catostomid fishes. *J. Fish. Res. Bd. Can.* **23**, 1761–1788.

HOGMAN W.J. (1968) Annulus formation on scales of four species of Coregonids reared under artificial conditions. *J. Fish. Res. Bd. Can.* **25**, 2111–2122.

HUBBY J.L. & THROCKMORTON L.H. (1965) Protein differences in Drosophila II. Comparative species of genetics and evolutionary problems. *Genetics* **52**, 205–215.

HUNTER R.L. & MARKERT C.L. (1957) Histochemical demonstration of enzymes separated by zone electrophoresis in starch gels. *Science* **125**, 1294–1295.

KOEHN R.K. (1969) Esterase heterogeneity: Dynamics of a polymorphism. *Science* **163**, 943.

KOEHN R.K. & RASMUSSEN D.I. (1967) Polymorphic and monomorphic serum esterase heterogeneity in Catastomid fish populations. *Biochem. Genet.* **1**, 131–145.

MACAN T.T. (1970) *Biological studies of English Lakes*, Chap. 11. Longman.

PRAKASH S., LEWONTIN R.C. & HUBBY J.L. (1969) Molecular approach to the study of genic heterozygosity IV. Variation in central, marginal, and isolated populations. *Genetics* **61**, 840–858.

SEMEONOFF R. & ROBERTSON F.W. (1968) A biochemical and ecological study of plasma esterase polymorphism in natural populations of the Field Vole, *Microtus agrestis* L. *Genetics* **1**, 205–227.

SMITHIES O. (1955) Zone electrophoresis in starch gels: group variation in the serum proteins of normal human adults. *Biochem. J.* **61**, 629–641.

TAMARIN R.H .& KREBS C.J. (1969) *Microtus* population biology II: Genetic changes at the transferin locus in fluctuating populations of two vole species. *Evolution* **23**, 183.

TESCH F.W. (1968) In *Methods for assessment of fish production in fresh waters* (W.E.Ricker, ed.), pp. 93–123.

15 ✤ Gene Duplication and Haemoglobin Polymorphism

J. B. CLEGG

The introduction of electrophoretic techniques in the early 1950s soon led to the discovery that the haemoglobins of many animal species are polymorphic. In man the crucial discovery of haemoglobin S in the red cells of persons with sickle-cell anaemia was a striking illustration of the concept of disease at the molecular level, and it subsequently stimulated a great deal of research into the genetic aspects of the control of protein synthesis. Early ideas about the genetics of human haemoglobin were developed largely through family studies of individuals with various abnormal haemoglobins; particularly noteworthy were the findings of Smith & Torbert (1958) who established the existence of individual α- and β-chain genes, most probably on separate chromosomes. Haemoglobin variants proved to be the products of alleles of either α- or β-chain genes, and Hbs F and A_2 were shown to be due to the existence of separate γ- and δ-chain genes. The determination of the amino acid sequences established considerable homologies between the α, β, γ and δ-chains and it was suggested that these could be most simply accounted for by assuming that the genes which determine them were originally derived from a common ancestral gene. Successive duplications followed by separate evolution of the resulting genes by point mutations would then give rise to the different but related genes that exist today (Ingram 1961). The very close homology of the β- and δ-chains and the fact that the two genes lie close together on the same chromosome was taken as an indication that they have existed as separate entities only recently in evolutionary history.

This concept of gene duplication, although not specifically applied to particular haemoglobin chains, had been used by Bangham & Lehmann (1958) in suggesting the non-allelic nature of the two haemoglobins commonly found in horses. Other examples followed: it was assumed, without much direct evidence, that multiple haemoglobins which were

298

not the result of alleles, owed their existence to non-allelic genes. This assumption was seriously questioned by von Ehrenstein (1966) who in the case of rabbit haemoglobin put forward the somewhat heretical suggestion that multiple haemoglobin chains could be derived from a single gene. This radical hypothesis threatened for a time to upset the long-cherished one-gene one-polypeptide chain idea and although it is not now in favour it had the effect of stimulating numerous experiments in attempts to prove or disprove it. As a result, genetic evidence in favour of duplicate structural genes for haemoglobin chains was soon forth-coming. Before describing some of these experiments it is worth consider-ing briefly the case of rabbit haemoglobin and the questions it raised.

During experiments in a cell free system on the coding properties of *E.coli* leucine transfer RNA, Weisblum *et al.* (1965) found evidence that one particular leucine position in the α-chain differed from all other leucine positions in rabbit haemoglobin. Von Ehrenstein (1966) later assigned this exceptional codon to position α48, and subsequent determin-ation of the amino acid sequence showed that not only was this position unusual in its response to a minor leucine transfer RNA but also that both leucine and phenylalanine were present; in haemoglobin from a single rabbit about equal amounts of the two amino acids were found at α48. Subsequently, another five positions in the α-chain were found to be occupied by more than one amino acid. Von Ehrenstein suggested that although a multigenic origin for the sequence variations could not be excluded, the facts were consistent with the idea that variations in the translation of messenger RNA could account for the multiple α-chains. This 'ambiguity hypothesis', as it has become known, was difficult to prove or disprove, particularly since all the variations involved neutral amino acids and chromatographic or electrophoretic fractionation of the α-chains was impossible. It was three years before the doubts about the fidelity of the genetic information were resolved, at least in the case of the rabbit, by the demonstration that the variants of the α-chain form two linked sets which segregate in a Mendelian fashion (Hunter & Munro 1969), thus indicating the existence of two alleles for the α-chain.

In the meantime, however, the ambiguity hypothesis had focused interest on to other species with multiple haemoglobins, in the hope that more favourable circumstances might enable a conclusive test of the hypothesis to be made. In the case of the mouse, the horse, the goat, and man, particularly the last two, evidence was obtained which not only disproved the ambiguity hypothesis, but suggested the existence of linked non-allelic genes as being responsible for the haemoglobin polymorphisms involved.

The haemoglobins of certain strains of mice show obvious differences in solubility in phosphate buffer when compared to the haemoglobin of the strain C57BL/6 (Popp & Amand 1960). Genetic studies indicated that the 'low solubility' locus segregated with the α-chain gene (Popp 1965, Hutton, Schweet, Wolfe & Russell 1963) although peptide 'fingerprints' of the α-chains of the strains involved showed no differences.

Detailed amino acid sequence studies, however, did reveal differences between the α-chains of C57BL/6 and SEC mice, and in particular the asparagine residue at position α68 in the C57BL/6 strain was replaced by a mixture of threonine and serine in SEC (Rifkin, Rifkin & Konigsberg 1966). The fact that this occurred in a strain which had been inbred for 82 generations made it unlikely to be due to heterozygosity. Similar results were obtained with the strain BALB/c (Popp 1967), and haemoglobins from the strain C3H.B and C3H were likewise found to be composed of mixtures of two forms of α-chains (Hilse & Popp 1968). In this latter case, however, some unique characteristics of the C3H α-chain made it possible to distinguish the genetic basis for the existence of the multiple α-chains. Determination of the relevant portions of the amino acid sequences of the α-chains of C3H.B and C3H haemoglobins showed that position α68 in C3H.B was occupied by serine and asparagine, α62 by valine and α25 by glycine, whereas in strain C3H, α25 had

TABLE 15.1. Relationships of the primary structures of α-chains of mouse haemoglobins.

Strain of origin of haemoglobin	Residue numbers showing amino acid differences		
	25	62	68
C57BL/6	Gly	Val	Asn
NB	Val	Ile	Ser
BALB/c	Gly	Val	Ser
	Gly	Val	Thr
C3H.B	Gly	Val	Ser
	Gly	Val	Asn
C3H	Val	Ile	Ser
	Gly	Val	Asn

both glycine and valine, $\alpha 62$ had both valine and isoleucine, and $\alpha 68$ had serine and asparagine. These relationships are shown in table 15.1. Positions $\alpha 62$ and $\alpha 68$ both occur in tryptic peptide $\alpha T9$, and chromatography of this peptide from the C3H α-chain revealed two forms, one having isoleucine at $\alpha 62$ and serine at $\alpha 68$, the other having valine at $\alpha 62$ and asparagine at $\alpha 68$.

This important observation showed that the amino acid mixtures at $\alpha 62$ and $\alpha 68$ were not due to translational ambiguities since four forms of the peptide $\alpha T9$ would have been expected if this had been the case (i.e. Val^{62}/Ser^{68}, $Val^{62}/Aspn^{68}$, Ile^{62}/Ser^{68} and $Ile^{62}/Aspn^{68}$). The presence of two α-chain structural genes was thus indicated and genetic studies confirmed this and established their close linkage.

The occurrence of haemoglobin polymorphism in the domestic goat has been well established for some time (Bernhardt 1964; Huisman *et al.* 1967, 1968). The two major haemoglobins, A and B, differ in their α-chains, while the respective β-chains are identical. Structural studies of the α^A-chains showed that, in fact, two forms existed which differed from each other in four positions as shown in table 15.2.

TABLE 15.2. Relationships of the primary structures of α-chains of goat haemoglobins.

	19	26	75	113	115
α^{IA}	Gly	Ala	Asp	Leu	Asx
α^{IIA}	Ser	Thr	Asp	His	Ser
α^B	Gly	Ala	Tyr	Leu	Asx

The possibility of ambiguous messenger RNA translation did not have to be considered here because peptides with different and unique (i.e. non-ambiguous) sequences were isolated from the two α-chains (Huisman *et al.* 1968). Furthermore, it was shown that the α-chain from Hb B (α^B) differed from the α^{IA}-chain by a single point mutation at $\alpha 75$ (thus incidentally providing the test that von Ehrenstein had suggested as a way of settling the ambiguity issue), and since in a goat with both Hb A and Hb B all three types of α-chain were present, the α^I and α^{II} chains could not be products of allelic genes but must derive from non-allelic genes. This was confirmed when a goat homozygous for Hb B was found to have both α^B- and α^{IIA}-chains, but no α^{IA}-chains (Garrick & Huisman 1968). The situation is summarized in fig. 15.1.

$$A\text{-}genotype \qquad B\text{-}genotype$$
$$\alpha^{IA} \qquad\qquad\quad \alpha^{B}$$
$$| \qquad\qquad\qquad | \qquad\qquad (\alpha^{B} = \alpha^{IA}\ 75\ Tyr)$$
$$\alpha^{IIA} \qquad\qquad\quad \alpha^{IIA}$$

FIGURE I5.I.   Goat haemoglobin α-chain genotypes.

Hb polymorphism was first described in horses in 1955. Bangham & Lehmann (1958) found that most horses had two distinct haemoglobin fractions on electrophoresis, the faster component comprising 60 per cent of the total Hb. A few horses were observed to have a higher proportion of the fast component, however, and Braend (1967) confirmed and extended these observations by the discovery in a population of Norwegian horses of a third phenotype in which only the fast Hb was present. The fast and slow haemoglobin ratios in the three phenotypes were 60/40, 80/20, and 100/0. Genetic studies indicated that the 80/20-type was the result of a mating between the 60/40- and 100/0-types.

Kilmartin & Clegg (1967) found that the difference between the fast and slow Hbs was due to a substitution of the glutamine residue at position 60 in the fast Hb α-chain (α_f) by lysine in the slow Hb α-chain (α_s). (The β-chains of the fast and slow Hbs were identical.) Furthermore, in a number of the 60/40-type horses studied, position $\alpha24$ in both the α_f- and α_s-chains was occupied by both tyrosine and phenylalanine. These particular animals thus had four haemoglobins. Other horses were later found in which $\alpha24$ in both fast and slow α-chains was occupied solely by either tyrosine or phenylalanine. Although the distribution of these different α-chains suggested that the genes controlling the structures of the $\alpha24$ Tyr-chains and $\alpha24$ Phe-chains might be alleles, it was not possible at that time to carry out any family studies.

These early observations have since been extended (Clegg 1970) to cover the Norwegian horses from the population studied by Braend. The 60/40-types were similar to those studied in England, but the 80/20- and 100/0-types (which appear to be rare outside Norway) showed marked differences. In all the 100/0-types examined, for example, α_f24 was occupied by tyrosine alone, while in the 80/20-types, the α_f- and α_s-chains had either tyrosine alone at both α_f24 and α_s24 or else had tyrosine and phenylalanine at α_f24 and *only* phenylalanine at α_s24. These findings are summarized in table 15.3.

It was extremely fortunate that pedigree data was readily available for almost all the horses in this interesting group, and it was thus possible

TABLE 15.3. Horse haemoglobin and α-chain phenotypes.

Haemoglobin phenotype	60/40	80/20	100/0
α-chains found	$\alpha_f 24$ Tyr $\alpha_s 24$ Tyr BI^*	$\alpha_f 24$ Tyr $\alpha_s 24$ Tyr A/BI	$\alpha_f 24$ Tyr A
	$\alpha_f 24$ Tyr/Phe $\alpha_s 24$ Tyr/Phe BI/BII	$\alpha_f 24$ Tyr/Phe $\alpha_s 24$ Phe A/BII	
	$\alpha_f 24$ Phe $\alpha_s 24$ Phe BII		

* The notations BI, A/BI etc. refer to the appropriate genotypes described in the text.

to amplify the simple chemical studies made originally by some family studies on horses in which the tyrosine/phenylalanine heterogeneity occurred at position $\alpha 24$. These experiments established that in the 60/40-type horses with tyrosine and phenylalanine at $\alpha_{f,s}24$ the Tyr and Phe loci behaved as alleles, one locus having only tyrosine and the other phenylalanine. Also, in the 80/20-type horses with tyrosine and phenylalanine at $\alpha_f 24$ and phenylalanine at $\alpha_s 24$ (table 15.3) the Tyr and Phe loci segregated as alleles in which the α_f-chain and α_s-chain loci were linked, the $\alpha_f 24$ Phe locus segregating with the $\alpha_s 24$ Phe locus and the $\alpha_f 24$ Tyr locus segregating independently (fig. 15.2).

These observations have been explained (Clegg 1970) by the assumption that two main genotypes (A and B, below) are present in the Norwegian horse population, each consisting of two linked α-chain genes, one gene being responsible for 60 per cent of the α-chain production and the other 40 per cent:

Type A

$60\alpha_f$
|
$40\alpha_f$

Type B

$60\alpha_f$
|
$40\alpha_s$

On this basis the 100/0-type horse is an AA homozygote, the 60/40-type a BB homozygote, and the 80/20-type an AB heterozygote.

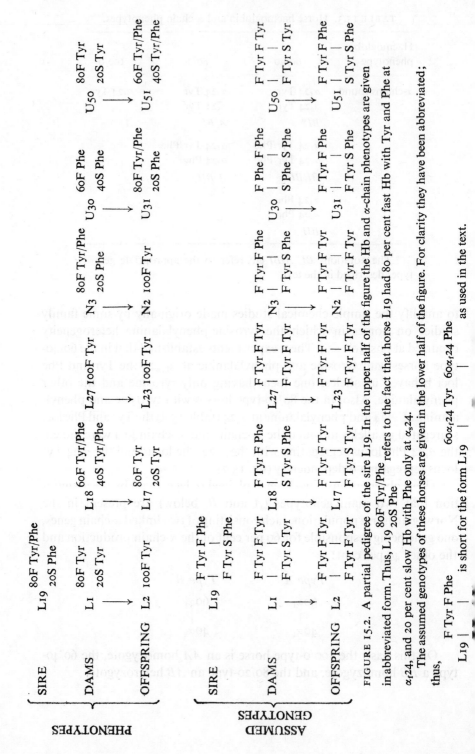

FIGURE 15.2. A partial pedigree of the sire L19. In the upper half of the figure the Hb and α-chain phenotypes are given in abbreviated form. Thus, L19 $\frac{80F\ Tyr/Phe}{20S\ Phe}$ refers to the fact that horse L19 had 80 per cent fast Hb with Tyr and Phe at $\alpha_f 24$, and 20 per cent slow Hb with Phe only at $\alpha_s 24$.

The assumed genotypes of these horses are given in the lower half of the figure. For clarity they have been abbreviated; thus,

L19 $\frac{F\ Tyr\ F\ Phe}{—\ —}$ is short for the form L19 $\frac{F\ Tyr\ F\ Phe}{60\alpha_f 24\ Tyr\quad 60\alpha_f 24\ Phe}$ as used in the text.

Inasmuch as all the 100/0-type horses had tyrosine alone at a $\alpha_f 24$ it was suggested that the A genotype is actually

$$60\alpha_f 24 \text{ Tyr}$$
$$|$$
$$40\alpha_f 24 \text{ Tyr}$$

and from the tyrosine/phenylalanine distribution at $\alpha 24$ in the fast and slow α-chains of the 80/20- and 60/40-type horses it was suggested that the B genotype is polymorphic, consisting of the two alleles BI and BII.

BI	BII
$60\alpha_f 24$ Tyr	$60\alpha_f 24$ Phe
$|$	$|$
$40\alpha_s 24$ Tyr	$40\alpha_s 24$ Phe

thus accounting for the various phenotypes shown in table 15.3.

How this complex situation arose is a mystery and seems likely to remain so until much more information about the number of possible phenotypes which exist becomes available. Thus, out of the 16 possible genotypes which might theoretically derive from combinations of pairs of the four genes $\alpha_f 24$ Tyr, $\alpha_s 24$ Tyr, $\alpha_f 24$ Phe, $\alpha_s 24$ Phe, only three have been observed so far, and until it is known whether others exist speculation as to their origin seems unlikely to prove very fruitful.

A comparison with the situation in other Equidae, though, is interesting, for although the donkey has a single haemoglobin (with, as far as one can tell, no heterogeneity involving neutral amino acid variations of the sort encountered in the rabbit) both the onager and some sub-species of the zebra have two haemoglobins. In each case the variant Hb has a different amino acid substitution in the α-chain from that which gives rise to the α_f- and α_s-chains in the horse. Furthermore, in the zebra (the onager data is still incomplete) *both* α-chains have additional *identical* mutations distinguishing them from horse Hb α-chains. Therefore if α-chain duplication is responsible for the two zebra haemoglobins, the duplication event itself must have occurred independently of that in the horse. It thus seems likely that haemoglobin gene duplication took place on a number of occasions during the evolution of the Equidae.

It had seemed for a time that the complexities manifest in the haemoglobin genetics of some animal species were not to be found in man. The discovery that human foetal haemoglobin is heterogeneous in the γ-chain (Schroeder *et al.* 1968) provided the first evidence to the contrary. The γ-chains of Hb F, isolated from the cord bloods of infants representing a variety of ethnic groups, consisted of two types, with either glycine or alanine in position γ136, and the world-wide occurrence of this phenomenon in all the normal subjects studied discounted heterozygosity at a single allele as an explanation. In infants who were heterozygous for Hb F and a foetal Hb variant, however, the abnormal γ-chains had only glycine *or* alanine at γ136, while in the normal γ-chains from these infants both amino acids were present. The interpretation of this is that the γ136 Gly- and γ136 Ala-chains are the products of non-allelic structural genes and the variants are mutants of one or the other allele. Further work on the nature of the Hb F present in infants and adults in normal and various abnormal conditions such as hereditary persistence of foetal haemoglobin and β-thalassaemia (Huisman *et al.* 1969, 1970; Schroeder *et al.* 1970) has shown a remarkable heterogeneity at the molecular level suggesting, in the latter condition at least, an even more complicated situation than hitherto assumed.

Is there any evidence that any of the other human haemoglobin chains are produced by non-allelic genes? For the β- and δ-chains the answer seems to be no, for there are many examples of homozygotes and double heterozygotes for various β- and δ-chain variants. α-chains present a more confused picture. The low percentage of most (but not all) α-chain variants has led to the suggestion that non-allelic α-chain genes might exist in humans (Lehmann & Carrell 1968) and this speculation has been supported by the discovery of two different families with members having *two* α-chain haemoglobin variants in addition to Hb A (Abramson *et al.* 1970; Brimhall *et al.* 1970). In apparent contradiction, however, is the case of the New Britain population where examples of homozygous Hb J (Tongariki), an α-chain variant, have been found (Abramson *et al.* 1970). Thus the evidence for non-allelic α-chain genes is at present equivocal but it may be that examples of both single and duplicated α-chain genes exist in the human population. The New Britain situation is in some ways very similar to that found with horses in Norway. Both represent inbred groups. In the Norwegian horses, animals homozygous for the fast Hb alone are found, whereas the usual situation in most of the world is the occurrence of two Hbs which are presumed to be the products of non-allelic genes. Horses having only the fast Hb are assumed to have duplicated α-chain genes also, though in fact it is difficult to prove this conclusively; they could equally well have only one.

Many generations of localized inbreeding in the New Britain population might have produced a similar situation in which two α^J-genes have become located on the same chromosome.

Proof of the existence of duplicated α-chain genes in the human population would have important consequences for genetics and medicine, for example, in the study of α-thalassaemia. The mechanisms which control the expression of mammalian genes are unknown, but a prerequisite to their understanding is a thorough knowledge of the genetic system involved. It seems likely that haemoglobin, for long a model protein in structural studies, will continue to play an important role in the field of gene control in mammalian systems.

REFERENCES

ABRAMSON R.K., RUCKNAGEL D.L., SHREFFLER D.C. & SAAVE J.J. (1970) Homozygous HbJ Tongariki: Evidence for only one alpha-chain structural locus in Melanesians. *Science* 169, 194.

BANGHAM A.D. & LEHMANN H. (1958) Multiple haemoglobins in the horse. *Nature* 181, 267.

BERNHARDT D. (1964) Elektrophoretische Untersuchungen von hämoglobinen bei Ziegen, Hunden, Katen und Nerzen. *Dt. Tierärztebl. Wschr.* 71, 461.

BRAEND M. (1967) Genetic variation of horse haemoglobin. *Hereditas* 58, 385.

BRIMHALL R., HOLLAN S., JONES R.T., KOLER R.D. & SZELENYI J.G. (1970) Multiple α-chain loci for human haemoglobin. *Clin. Res.* 18, 184.

CLEGG J.B. (1970) Horse haemoglobin polymorphism: Evidence for two linked non-allelic α-chain genes. *Proc. R. Soc. B.* 176, 235.

VON EHRENSTEIN G. (1966) Translational variations in the amino acid sequence of the α-chain of rabbit haemoglobin. *Cold Spring Harb. Symp. quant. Biol.* 31, 705.

GARRICK M.D. & HUISMAN T.H.J. (1968) Gene duplication of the α-chain of goat haemoglobin: Evidence from a homozygous mutant. *Biochim. biophys. Acta*, 168, 585.

HILSE K. & POPP R.A. (1968) Gene duplication as the basis for amino acid ambiguity in the α-chain polypeptides of mouse haemoglobin. *Proc. natn. Acad. Sci.* 61, 930.

HUISMAN T.H.J., WILSON J.B. & ADAMS H.R. (1967) The heterogeneity of goat haemoglobin: Evidence for the existence of two non-allelic and one allelic α-chain structural genes. *Arch. Biochem.* 121, 528.

HUISMAN T.H.J., BRANDT G. & WILSON J.B. (1968) The structure of goat haemoglobins. II Structural studies of the α-chains of the haemoglobins A and B. *J. Biol. Chem.* 243, 3675.

HUISMAN T.H.J., SCHROEDER W.A., DOZY A.M., SHELTON J.R., SHELTON J.B., BOYD E.M. & APELL G. (1969) Evidence for multiple structural genes for the gamma chain of human foetal haemoglobin in hereditary persistence of foetal haemoglobin. *Ann. N.Y. Acad. Sci.* 165, 320.

HUISMAN T.H.J., SCHROEDER W.A., STAMATOYANNOPOULOS G., BOUVER N., SHELTON J.R., SHELTON J.B. & APELL G. (1970) Nature of foetal haemoglobin in the Greek type of hereditary persistence of foetal haemoglobin with and without concurrent β-thalassaemia. *J. clin. Invest.* **49**, 1035.

HUTTON J.J., SCHWEET R.S., WOLFE H.G. & RUSSELL E.S. (1963) Haemoglobin solubility and α-chain structure in crosses between two inbred mouse strains. *Science* **143**, 252.

HUNTER T. & MUNRO A. (1969) Allelic variants and the amino acid sequence of the α-chain of rabbit haemoglobin. *Nature* **223**, 1270.

INGRAM V.M. (1961) Gene evolution and the haemoglobins. *Nature* **189**, 704.

KILMARTIN J.V. & CLEGG J.B. (1967) Amino-acid replacements in horse haemoglobin. *Nature* **213**, 269.

LEHMANN H. & CARRELL R.W. (1968) Differences between α- and β-chain mutants of human haemoglobin and between α- and β-thalassaemia. Possible duplication of the α-chain gene. *Br. med. J.* **4**, 748.

POPP R.A. (1965) Haemoglobin variants in mice. *Fedn Proc.* **24**, 1252.

POPP R.A. (1867) Haemoglobins of mice: Sequence and possible ambiguity at one position of the α-chain. *J. molec. Biol.* **27**, 9.

POPP R.A. & AMAND W.S. (1960) Studies on the mouse haemoglobin locus. I. Identification of haemoglobin types and linkage of haemoglobin with albinism. *J. Hered.* **51**, 141.

RIFKIN D.B., RIFKIN M.R. & KONIGSBERG W. (1966) The presence of two major Hb components in an inbred strain of mice. *Proc. natn. Acad. Sci.* **55**, 586.

SCHROEDER W.A., HUISMAN T.H.J., SHELTON J.R., SHELTON J.B., KLEIHAUER E.F., DOZY A.M. & ROBBERSON B. (1968) Evidence for multiple structural genes for the γ-chain of human fetal hemoglobin. *Proc. natn Acad. Sci.* **60**, 537.

SCHROEDER W.A., HUISMAN T.H.J., SHELTON J.R., SHELTON J.B., APELL G. & BOUVER N. (1970) Heterogeneity of foetal haemoglobin in β-thalassaemia of the negro. *Am. J. hum. Genet.* **22**, 505.

SMITH E.W. & TORBERT J.V. (1958) Study of two abnormal haemoglobins with evidence for a new genetic locus for haemoglobin formation. *Bull. Johns Hopkins Hosp.* **102**, 38.

WEISBLUM B., GONANO F., VON EHRENSTEIN G. & BENZER S. (1965) A demonstration of coding degeneracy for leucine in the synthesis of protein. *Proc. natn. Acad. Sci.* **53**, 328.

16 ❊ The Molecular Basis of Thalassaemia

D. J. WEATHERALL

In his monograph *Genetic Polymorphism*, E.B.Ford (1965) discussed the sickle-cell mutation as a model system for examining the phenomena of balanced polymorphism in human genetics. The haemoglobinopathies remain a valuable tool for the study of gene action in Man but in the last few years the emphasis in this field has changed from population genetics to the study of abnormal haemoglobin synthesis at a molecular level. This trend is particularly noticeable in recent work on thalassaemia.

The thalassaemia syndromes are a group of genetic disorders of haemoglobin synthesis characterized by defective production of one or more of the globin chains of haemoglobin. The anaemias which result from the thalassaemia mutations constitute by far the commonest disorders in Man due to single autosomal genetic defects. For this reason the elucidation of the basic defect in protein synthesis in these conditions is of considerable practical importance.

In this paper some theoretical and experimental aspects of recent work on the molecular basis of thalassaemia are examined.

INTRODUCTION

The human haemoglobins all have a similar structure consisting of four peptide chains, each of which is associated with a haem group. Each type of haemoglobin has one pair of chains in common, the α-chains. In haemoglobin F these are combined with γ-chains ($\alpha_2\gamma_2$) and in haemoglobins A and A_2 with β- and δ-chains respectively ($\alpha_2\beta_2$ and $\alpha_2\delta_2$). Inherited disorders of haemoglobin production fall into two groups. First there are those like sickle-cell anaemia in which there is a structural abnormality in one of the globin chains. In the other group, the thalassaemias, there is a defect in the rate of production of one or more of the globin chains.

Before 1966 the genetic basis of the thalassaemias was deduced from the associated haemoglobin constitution and pattern of inheritance (Itano 1957, Ingram & Stretton 1959). In one form of thalassaemia the anaemia is associated with persistance of foetal haemoglobin production beyond the neonatal period. Ingram and Stretton suggested that this form of thalassaemia follows defective β-chain synthesis and that γ-chain synthesis persists beyond the neonatal period in an attempt to compensate for the lack of β-chains. However, some patients with thalassaemia do not have increased foetal haemoglobin levels but small quantities of haemoglobins consisting entirely of the β-chains of haemoglobin A (β_4 or haemoglobin H) or the γ-chains of haemoglobin F (γ_4 or haemoglobin Bart's). It was suggested that these patients have a genetic defect in the production of α-chains, α-thalassaemia, the excess β-chains or γ-chains combining to form haemoglobins H or Bart's respectively. Available genetic data indicate that the β-thalassaemia gene is an allele or closely linked to the locus controlling the structure of the β-chain (Weatherall 1965).

In the period since 1960 extensive genetic studies have indicated that both α- and β-thalassaemia represent a heterogeneous group of disorders and a variety of clinical syndromes has been defined, all associated with persistance of foetal haemoglobin into adult life or the presence of haemoglobins H or Bart's. Furthermore it is now clear that patients with the clinical picture of thalassaemia may be heterozygous for one or more genes for both thalassaemia and structural haemoglobin variants. In populations where both these types of genetic disorder occur with a high frequency a complex series of clinical conditions result from their interaction. In Thailand, for example, it has been estimated that out of a population of approximately 30 million some 323,000 individuals suffer chronic ill-health due to the deleterious effects of combinations of the genes for different types of thalassaemia and structural haemoglobin variants (Wasi *et al.* 1969). These figures, which are summarized in table 16.1, give some indication of the immense public health problem caused by thalassaemia in Southeast Asia.

In 1966 a method was devised for the separation of the α- and β-chains of human haemoglobin which gave a quantitative yield of each chain with small amounts of starting material (Clegg, Naughton & Weatherall 1966). Using this technique combined with *in vitro* labelling of haemoglobin following short term incubation of reticulocytes or bone marrow with ^{14}C- or ^{3}H-labelled amino acids, it has been possible to compare the

TABLE 16.1. Major haemoglobin types in the more common thalassaemic diseases seen in Thailand (from Wasi *et al.* 1969).

Diseases	Major Hb types	Estimated No. of patients*
β-thalassaemia homozygosity	F or F + A	18,750
β-thalassaemia/Hb E	E + F	48,750
Hb H	A + H	151,986
Hb H/Hb E	A + E + Bart's	22,709
Hb Bart's hydrops foetalis	Bart's	80,790
	Total =	322,985

* Estimated for the population of 30 million.

rate of production of the individual peptide chains of haemoglobin in thalassaemic and non-thalassaemic individuals (fig. 16.1).

In reticulocytes obtained from patients with homozygous β-thalassaemia α-chains are always synthesized at a greater rate than β- and γ-chains. This imbalance of globin chain synthesis is also found in heterozygous β-thalassaemia and in the heterozygous states for both β-thalassaemia and β-chain structural haemoglobin variants. The imbalance of α- and β-chain production results in a large intracellular pool of α-chains which are unstable and precipitate, causing rigid inclusion bodies and damage to the red cell membrane. These results have now been confirmed in several laboratories and there is no doubt whatever that the basic lesion in β-thalassaemia is an imbalance of globin chain synthesis (Weatherall, Clegg & Naughton 1965, Heywood, Karon & Weissman 1965, Bank & Marks 1966, Bargellesi, Pontremoli & Conconi 1967, Weatherall *et al.* 1969, Modell *et al.* 1969).

The pattern of haemoglobin synthesis in β-thalassaemia falls into two main groups. There are patients in whom there is a relative deficiency of β-chains and another group in which there is a total deficiency of β-chain synthesis (Bargellesi *et al.* 1967, Weatherall *et al.* 1969). These types of β-thalassaemia run true within families and probably represent distinct genetic varieties of the disorder.

A similar approach to the study of haemoglobin synthesis has been used to examine patients with various types of α-thalassaemia. Since α-chains are shared by both foetal and adult haemoglobin, severe homozygous α-thalassaemia causes intrauterine anaemia and stillbirth.

CHAPTER 16

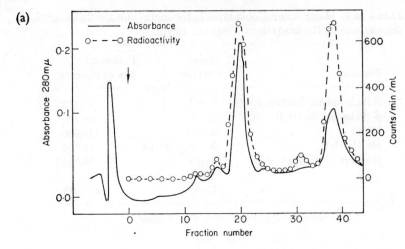

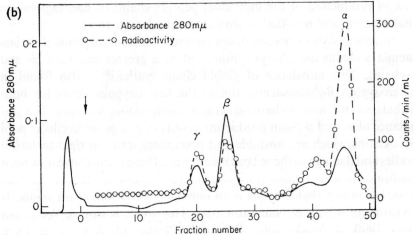

FIGURE 16.1. The patterns of globin chain synthesis in normal and thalassaemic individuals. Reticulocytes were incubated with ³H-leucine for 45 minutes and the cells washed, lysed and converted to 'globin'. The globin was dissolved in 8M urea and run on a CMC column. The order of elution of the globin chains is γ, β and α. (a) Cells from a normal individual showing equal labelling on α- and β-chains. (b) Cells from an individual with β-thalassaemia showing large excess of α-chain synthesis.

It has been possible to study haemoglobin synthesis in blood from one such infant and a total absence of α-chain synthesis has been demonstrated; the haemoglobin consisted of Bart's and H only (Weatherall, Clegg & Wong 1970). In the milder form of α-thalassaemia, haemoglobin

H disease, there is a moderate imbalance of globin chain synthesis such that β-chain synthesis exceeds that of α-chain synthesis by a factor of about three (Clegg & Weatherall 1967). This imbalance of chain synthesis results in a large intracellular pool of β-chains, some of which combine with newly made α-chains to make haemoglobin A, while the remainder form β_4 tetramers or haemoglobin H. It is clear from these experiments that haemoglobin H precipitates as red cells age. The resulting inclusion bodies cause the red cells to be sequestered in the spleen.

It is clear from these biosynthetic studies that the basic defect in all forms of thalassaemia is a reduced rate of synthesis of a structurally normal globin chain. The deficit of peptide chains may be partial or total. Genetic data indicate that the gene for β-thalassaemia is an allele of, or closely linked to, the β-structural locus. Although these observations provide a fairly clear picture of the genetic and pathophysiological aspects of thalassaemia, less progress has been made in elucidating the molecular defect in protein synthesis which results in globin chain imbalance.

THEORETICAL MODELS FOR THE MOLECULAR DEFECT IN THALASSAEMIA

A major problem in designing an experimental system to examine the molecular defect in thalassaemia is that of ignorance about the control mechanisms involved in Mammalian protein synthesis. Indeed very little is known about the factors which control the rate of globin synthesis and synchronize the rate of haem and globin synthesis during red cell maturation. Presumably control of haemoglobin synthesis can occur at both nuclear and cytoplasmic levels. The erythrocyte is a highly specialized cell which is concerned with the synthesis and maintenance of a single protein, haemoglobin. Since it loses its nucleus and hence the capacity to synthesize new messenger RNA during development, control at the cytoplasmic level may be of considerable importance.

Clearly the final amount of any peptide chain must be related to the rate of production of the messenger RNA which directs its synthesis. While it is tempting to speculate about the existence of regulator or operator genes in Man it seems likely that control mechanisms for messenger RNA production in mammals will be much more complex than in the microbial system described by Jacob & Monod (1961). In fact there is evidence (reviewed by Anderson 1968) that, in eucaryocytic cells, the level of gene product is more dependent on the rate of messenger RNA translation than transcription and that cytoplasmic control assumes greater importance in higher organisms.

Another possible mechanism in the nuclear control of Mammalian protein synthesis is that of gene duplication with the production of a multicistronic segment of DNA. This possibility, which has been discussed by Ford (1965), is just starting to have some experimental verification in both human and animal haemoglobin genetics (reviewed by Schroeder & Huisman 1970). There is now good evidence for multiple α-chain loci in many animals and in at least some human populations. There are at least two and possibly more, γ-chain loci directing the production of foetal haemoglobin in Man. It seems likely that in such multicistronic systems the final level of gene product will be modified by the number of active loci and hence copies of messenger RNA for a particular gene product.

It is certain that at least some control of the rate of globin-chain synthesis occurs at a cytoplasmic level. Quite clearly the rate of chain initiation, translation and termination may all be rate limiting steps in globin synthesis. It is quite conceivable that the rate of chain folding to assume secondary and tertiary configurations and the rate of subunit association may also be rate limiting. Haem appears to be involved in globin chain initiation (Rabinovitz *et al.* 1969) and the rate of haem production may be a further rate-limiting factor. The translation time will be dependent on the availability of the various transfer RNAs which transport amino acids to the growing peptide chain. Since the genetic code is known to be degenerate the differential availability of various transfer RNAs may well be a rate limiting factor in globin chain synthesis. From a consideration of these hypothetical rate-limiting steps it is possible to suggest a series of levels at which protein synthesis might be defective in thalassaemia, some of which are open to experimental study.

The thalassaemic defect could be in the production of messenger RNA, i.e. a primary defect in transcription. This could follow an RNA polymerase mutation, controller gene mutation or simple deletion of a whole gene or a portion of a multicistronic segment. Unequal crossing over at the closely linked δ- and β-structural genes might produce a series of slowly synthesized composite δ-β-chains. In fact there is evidence that this occurs in one form of thalassaemia, haemoglobin Lepore thalassaemia, in which a composite δ-β-gene results from unequal crossing over at the δ-β-gene complex (Baglioni 1962). The product of this locus is a composite δ-β-chain which is synthesized at a slow rate so producing the clinical picture of thalassaemia. However, there is no evidence that this type of phenomenon is the cause of the majority of forms of thalassaemia.

There are several different ways in which a breakdown of cytoplasmic control might result in the clinical picture of thalassaemia. There might be a reduced rate of chain initiation, translation or termination. A defect in translation could result from an altered codon in the messenger RNA which codes for a transfer RNA present in rate-limiting amounts, so reducing the rate of α- or β-chain synthesis. A transfer RNA deficiency might also follow a mutation in the gene controlling its synthesis or in the enzyme system responsible for its acylation. Examples of this type of mutation have been found in the histidine-synthesizing system in *Salmonella*. There might be a mutation in the terminating codon such that there is a 'pile up' of finished globin chains on the ribosomes. Any mutation involving a delay in translation near the N-terminal end of a globin chain could result in the affected messenger RNA carrying fewer ribosomes and hence being open to attack by ribonucleases with premature destruction. Since ribosomes play an important role in codon-anticodon recognition, a mutation of the ribosomal RNA or protein might also result in a prolonged translation time. Finally a mutation of the 'Amber' type, which produces premature chain termination and the synthesis of fragments of peptide chain, might be responsible for forms of thalassaemia in which no gene product is formed.

These possibilities form the background to the experimental work summarized in the next section.

EXPERIMENTAL APPROACH TO THE
MOLECULAR DEFECT IN THALASSAEMIA

Faced with such a plethora of molecular possibilities, it first seemed necessary to try to determine the level at which protein synthesis is defective in various types of thalassaemia, i.e. is the defect one of transcription or translation? Experiments have been designed to try to distinguish between these possibilities based on the following assumptions.

1 If the translation time is normal for the deficient globin chain a cytoplasmic defect other than chain initiation can be excluded at least in those forms of thalassaemia in which *some* globin chain is synthesized.

2 If no fragments of globin chain can be isolated from thalassaemic cells an 'Amber' type of mutation can be excluded as a causal mechanism at translational level in those forms of thalassaemia in which *no* globin chain is synthesized.

The results of experiments designed to test these possibilities have been reported in detail elsewhere (Clegg *et al.* 1968, Weatherall & Clegg 1969, Weatherall *et al.* 1970) and will only be briefly outlined here.

In order to measure the translation time for β-chains in thalassaemic reticulocytes and to compare it with that in normal red cell precursors the principle used by Dintzis (1961) to study the assembly of globin

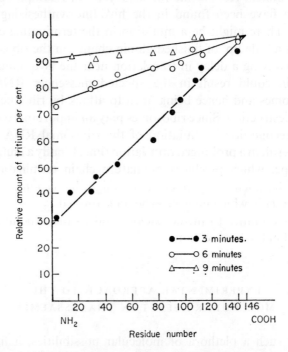

FIGURE 16.2. The assembly patterns of normal human β-chains. Reticulocytes were incubated for 3, 6 and 9 minutes with ^{3}H-leucine and then the cells were lysed and the α- and β-chains separated. The ^{3}H-labelled β-chains were mixed with uniformly labelled ^{14}C-β-chains as internal standards. The chains were then fingerprinted and the specific activity i.e. ^{3}H/^{14}C ratio of constituent peptides determined.

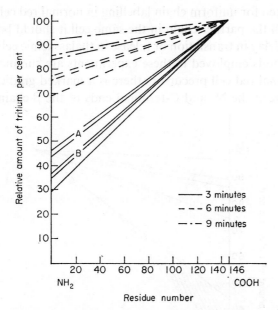

FIGURE 16.3. The assembly of human β-chains using a similar experimental method to fig. 16.2. Each slope was derived from a separate experiment and the difference in slopes indicates the variability of assembly times in normal reticulocytes. The experiments labelled A and B were duplicated, assembly times measured with and without preincubation with unlabelled leucine and with and without removal of the ribosomes to examine the effects of altered experimental conditions on the assembly times.

chains in the rabbit was applied. Peptide chains are synthesized by step-wise addition of amino acids from the N- to the C-terminal ends. If red cell precursors are exposed to radioactive amino acids for a short period of time the first part of the growing chain to become labelled is the C-terminal end. As the incubation continues radioactivity will gradually spread down the chain and finally the N-terminal residues will be labelled. Thus at very early incubation times there will be a gradient of radioactivity between N- and C-terminal ends of the chains. When the N-terminal peptides become labelled chains which are made subsequently will all be uniformly labelled and will produce an increasing background of radioactivity in the cell such that the initial gradient is lost. Thus the time it takes to uniformly label the α- and β-chains reflects the time it takes for ribosomes to move across the messenger RNA, reach the terminating codon and release the completed peptide chain. If this time is compared

with that taken for uniform chain labelling in normal red cell precursors
and also with the partner chain in the same cell it should be possible to
recognize a delay in translation or release in thalassaemic cells.

The methods employed in these experiments are summarized in fig.
16.2. In normal red cell precursors there is a marked gradient of radioactivity between the N- and C-terminal ends of the β-chain after three

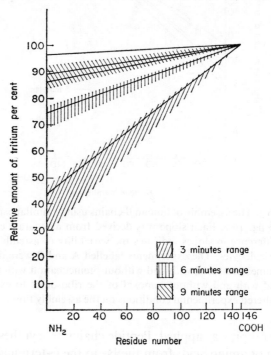

FIGURE 16.4. Identical experiments to those shown in figs. 16.2 and 16.3
utilizing cells from patients with β-thalassaemia. The uppermost slope
represents a 12-minute cell incubation profile. The shaded area represents
the range of normal values from fig. 16.3.

minutes of incubation with ³H-leucine and this gradually becomes less
marked; at 9 minutes there is virtually no difference in the specific
activities of the two ends of the chain. This 'uniform labelling time' is
consistent within a given experiment but there is some variation between
the reticulocytes of different individuals as shown in fig. 16.3. Since these
experiments were performed with peripheral reticulocytes it was
obviously important to decide whether the translation time was the

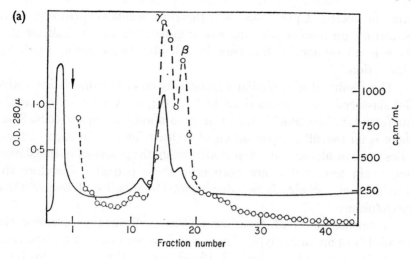

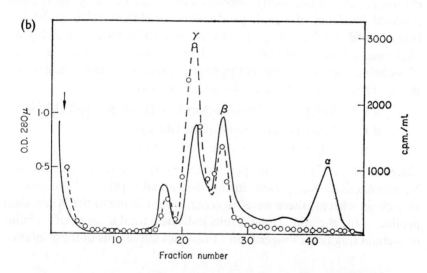

FIGURE 16.5. Haemoglobin synthesis in homozygous α-thalassaemia. Reticulocytes were incubated with radioactive leucine as indicated in fig. 16.1 and the cells washed, lysed and converted to 'globin'. The elution pattern of the globin chains shows a total absence of α-chains. (a) untreated haemolysate, (b) haemolysate with human haemoglobin A added as carrier. This lysate was also subjected to finger printing and the N-terminal peptide radioactivity counted.

same in nucleated precursors and therefore similar experiments were performed on human bone marrow samples. The results indicate that there is no significant difference in assembly times during erythroid maturation.

The results of nine similar experiments on cells from patients with β-thalassaemia are summarized in fig. 16.4. It will be seen that the 'uniform labelling time' is the same as in non-thalassaemic cells. Since there is an overall β-chain deficit of at least fourfold in each of these cases these results provide clear evidence that the processes of translation and chain termination are normal in these patients. Therefore the deficiency of β-chains is not due to an abnormality of one of these mechanisms.

A further series of experiments was performed to examine the possibility of an amber type of mutation in those forms of thalassaemia in which no chain is synthesized. Blood was obtained from a hydropic baby, homozygous for α-thalassaemia, who lived for a few minutes after delivery. The cells were incubated for four hours with ^{3}H-leucine. The elution pattern of radioactive globin chains is shown in fig. 16.5. There is a total deficiency of α-chain production, the globin chains synthesized being γ- and β-only. A fingerprint was prepared from a total cell lysate after incubation, unlabelled 'carrier' haemoglobin A added as a source of α-chains and the N-terminal peptides of the α-, β- and γ-chains prepared. These peptides have the following constitution.

α 1–2   Val-*Leu*-Ser-Pro-Ala-Asp-Lys-Thr-AspNH$_2$-Val-Lys^{11}

β 1   Val-His-*Leu*-Thr-Pro-Glu-Glu-Lys^8

γ 1   Gly-His-Phe-Thr-Glu-Glu-Asp-Lys^8

Since the N-terminal γ-chain peptide has no leucine it acts as a control. No radioactivity was present in the N-terminal peptides of the α-chain or γ-chain whereas there were 300 counts per minute in the N terminal peptide of the β-chain. These results indicate a total absence of α-chains or α-chain fragments longer than 11 residues in the cells of these infants.

DISCUSSION

The results of these experiments indicate that, in homozygous α-thalassaemia, there is a total deficiency of α-chain synthesis. Furthermore the absence of α-chain fragments rules out an amber type of mutation unless small peptide fragments can be removed from red cells almost instantaneously. Little is known about proteolytic activity in red cell precursors but Baglioni, Colombo & Jacobs-Lorena (1969) have some evidence that chains synthesized from the N-terminus and prematurely released are not as susceptible to degradation as chains initiated at sites other than the N-terminal end.

The pattern of β-chain assembly in reticulocytes of patients with β-thalassaemia who synthesize β-chains at a markedly reduced rate indicates that the translation time for β-messenger RNA is not delayed. This finding rules out a defect in translation or chain termination and hence most of the transfer RNA or ribosomal mutations which were suggested as possible molecular defects in thalassaemia.

The finding of a gross reduction in the rate of peptide chain production with a normal translation time suggests that the basic defect in this form of β-thalassaemia is that of transcription or chain initiation. The problem of chain initiation is currently being studied by examining the distribution of nascent α- and β-chains on polysomes of different sizes. However the process of transcription has received little study in Mammalian cells. Recently a 9S RNA fraction with many of the properties of messenger RNA has been isolated in both rabbit (Labrie 1969) and mouse (Williamson, personal communication) reticulocytes. Should it be possible to isolate a similar fraction from human reticulocytes and to compare the electrophoretic properties of messenger RNA of normal reticulocytes with that of thalassaemic cells it should be possible to characterize a quantitative defect in messenger RNA production.

At the beginning of this paper it was pointed out that the present tendency in the human haemoglobin field is to move from population to molecular genetics. If the defect in thalassaemia turns out to be one of transcription this trend may have to be reversed. Linkage data suggest that the β-thalassaemia and β-structural loci may be separable, provided that sufficient critical families can be found (Weatherall 1967). Recent surveys in West Africa have shown that populations exist with a high enough incidence of both β- and δ-structural haemoglobin variants and β-thalassaemia to make this a distinct possibility (Weatherall et al. 1970).

REFERENCES

ANDERSON W. FRENCH (1968) Current Potential for Modification of Genetic Defects. *Joseph P. Kennedy, Jr. Foundation Fourth International Scientific Symposium.*

BAGLIONI C. (1962) The fusion of two peptide chains in hemoglobin Lepore and its interpretation as a genetic deletion. *Proc. natn. Acad. Sci.* **48**, 1880.

BAGLIONI C., COLOMBO B. & JACOBS-LORENA M. (1969) Chain termination: A test for a possible explanation of thalassemia. *Ann. N.Y. Acad. Sci.* **165**, 212.

BANK A. & MARKS P.A. (1966) Excess alpha chain synthesis relative to beta chain synthesis in thalassaemia major and minor. *Nature* **212**, 1198.

BARGELLESI A., PONTREMOLI S. & CONCONI F. (1967) Absence of beta-globin synthesis and excess of alpha-globin synthesis in homozygous beta-thalassaemia. *Europ. J. Biochem.* **1**, 73.

CLEGG J.B., NAUGHTON M.A. & WEATHERALL D.J. (1966) Abnormal human haemoglobins: Separation and characterization of the α- and β-chains by chromatography, and the determination of two new variants Hb Chesapeake and Hb J (Bangkok). *J. molec. Biol.* **19**, 91.

CLEGG J.B. & WEATHERALL D.J. (1967) Haemoglobin synthesis in alpha-thalassaemia (Haemoglobin H disease). *Nature*, **215**, 1241.

CLEGG J.B., WEATHERALL D.J., NA-NAKORN S. & WASI P. (1968) Haemoglobin synthesis in β-thalassaemia. *Nature* **220**, 664.

DINTZIS H.M. (1961) Assembly of the peptide chains of hemoglobin. *Proc. natn. Acad. Sci.* **47**, 247.

FORD E.B. (1965) *Genetic polymorphism.* Faber and Faber.

HEYWOOD J.D., KARON M. & WEISSMAN S. (1965) Asymmetrical incorporation of amino acids into the alpha and beta chains of hemoglobin synthesised in thalassemic reticulocytes. *J. Lab. clin. Med.* **66**, 476.

INGRAM V.M. & STRETTON A.O.W. (1959) Genetic basis of the thalassaemia diseases. *Nature* **184**, 1903.

ITANO H.A. (1957) The human hemoglobins: their properties and genetic control. *Adv. Protein Chem.* **12**, 215–268.

JACOB F. & MONOD J. (1961) Genetic regulatory mechanisms in the synthesis of proteins. *J. molec. Biol.* **3**, 318.

LABRIE F. (1969) Isolation of an RNA with the properties of haemoglobin messenger. *Nature* **221**, 1217.

MODELL C.B., LATTER A., STEADMAN J.H. & HUEHNS E.R. (1969) Haemoglobin synthesis in β-thalassaemia. *Br. J. Haemat.* **17**, 485.

RABINOVITZ M., FREEDMAN M.L., FISHER J.M. & MAXWELL C.R. (1969) Translational control in hemoglobin synthesis. *Cold Spring Harb. Symp. quant. Biol.* **34**, 567.

SCHROEDER W.A. & HUISMAN T.J.H. (1970) Nonallelic structural genes and hemoglobin synthesis. *XIII Internat. Cong. Haemat.* Plenary Sessions 26, J.F. Lehmanns Verlag, Munchen.

WASI P., NA-NAKORN S., POOTRAKUL S., SOOKANEK M., DISTHASONGCHAN P., PORNPATKUL M. & PANICH V. (1969) Alpha- and beta-thalassemia in Thailand. *Ann. N.Y. Acad. Sci.* **165**, 60.

WEATHERALL D.J. (1965) *The Thalassaemia Syndromes.* Blackwell Scientific Publications, Oxford.

WEATHERALL D.J. (1967) The Thalassemias. In A.G.Steinberg & A.G.Bearn (Eds.), *Progress in Medical Genetics, V,* p. 8. Grune and Stratton, New York.

WEATHERALL D.J. & CLEGG J.B. (1969) Disordered Globin synthesis in thalassaemia. *Ann. N.Y. Acad. Sci.* **165**, 242.

WEATHERALL D.J., CLEGG J.B., NA-NAKORN S. & WASI P. (1969) The pattern of disordered haemoglobin synthesis in homozygous and heterozygous β-thalassaemia. *Br. J. Haemat.* **16**, 251.

WEATHERALL D.J., CLEGG J.B. & NAUGHTON M.A. (1965) Globin synthesis in thalassaemia: an in vitro study. *Nature* **208**, 1061.

WEATHERALL D.J., CLEGG J.B. & WONG H.B. (1970) The haemoglobin constitution of infants with the haemoglobin Bart's hydrops foetalis syndrome. *Br. J. Haemat.* **18**, 357.

WEATHERALL D.J., GILLES H.M., CLEGG J.B., BLANKSON J.A., MUSTAFA D. & BOI-DOKU F.S. (1970) *Preliminary surveys for the incidence of the thalassaemia genes in some African populations.* (In press.)

17 ❀ Blood Group Interactions between Mother and Foetus

C. A. CLARKE

It was Ford (1942) who first regarded the human blood groups as examples of balanced polymorphism and suggested that there would be found physical advantages and disadvantages associated with them (such as susceptibility to certain diseases) which would maintain the genes in the population. It was however many years before practical work was begun on the subject and only when Aird, Bentall & Fraser Roberts (1953) had shown that a significantly high proportion of those suffering from cancer of the stomach belong to group A and of those with peptic ulcer to group O did the idea become generally accepted. Though of great theoretical interest, such associations between blood groups and disease are not of help to particular patients, and the purpose of this paper is to draw attention to other situations which have medical (or veterinary) implications to individuals and yet which also play a part in maintaining the polymorphisms. Such for example are disorders due to interactions between the blood groups of mother and foetus which may result in haemolytic disease of the newborn, conditions where there is maternal immunization by foetal white cells and platelets, and, more speculatively, where blood group compatibility may determine the course of certain types of cancer.

Some of these conditions are now described and it will be seen that as far as selection is concerned the matter is extremely complex, and account has to be taken not only of naturally occurring interactions but also of the effects of modern medical procedures.

I. BLOOD GROUP INCOMPATIBILITIES AFFECTING THE RED CELLS

I. ABO HAEMOLYTIC DISEASE OF THE NEWBORN

This is usually a mild disease characterized by haemolytic jaundice which clears up spontaneously. It is clinically apparent in only about 3 per cent

324

of newborn infants, whereas in about 20 per cent of births mother/child ABO incompatibility is present. There must, therefore, be protective mechanisms and some of these (as will be seen) are well recognized, but there are many problems remaining, not the least being that there is no uniformity of opinion as to what constitutes the disease. Some facts about the illness are as follows.

(i) Group O mothers are almost the only ones concerned, and in contrast to Rh (see below) it is often the first ABO incompatible child which shows signs of the disease. The reason for group O mothers usually being involved is that they form 7S anti-A or anti-B more readily than do those who are group A or group B, though why this should be so is unknown. The fact that the first baby is often affected may sometimes be explained by the mother having been immunized by some other (non-human) antigen with A specificity—for instance by an earlier injection containing horse serum. The formation by the mother of 7S antibody is very important, since it is only this which will cross the placenta and thus damage the baby; the naturally occurring anti-A or anti-B in not-O women is nearly all of the 19S type, which having a larger molecule remains in the maternal circulation.

(ii) There is no certain diagnostic test for ABO haemolytic disease, in contrast to Rhesus haemolytic disease of the newborn where all clinically affected babies are Coombs positive; that is, the antibody can be demonstrated on their red cells by the antiglobulin test. In ABO disease, on the other hand, badly affected children may not show a positive Coombs test and in some cases where this test is positive the children are not affected at all (Gell & Coombs 1968). The reason for this is uncertain but possibly a subtype of 7S antibody is involved requiring special techniques for its demonstration.

(iii) Babies who are secretors of ABH substances are more likely than non-secretors to cause a rise in titre of 7S immune anti-A or anti-B, the blood group substance being known to cross the placenta at delivery (Voak 1969), and possibly earlier.

(iv) The foetus has a varying amount of blood group substance in its plasma and the more there is present the more likely is any 7S maternal antibody to be neutralized, thereby preventing it from attacking the foetal red cells.

(v) The antigens on the foetal red cells being poorly developed in some cases may not immunize normally, though if this is the case at all

often it raises difficulties in relation to protection against Rh immunization (see below).

(vi) Group A_2 infants are rarely affected because A_2 cells have fewer
antigen sites than those of A_1.

2. Rh HAEMOLYTIC DISEASE

This is a much more serious condition, the babies not infrequently
requiring exchange or intrauterine transfusion, and the disease is also
responsible for a proportion of stillbirths and brain damage. It is, therefore, a cause of considerable anxiety to many parents. Nevertheless, in
spite of the fact that 15 per cent of the population lack the Rh factor, only
a small proportion of such women who bear Rh positive babies actually
become immunized. Again, therefore, protective mechanisms have to be
considered, and there are many factors which influence the production
of Rh antibodies (see McConnell 1969, quoting Woodrow 1969). Some
of these will now be considered:

(i) *ABO incompatibility between mother and foetus*

It has long been known that incompatibility between mother and baby
(e.g. mother: Rh negative, group O; foetus: Rh positive, group A) gives
good protection against Rh immunization. Levine (1943) was the first
to recognize this when he studied the ABO and Rh groups of the parents
of erythroblastotic infants and noted an excess of matings in which the
father's ABO blood group was compatible with that of the mother.
Nevanlinna & Vainio (1956) made a great contribution to the subject by
studying families in which the mother developed anti-D even though the
father was ABO incompatible (in this situation a baby could, of course,
be compatible if the father was heterozygous). They concluded that an
ABO compatible foetus is much more likely to immunize its mother than
one which is incompatible. Clarke, Finn, McConnell & Sheppard (1958)
confirmed this finding in an investigation of 91 families, in all of which
Rh haemolytic disease had occurred. No less than 23 of them were from
ABO incompatible matings as between mother and father and in 14 of
these families the immunizing foetus could be determined with certainty.
They found that its ABO group was always compatible with that of the
mother, the probability of finding this by chance being less than 1 in 500.
Most people think that the protection is limited to pre-immunization

(affected children being equally commonly ABO compatible and in-compatible), but some authorities are of the opinion that a minor degree of protection may be conferred on an ABO incompatible Rh positive foetus even after the mother has been immunized to Rh (Murray, Knox & Walker 1965b).

Experimental confirmation of protection by ABO incompatibility has come from Stern, Goodman & Berger (1961) who injected Rh positive blood into Rh negative subjects; when the injected cells were ABO compatible, 17 out of 24 developed anti-D, while injection of ABO incompatible cells resulted in only 5 out of 32 being immunized. It is of interest that protection is much more effective when the mother is group O than when she is A or B, and this is because such mothers are more likely to have both 7S and 19S types of A (or B) antibody (see above).

The mechanism by which ABO incompatibility protects against Rhesus immunization is uncertain. Race and Sanger (1968) suggest that incompatible cells would be rapidly destroyed and thus not able to provide an antigenic stimulus. The destruction might be either intra- or

TABLE 17.1. Comparison of foetal cell counts in foeto-maternal ABO compatibility and incompatibility (Bowley, personal communication). From Clarke 1968a, by kind permission of the editor of the *Brit. med. Bull.*

Foetal blood	'Significant' foeto-maternal bleeding	'Insignificant' foeto-maternal bleeding	No bleeding	Total
ABO-compatible D-positive	85 (18·6%)	184 (40·2%)	189 (41·2%)	458 (100%)
ABO-compatible D-negative	70 (20·2%)	147 (42·2%)	131 (37·6%)	348 (100%)
ABO-incompatible D-positive	5 (4·2%)	28 (23·5%)	86 (72·3%)	119 (100%)
ABO-incompatible D-negative	3 (4·0%)	20 (26·7%)	52 (69·3%)	75 (100%)
Total	163	379	458	1,000

Foeto-maternal bleeding which gives foetal cell counts of 5 or more is called 'significant', while bleeding giving counts of from 1 to 4 inclusive is called 'insignificant'. It will be seen that, when the foetal blood is ABO-incompatible with the mother, there is a marked decrease in the proportion of both 'significant' and 'insignificant' foeto-maternal bleeding.

extravascular, the latter taking place predominantly in the liver (Mollison 1967), and foetal cells sequestrated here, rather than in the spleen, are probably less well placed for antibody stimulation. This explanation seems more likely than that of clonal competition, where it is postulated that a group O woman, having many anti-A or anti-B forming cells available would taken up A or B erythrocytes and therefore little Rh antigen would reach the few anti-Rh forming cells (Stern *et al.* 1961) Against this, however, is the fact that the anti-A or B titre does not usually rise in these circumstances.

Whatever the truth of the matter, protection by ABO incompatibility seems to be non-specific, i.e. antibody formation against all the antigenic determinants on the erythrocytes is suppressed, a point of great import- ance when considering the mechanism by which prophylactic anti-D gammaglobulin exerts its effect (see McConnell 1969).

A study of the figures in table 17.1 shows how few foetal cells remain in the maternal circulation when there is foeto-maternal ABO incompati- bility.

(ii) *Rh immunization in relation to the type of ABO compatibility between mother and child*

It is clear that mother and child may belong to the same ABO group (e.g. mother O, baby O), or different ones (e.g. mother A, baby O) though in both cases the baby is ABO compatible, its cells not being agglutinated by the maternal serum. Renkonen, Seppälä & Timonen (1968) found that immunization was more likely to have occurred in cases where the blood groups of the mother and of the first affected child were identical. Why this should be so is unknown, but the matter is mentioned here because of a similar ABO finding in choriocarcinoma (see pp. 339–340).

(iii) *The Rh genotype of the baby*

The number of D antigen sites on cells of different Rh types was estimated by Rochna and Hughes-Jones (1965) using purified ^{125}I-labelled anti-D IgG. Their findings were in keeping with the serological observations that cDE cells react more strongly with antibody *in vitro* than those with CDe and that on the whole homozygous cDE/cDE cells are agglutinated more strongly than are those of any other Rh type. Murray (1957) studied the Rh types of children whose mothers had become immunized and concluded that a *cDE/cde* foetus was more likely to immunize than a *CDe/cde* one. Also, it was later shown that in first-affected infants those of *cDE/cde* were more likely to need treatment for haemolytic disease (Murray, Knox & Walker, 1965a).

(iv) *Sex of the baby*

Renkonen & Timonen (1967) published data suggesting that Rh negative mothers are more likely to be immunized by male than by female foetuses. The sex ratios of the immunizing foetuses in their two series were 1·44 and 1·74 respectively. Support has come from Woodrow & Donohoe (1968) in Liverpool, for they found that in 63 first pregnancies immunized six months after delivery the sex ratio was 1·5, but the reason for this is unknown. Since immunization is caused by foeto-maternal haemorrhage one would have expected larger bleeds in the instances where the foetus was male, but in fact Woodrow and Donohoe found that the distribution of the post-delivery foetal cell scores was the same for both sexes.

(v) *Other genetic factors*

The variability of the immune response to the D antigen both by male volunteers injected with Rh positive blood (Clarke *et al.* 1963, Gunson, Stratton, Cooper & Rawlinson 1970) and by women following pregnancy, may be because the ability to make antibodies is partly inherited. If so, one might expect Rh haemolytic disease to run in families, and this could be tested by investigating Rh negative sisters of immunized women and comparing the immunization rate with that in sisters of non-immunized controls.

(vi) *Size of transplacental haemorrhage in relation to immunization*

A survey of Rh negative primiparae was carried out in Liverpool and 760 of these women who had had Rh positive, ABO compatible babies had their blood tested for foetal cells soon after delivery by the method of Kleihauer, Braun & Betke (1957) (see plate 17.1), and they were also investigated for Rh antibodies six months later (Woodrow & Donohoe

TABLE 17.2. Relationship of foeto-maternal haemorrhage after delivery to Rh-immunization six months later (first delivery: Rh-positive, ABO-compatible baby). (From Woodrow & Donohoe 1968, by courtesy of the authors and the Editor of the *Brit. Med. J.*)

	Foetal cell score							
	0	1	2	3–4	5–10	11–39	40+	Total
Anti-D present	13 (4)	6 (2)	8 (2)	10 (3)	11 (2)		4	61 (14)
Anti-D absent	346	141	73	50	42	33	14	699
Total	359	147	81	60	53	42	18	760
% with anti-D	3·6	4·1	9·9	16·7	20·8	21·4	22·2	8·0

Figures in parentheses denote anti-D detected by papain technique only.

1968). The results are shown in table 17.2 and it will be seen that there is a good correlation between the size of the transplacental haemorrhage and the chances of Rh immunization—the larger the bleed the greater the likelihood.

Nevertheless this is not the whole story, because as many as 13 of 61 immunized women had had no foetal cells detected at delivery, and it may well be that there are two mechanisms involved in immunization. In the first a sizeable transplacental haemorrhage either immunizes at six months or not at all, whereas in the second, tiny TPHs may result only in 'sensibilization', and the stimulus of a second pregnancy is necessary to produce overt antibodies (Cohen, Gustafson & Evans 1964, Zipursky, Pollack, Chown & Israels 1965, Woodrow & Donohoe 1968, Spensieri, Carnevale-Arella & Caldana 1970).

In animals also it has been shown that large antigenic doses may result in good antibody production but poor preparation for a second response, while small antigenic doses can produce a poor or undetectable primary antibody response but good preparation for a secondary one (Siskind, Dunn & Walker 1968).

(vii) *Immunological tolerance*

One possible reason for an Rh negative woman not becoming immunized by the red cells of her Rh positive foetus is that she may be tolerant of foetal antigens.

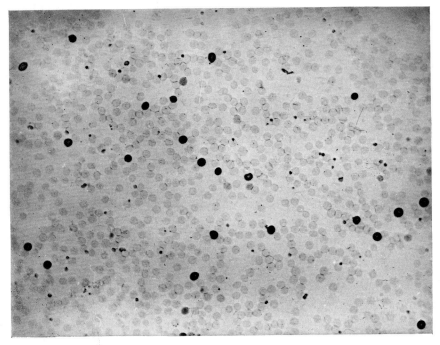

PLATE 17.1. *Foetal cells in the maternal circulation.* By means of an acid elution technique the foetal cells can be stained and counted, the adult cells remaining as ghosts while the foetal cells stand out as dark refractile bodies (Kleihauer, Braun & Betke 1957).

[to face p. 330

Jakobowicz, Crawford, Graydon & Pinder (1959) measured the strength of anti-A in the serum of 230 group O and group B recruits before and after anti-tetanus injections (which contain A specificity). The average rise in antibody was significantly less if the mother of the recruit was group A than if she was group O or B.

In theory, an Rh positive woman might make her female Rh negative foetus tolerant of the Rh antigen, and thus less likely to have an immune response to Rh (in pregnancy) than an Rh negative woman whose mother had been Rh negative. Race & Sanger (1968) give an account of this attractive 'grandmother' hypothesis (originally put forward by Rogers Brambell), but in the several series of grandmothers' bloods which have been collected no support for it has been found (Owen, Wood, Ford, Sturgeon & Baldwin 1954, Mayeda 1962). However, the actual Rh genotypes concerned have not been considered, and this might be an important point. On the other hand, Konugres (1964) tested the grand-mothers of children with haemolytic disease due to anti-A and anti-B and his results suggested that some degree of tolerance to A and B can be developed *in utero*. Rogers Brambell (1970) provides a comprehensive review of the transmission of passive immunity from mother to young.

In animals the work of Anderson & Benirschke (1964) on armadillos, which have monozygotic quadruplets, supports the idea of tolerance in pregnancy. One of the foetuses was removed from the uterus of an animal and its skin grafted on to the mother as well as on to pregnant and non-pregnant controls. The graft to the mother survived for 35 days (at which time she died) while those both to the pregnant and non-pregnant controls were rejected at about two weeks. In another instance a graft from a week-old offspring survived for 329 days on the mother.

(viii) *Immunological inertia during pregnancy*

Occasionally transplacental haemorrhage occurs during pregnancy, sometimes in the second trimester, leaving ample time for antibodies to develop before the end of that pregnancy. In fact, however, it is very rare to find anti-D in primiparae at delivery. Nevanlinna & Vainio (1962) found only 4 in 4,153 of such cases and Woodrow & Donohoe (1968) only 6 in 2,000, but Chown gives a higher incidence, about 2 per cent (Chown 1968, 1969).

To explain this lack of immunization by foetal bleeds occurring during pregnancy, it may be that there is a state of maternal anergy

12

during this time (Billingham 1964), but if so it only applies to primary immunization because women primed by one Rh positive pregnancy can rapidly develop antibodies quite early in a second one. Perhaps the transplacental migration of foetal lymphoid cells detected by Walknowska, Conte & Grumbach (1959) is relevant, for if they become temporarily grafted into her reticulo-endothelial system they might prevent immunization by transplacental haemorrhage occurring at delivery—though if this were so the effect would be expected to last longer than it in fact does. In this connexion it is of great interest that Hellström, Hellström, Inge & Brawn (1969) found evidence in mice which suggested that a specific serum factor in the mother, when pregnant with antigenically foreign foetuses, might protect these foetuses from the potentially destructive effects of maternal lymphocytes which had penetrated the placenta. Relevant to this in Man may be the finding of Leventhal *et al.* (1970) that some maternal plasmas inhibit the mixed leucocyte response when the mother's lymphocytes are tested against their children's cells.

Furthermore, pregnancy has an unfavourable effect on the prognosis of smallpox in an unvaccinated woman (Rao, Prahlad, Swaninatham & Lakshini 1963).

3. PREVENTION OF Rh ISOIMMUNIZATION

When Ford (1936) analysed the available data on the genetics of the African mimetic butterfly *Papilio dardanus* he started off a rather unpredictable chain of events. The first link was our investigation of the evolution of mimicry in *P. dardanus* (Clarke & Sheppard 1960) and later in *P. memnon* (Clarke, Sheppard & Thornton 1968), and this led to the study of human polymorphic (mainly blood group) systems. Our principal conclusions in the butterflies were that the mimetic patterns are determined by a series of closely linked genes (or a supergene) and that crossovers account for some of the unusual forms. Furthermore, there are many examples of genic interaction, notably the sex-control of the mimicry to the female and the marked modification of the expression of the major genes observed in race crosses. The work generally had parallels with the Rhesus blood groups and we first became particularly interested in their interaction with the ABO system (see above). Later it occurred to us that the protection conferred by naturally occurring anti-A or B might be mimicked in ABO compatible cases by injecting anti-D antibody, which would inactivate any Rh positive cells which had crossed the placental barrier, and Finn (1960) was the first to put forward this idea.

After five years of experimental work on Rh negative male volunteers and post-menopausal women, we showed that it was possible to prevent most cases of immunization and we then designed two clinical trials to test protection in Rh negative mothers.

(i) *The First Liverpool Group Trial*
(Combined Study 1966).

This began in 1964 and ended in 1968 except for the follow-up of the subsequent pregnancies. The figures to date are given in tables 17.3 and 17.4. The trial was designed to give the greatest amount of information on the efficacy of the treatment while injecting as few women as possible, and each centre adopted the same protocol. Since the likelihood of immunization was found to be related to the size of the transplacental haemorrhage (TPH) at delivery (see p. 329), only previously untransfused primiparae who had had a TPH with a Kleihauer score of 5 or over (see plate 17.1), and who had had an Rh positive, ABO compatible baby were included (presumed high risk cases) and alternate cases were kept as controls. The dose was 5 ml of anti-D gammaglobulin containing about 1,000 μg of anti-D given intramuscularly within 48 hours of delivery. The gammaglobulin was prepared by Dr W.d'A.Maycock and his staff at the Lister Institute, London. It will be seen from tables 17.3 and 17.4 that the results were successful, there being only two failures in the series (see below). The trial demonstrated that we had in fact picked a high risk group (immunization rate in the controls 22 per cent) and that here the subsequent pregnancies are protected (table 17.4). The fact that the subsequent babies are normal is the real proof of the efficacy of the treatment, since 'no antibodies at six months' might simply mean that overt immunization had been postponed.

(ii) *The Second Liverpool Trial*

Here the dose of anti-D gammaglobulin was 1 ml (about 200 μg), other centres were not involved, and the injection was given to those primiparae with a foetal cell score of from 0 to 4 in 50 low power fields. Table 17.5 shows that the six months figures are again very satisfactory, but in the second pregnancies (table 17.6) there are three failures in the treated as

TABLE 17.3. Liverpool group 5 ml high risk trial. (Rh antibodies six months post-delivery.)

Centre	No.	Not immunized	Immunized
CONTROLS			
Liverpool	92 (40)*	72 (30)	20 (10)
Baltimore	31 (14)	23 (10)	8 (4)
Sheffield	41 (15)	31 (10)	10 (5)
Leeds and Bradford	12 (9)	12 (9)	0 (0)
Totals	176 (78)	138 (59)	38 (19)
TREATED			
Liverpool	94 (40)	93 (40)	1 (0)
Baltimore	26 (13)	26 (13)	0 (0)
Sheffield	39 (14)	39 (14)	0 (0)
Leeds and Bradford	14 (11)	14 (11)	0 (0)
Totals	173 (78)	172 (78)	1 (0)

* The figures in parentheses are those when the trial was published (A Combined Study 1966). They are included in the larger figures.

TABLE 17.4. Liverpool group 5 ml trial. (Rh antibodies at the end of second Rh positive pregnancy.)

Centre	No.	Not immunized	Immunized
CONTROLS			
Liverpool	38	30	8 (7)*
Baltimore	4	3	1
Sheffield	19	10	9 (5)
Leeds and Bradford	4	2	2
Totals	65	45	20 (12)
TREATED			
Liverpool	59	57	2 (1)
Baltimore	12	12	0
Sheffield	10	10	0
Leeds and Bradford	7	7	0
Totals	88	86	2 (1)

* Numbers in parentheses are mothers who had developed antibodies after the first pregnancy. They are included in the larger number.

TABLE 17.5. Liverpool 1 ml trial. (Rh antibodies six months post-delivery.)

	No.	Not immunized	Immunized
Controls	362	349	13
Treated	353	353	0
Totals	715	702	13

TABLE 17.6. Liverpool 1 ml trial. (Rh antibodies at the end of second Rh positive pregnancy.)

	No.	Not immunized	Immunized
Controls	127	114	13 (4)*
Treated	128	125	3
Totals	255	239	16 (4)

* The number in parentheses refers to mothers who had developed antibodies after the first pregnancy. They are included in the larger number.

against five new immunizations in the controls. This appears paradoxical compared with the first trial but it seems very likely that these failures are women who had been 'sensitized', i.e. immunized by a bleed early in the first pregnancy but with no demonstrable antibodies (NDA), and that their hidden antibody had removed the foetal cells which had crossed the placenta during the pregnancy. If this is the case, then the anti-D gammaglobulin would have been given too late.

(iii) *Results in the U.S.A.*

Here multiparae as well as primiparae were included (Freda, Gorman & Pollack 1964) and the Kleihauer test was not carried out as a routine. It will be seen (table 17.7) that the latest results (Pollack, personal communication) are slightly better than ours, a fact which is remarkable since multiparae (some of whom might be expected to be 'sensitized') were excluded in Liverpool. The explanation is uncertain, but it may be that the American commercial anti-D (RhoGAM) has greater affinity; and it is also true that in the earlier U.S.A. cases a larger dose (about

1,000 μg) was given—though for some time now about 300 μg has been given with equally good results.

It was Hamilton (in St Louis, Missouri) who appreciated the importance of the second pregnancy on reading our 1961 paper (Finn *et al.* 1961) and he never troubled to test at six months at all. He treated women with plasma, not gammaglobulin, and in 149 subsequent babies has found no evidence of sensitization using the direct Coombs test on the babies' cord blood (Hamilton 1967 and personal communication).

TABLE 17.7. U.S.A. Ortho Trials (RhoGAM) tested six months post-delivery. (Pollack, personal communication.)

	Total	Immunized	Not immunized	Per cent immunized
Controls	1,476	102	1,374	6·9
Treated	3,389	6	3,383	0·2
Subsequent pregnancies				
Controls	155	23	132	12·9
Treated	390	5	385	1·3

(iv) *Other centres*

In the past few years the prophylaxis has been taken up all over the world where Rh negative women are found (see Clarke 1970, for details), and it will not be long before the exact degree of protection is determined.

II. LEUCOCYTE ANTIBODIES AND Rh

THE SEVERITY OF Rh HAEMOLYTIC DISEASE IN RELATION TO LEUCOCYTE ANTIBODIES

Parous women frequently have leucocyte or platelet (HL-A) antibodies in their serum in response to leucocytes which have crossed the placenta in earlier pregnancies (see Engelfreit 1966). Such antibodies, which are tissue antibodies, have generally been thought to be harmless to the foetus. The explanation of this could be that (a) they are absorbed by foetal tissues, (b) they need complement to act and this is not present to any great extent in newborn babies, or (c) that the mother and a particular baby happen to be compatible on the HL-A system.

Moulinier & Mesnier (1969) have, however, produced evidence to suggest that lymphocytotoxic antibodies may reinforce the destructive effects of Rh antibodies in Rh haemolytic disease. They found that severe forms of the disorder, leading to still-birth, were more common when an Rh negative mother had both Rh and HL-A immune antibodies, and this could parallel the observation that HL-A antibodies, present *before* transplantation in renal allograft recipients, can lead to hyperacute rejection. (Harris, Macaulay & Wentzel 1970) suggest various possibilities with regard to the Moulinier hypothesis.

(i) The association may be fortuitous.

Some women may be good antibody producers, HL-A antibodies implying no more than the ability to produce abundant Rh antibodies which alone are responsible for the death of the foetus. Alternatively, the known association of Rh haemolytic disease with placental pathology may result in a greater exposure of the mother to foetal HL-A antigen, particularly perhaps when the foetus has died. Thus it is possible that HL-A antibodies do not contribute to the destructive process of Rh haemolytic disease although they may fortuitously be associated with it.

(ii) There may be increased susceptibility of the foetus to maternal HL-A antibodies as a consequence of Rh haemolytic disease.

A placental barrier is thought to be an important factor in the invulnerability of the normal foetus to immunological rejection, perhaps preventing the invasion of the foetus by maternal immunocytes (Kirby 1969). This may be broken down in Rh haemolytic disease.

(iii) There may be an additive effect between anti-HL-A and anti-Rh antibodies.

Red cell precursors, unlike mature erythrocytes, have demonstrable HL-A antigens (Harris & Zervas 1969), and erythroblastosis foetalis due to Rh haemolytic disease, therefore, guarantees the presence of large numbers of red cell precursors, which, since they have HL-A antigens, may be susceptible to the additional effects of immune HL-A antibodies. These red cell precursors are the product of a bone marrow probably already at maximum production and without remaining reserves.

Moulinier's views so far have not been confirmed. In a very small series of women with extremely bad obstetric histories, double immunization did not seem to be a factor (Clarke *et al.* 1970), and the same is true of a larger series by Tovey, Darke & Fraser (1970) who arbitrarily took a titre of 1 in 20 as being critical for the diagnosis of severe haemolytic disease. They investigated 72 mothers, and there was no suggestion that the presence of cytotoxic HL-A antibodies, in addition to high anti-D, had been of any clinical significance.

Though the original finding, therefore, is doubtful, the matter remains interesting particularly as the anti-D gammaglobulin sometimes contains anti-leucocyte antibodies and this might have relevance to the prevention of the rejection response in transplantation work (see Clarke 1968b, Harris, Macaulay & Wentzel 1970, Clarke 1970).

III. HAEMOLYTIC DISEASE OF THE NEWBORN IN THE FOAL

This disorder has many parallels with Rh. Thus the first foal is rarely affected and immunization is thought to take place by transplacental passage of cells at delivery. Unfortunately, however, the erythrocytes of the foal are mature at birth and it is, therefore, not possible to score foetal cells in the maternal circulation. Also in contrast to Man, the immune antibody formed against the foal's incompatible antigens is usually of the agglutinating type (Roberts 1957), and this may be why it does not cross the placenta though it is secreted in the colostrum where for a short time it is extremely toxic. A foal normal at birth rapidly becomes ill on suckling and may be dead within 36 hours. Since after this time the milk contains no more antibody, the disease can be effectively prevented by weaning (Franks 1962), and this being so it is perplexing to read that the disease is of considerable economic importance in the bloodstock industry and that new methods of exchange transfusion were being devised (Cambridge *Daily News* 20 August 1966). The reason may be that weaning, though easy in theory, is difficult in practice (since a little colostrum may be taken) or it may even not be known that the foal is at risk. Also, a foster mother may be hard to come by, and colostrum appears to be necessary to ensure that the foal thrives.

By analogy with Man the disease ought to be preventable by giving the mare gammaglobulin against the incompatible antigens, but, bloodstock breeders are very sensitive to any suggestion that there is an inherited disease in their animals.

IV. HAEMOLYTIC DISEASE OF THE NEWBORN IN THE PIGLET (Roberts 1957)

Here, too, the antibody is transmitted to the offspring in the colostrum. The sow can be immunized by certain foetal red cell antigens during pregnancy (though the pig A antigen is not responsible since it is only formed after birth), and also (formerly) by the administration of swine fever vaccine, which was made from pooled pigs' blood. The disease has been produced experimentally by the injection of the sow during pregnancy with repeated doses of the incompatible (heterozygous) male's blood. Those piglets in the litter which were incompatible were all killed by HDN, while those which were compatible were unaffected.

V. CHORIOCARCINOMA AS A MODEL FOR THE IMPORTANCE OF IMMUNOLOGICAL FACTORS IN CANCER

A stimulating though unproven suggestion regarding cancer is that most of us at some time in our lives develop neoplasia but that usually we eliminate the rogue cells, possibly by a mechanism similar to that of graft rejection. Modest support for the hypothesis is afforded by the observation that women with a history of allergy may be less at risk for subsequent cancer than controls (Mackay 1966). However, this survey was retrospective and the sex difference remains unexplained.

A more profitable approach concerns choriocarcinomata. These malignant tumours are derived from foetal tissue, and, therefore, one of the chromosomes in each pair is paternal in origin and the growth must be considered as a partially allogeneic graft. This might have some relation to the progress or halting of the disease (see Clarke 1962, 1968c), which as is well known, runs an extremely unpredictable course. A priori, it would seem reasonable that where the mother and the cancer had similar histocompatibility genes the disease would take a more sinister course, and conversely, where the blood groups on the cancer cells were incompatible it might be kept in check or rejected. There is some support for these views. Scott (1962, 1968) found that the maternal ABO blood group distribution showed a shift away from O to A, B and AB which was in keeping with the idea that foeto-maternal compatibility is an aetiological factor, and furthermore the growth is commoner in in-breeding communities (Azar 1962, Iliya, Williamson & Azar 1967).

Ben-Hur, Robinson & Neuman (1965) found that in three patients with choriocarcinoma skin homografts from the husband were accepted, whereas those from other donors were rejected. The authors suggest that since some hydatidiform moles undergo transformation to choriocarcinomata every woman with a mole should be grafted with her husband's skin. Where this homograft is not rejected they advise against further pregnancies (presumably for about two years) since the risk of development of choriocarcinoma would be high and pregnancy would confuse the diagnosis.

Mogensen & Kissmeyer-Nielsen (1967, 1968) carried out family studies on choriocarcinoma cases in order to establish whether the matings could produce zygotes which were histocompatible with the mothers, the antigens investigated being the important histocompatibility antigens, i.e. the ABO and those at the HLA locus, and the authors found a high degree of histocompatibility. Since there is about a 50 per cent chance of any donor and recipient being compatible or incompatible for for the major histocompatibility genes, this could mean that in 50 per cent of those cases where the mole began to become malignant the cancer might be checked.

On the other hand, not all the evidence points to compatibility being important. While the observations of Robinson and his colleagues (1963, 1967) are in agreement about the skin grafting results, yet when they typed the leucocyte antigens of the patient and her husband they did not find compatibility. The explanation put forward was that the tumour induces maternal tolerance to the paternal antigens it carries. Something like this may occur in normal human pregnancy for it has been shown that in the mouse, post-partum females exhibit specific tolerance to the transplantation antigens of the mating male (Breyere & Barrett 1960). The mechanism for the induction of the tolerance is not clear, but the antigens of the sperm are not involved (Porter & Breyere 1964). See note on p. 344.

Further support for the view that immunity is important in cancer comes from the fact that patients with renal transplants appear prone to certain forms of cancer (lymphoma and lymphosarcoma) probably due to prolonged immunosuppressive therapy. Conversely, Mathé (1969) has shown that remissions in acute leukaemia can be greatly prolonged if immunity is stimulated by repeated scarification of the skin by BCG.

REFERENCES

AIRD, I., BENTALL H.H. & ROBERTS J.A.F. (1953) A relationship between cancer of the stomach and the ABO blood groups. *Br. med. J.* I, 799.

ANDERSON J.M. & BENIRSCHKE K. (1964) Maternal tolerance of foetal tissue. *Br. med. J,* I, 1534.

AZAR H.A. (1962) Cancer in Lebanon and the Near East, *Cancer* 15, 66.

BEN-HUR N., ROBINSON E. & NEUMAN Z. (1965) Possible screening method for prevention of chorionepithelioma in women with hydatidiform mole. *Lancet* 1, 611.

BILLINGHAM, R.E. (1964) Transplantation immunity and the maternal-foetal relation. *New Engl. J. Med.* 270, 667.

BILLINGTON W.D. (1969) Immunological processes in mammalian reproduction. In Matteo Adinolfi (ed.), *Immunology and Development*. Spastics International Publications, Heinemann Medical Books Ltd., London.

BREYERE E.J. & BARRETT M.K. (1960) Prolonged survival of skin homografts in parous female mice. *J. natn. Cancer Inst.* 25, 1405.

CHOWN B. (1968) On Rh immunisation and its prevention: Observations and thoughts. *Vox Sang.* 15, 249.

CHOWN B. and coauthors (1969) Prevention of primary Rh immunization. First report of the western Canadian trial. *J. Canad. med. Assoc.* 100, 1021–1024.

CLARKE C.A. (1962) *Genetics for the Clinician*, 2nd edn. Blackwell Scientific Publications, Oxford.

CLARKE C.A. (1968a) Prophylaxis of Rhesus isoimmunisation. *Br. med. Bull.* 24, 3.

CLARKE C.A. (1968b) Prevention of Rhesus isoimmunisation. *Lancet* 2, 1.

CLARKE C.A. (1968c) Immunology of pregnancy: significance of blood group incompatibility between mother and foetus. *Proc. R. Soc. Med.* 61, 1213.

CLARKE C.A. (1970) The prevention of Rh isoimmunisation. *Clin. Genet.* 1, 1.

CLARKE C.A., FINN R., MCCONNELL R.B. & SHEPPARD P.M. (1958) The protection afforded by ABO incompatibility against erythroblastosis due to Rhesus anti-D. *Int. Archs. Allergy appl. Immun.* 13, 380.

CLARKE C.A., DONOHOE W.T.A., MCCONNELL R.B., WOODROW J.C., FINN R., KREVANS J.R., KULKE W., LEHANE D. & SHEPPARD P.M. (1963) Further experimental studies on the prevention of Rh haemolytic disease. *Br. med. J.* 1, 979.

CLARKE C.A. & SHEPPARD P.M. (1960) The evolution of mimicry in *Papilio dardanus*. *Heredity* 14, 163.

CLARKE C.A., SHEPPARD P.M. & THORNTON I.W.B. (1968) The genetics of the mimetic butterfly, *Papilio memnon L. Phil. Trans. R. Soc.* B. 254, 37.

CLARKE, C.A., BRADLEY J., ELSON C.J., DONOHOE W.T.A., LEHANE D. & HUGHES-JONES N.C. (1970) Intensive plasmapheresis as a therapeutic measure in Rhesus-immunised women. *Lancet* 1, 793.

COHEN F., GUSTAFSON D.C. & EVANS M.M. (1964) Mechanisms of isoimmunisation I. The transplacental passage of foetal erythrocytes in homospecific pregnancies. *Blood* 23, 621.

Combined Study from centres in England and Baltimore (1966) Prevention of Rh haemolytic disease: Results of the Clinical Trial. *Br. med. J.* 2, 907.

ENGELFREIT C.R. (1966) Cytotoxic iso-antibodies against leucocytes. Thesis Amsterdam University.

FINN R. (1960) Erythroblastosis. *Br. med. J.* 1, 526.

FINN R., CLARKE C.A., DONOHOE W.T.A., McCONNELL R.B., SHEPPARD P.M., LEHANE D. & KULKE W. (1961) Experimental studies on the prevention of Rh haemolytic disease. *Br. med. J.* **1**, 1486.

FORD E.B. (1936) The genetics of *Papilio dardanus* Brown (Lep.). *Trans. R. ent. Soc. Lond.* **85**, 435.

FORD E.B. (1942) *Genetics for Medical Students*, 1st edn. Methuen & Co., London.

FRANKS D. (1962) Horse blood groups and haemolytic disease of the newborn foal. *Ann. N.Y. Acad. Sci.* **97**, 235.

FREDA V.J., GORMAN J.G. & POLLACK W. (1964) Successful prevention of experimental Rh sensitisation in Man with anti-Rh gammaglobulin antibody preparation (preliminary report). *Transfusion* **4**, 26.

GELL P.G.H. & COOMBS R.R.A. (1968) *Clinical Aspects of Immunology*, 2nd edn. Blackwell Scientific Publications, Oxford.

GUNSON H.H., STRATTON F., COOPER D.G. & RAWLINSON I. (1970) Primary immunisation of Rh negative volunteers. *Br. med. J.* **1**, 593.

HAMILTON E.G. (1967) Prevention of Rh isoimmunisation by injection of anti-D antibody. *Obstet. Gynec.* **30**, 812.

HARRIS R., MACAULAY M.B. & WENTZEL J. (1970) Prevention of leucocyte antibody formation. *Lancet* **1**, 544.

HARRIS R. & ZERVAS J.D. (1969) Reticulocyte HL-A antigens. *Nature, Lond.* **221**, 1062.

HELLSTRÖM K.E., HELLSTRÖM I. & BRAWN J. (1969) Abrogation of cellular immunity to antigenically foreign mouse embryonic cells by a serum factor. *Nature, Lond.* **224**, 914.

ILIYA F.A., WILLIAMSON S. & AZAR H.A. (1967) Choriocarcinoma in the Near East. *Cancer* **20**, 144.

JAKOBOWICZ R., CRAWFORD H., GRAYDON J.J. & PINDER M. (1959) Immunological tolerance within the ABO blood group system. *Br. J. Haemat.* **5**, 232.

KIRBY D.R.S. (1969) Is the trophoblast antigenic? *Transplant. Proc.* **1**, 53.

KLEIHAUER E., BRAUN H. & BETKE K. (1957) Demonstration von fetalem Hämoglobin in Erythrozyten eines Blutausstriches. *Klin. Wschr.* **35**, 637.

KONUGRES A.A. (1964) Immunological tolerance of the A and B antigens. *Proc. of the 9th Congress of the International Society for Blood Transfusion*, Mexico, 1962, p. 746. Grune and Stratton, New York.

LEVENTHAL BRIGID G., BUELL D.N., TERASAKI P., YANKEE R. & ROGENTINE G.N., JR. (1970) The mixed leucocyte response: effect of maternal plasma. Paper given at the Fifth Leukocyte Culture Conference, Ottawa, 25–27 June.

LEVINE P. (1943) Serological factors as possible causes in spontaneous abortion. *J. Hered.* **34**, 71.

McCONNELL R.B. (1969) The immunological relationship between mother and fetus. In Matteo Adolfini (ed.), *Immunology and Development*, Spastics International Publications, Heinemann Medical Books Ltd., London.

MACKAY W.D. (1966) The incidence of allergic disorders and cancer. *Br. J. Cancer* **20**, 434.

MATHÉ G. (1969) Approaches to the immunological treatment of cancer in Man. *Br. med. J.* **4**, 7.

MAYEDA L. (1962) The self-marker concept as applied to the Rh blood group system. *Am. J. hum. Genet.* **14**, 281.

MOGENSEN B. & KISSMEYER-NIELSEN F. (1967) Histocompatibility in generalised choriocarcinoma: a family study. *Bull. Europ. Soc. hum. Genetics* **1**, 64.

MOGENSEN B. & KISSMEYER-NIELSEN F. (1968) Histocompatibility antigens on the HL-A locus in generalised gestational choriocarcinoma: a family study. *Lancet* **1**, 721.

MOLLISON P.L. (1967) *Blood Transfusion in Clinical Medicine*, 4th edn. Blackwell Scientific Publications, Oxford.

MOULINIER J. & MESNIER F. (1969) *Proc. 12th Congress of the International Society for Blood Transfusion*, Moscow.

MURRAY S. (1957) The effect of Rh genotypes on severity in haemolytic disease of the newborn. *Br. J. Haemat.* **3**, 143.

MURRAY S., KNOX E.G. & WALKER W. (1965a) Haemolytic disease and the Rhesus genotypes. *Vox Sang.* **10**, 257.

MURRAY S., KNOX E.G. & WALKER W. (1965b) Rhesus haemolytic disease of the newborn and the ABO groups. *Vox Sang.* **10**, 6.

NEVANLINNA H.R. & VAINIO T. (1956) The influence of mother-child ABO incompatibility on Rh immunisation. *Vox Sang.* **1**, 26.

NEVANLINNA H.R. & VAINIO T. (1962) An attempt to calculate the probability of Rh immunisation during pregnancy. *Proc. 8th Congress of the International Society of Blood Transfusion*, Tokyo, 1960. L. Holland (ed.), p. 281. Karger, Basel.

OWEN R.D., WOOD H.R., FORD A.G., STURGEON P. & BALDWIN L.G. (1954) Evidence for actively acquired tolerance to Rh antigen. *Proc. natn. Acad. Sci. (Wash.)* **44**, 420.

PORTER J.B. & BREYERE E.J. (1964) Studies on the source of antigenic stimulation in the induction of tolerance by parity. *Transplantation* **2**, 246.

RACE R.R. & SANGER R. (1950) *Blood Groups in Man*, 1st edn. Blackwell Scientific Publications, Oxford.

RACE R.R. & SANGER R. (1968) *Blood Groups in Man*, 5th edn. Blackwell Scientific Publications, Oxford.

RAO A.R., PRAHLAD I., SWANINATHAM N. & LAKSHINI A. (1963). Pregnancy and smallpox. *J. Indian med. Assoc.* **40**, 353-63.

RENKONEN K.O. & TIMONEN S. (1967) Factors influencing the immunisation of Rh negative mothers. *J. med. Genet.* **4**, 166.

RENKONEN K.O., SEPPÄLÄ M. & TIMONEN S. (1968) Further factors influencing Rh immunisation: II. *J. med. Genet.* **5**, 123.

ROBERTS G. (1957) *Comparative aspects of haemolytic disease of the newborn.* Heinemann Medical Books Ltd., London.

ROBINSON E., SHULMAN J., BEN-HUR N., ZUCKERMAN H. & NEUMAN Z. (1963) Immunological studies and behaviour of husband and foreign homografts in patients with chorionepithelioma. *Lancet* **1**, 300.

ROBINSON E., BEN-HUR N., ZUCKERMAN H. & NEUMAN Z. (1967) Further immunologic studies in patients with chorionepithelioma and hydatidiform mole. *Cancer Res.* **27**, 1202.

Rochna E. & Hughes-Jones N.C. (1965) The use of purified ^{125}I-labelled anti-gammaglobulin in the determination of the number of D antigen sites on red cells of different phenotypes. *Vox Sang. (Basel)* **10**, 675.

Rogers Brambell F.W. (1970) *The transmission of passive immunity from mother to young.* North-Holland Publishing Co., Amsterdam and London.

Scott J.S. (1962) Choriocarcinoma; observations on the aetiology. *Am. J. Obstet. Gynec.* **83**, 185.

Scott J.S. (1968) Histocompatibility antigens in choriocarcinoma. *Lancet* **1**, 865.

Siskind G.W., Dunn P. & Walker J.G. (1968) Studies on the control of antibody synthesis. II Effect of antigen dose and suppression by passive antibody on the affinity of the antibody synthesised. *J. exp. Med.* **127**, 55.

Spensieri S., Carnevale-Arella E. & Caldana P.L. (1968) Rh isoimmunisation and transplacental transfer of foetal erythrocytes. *Monit. Obstet. Ginec. Endocr. Metabol.* 1968, 39/6 suppl., 889–906.

Stern K., Goodman H.S. & Berger M. (1961) Experimental isoimmunisation to hemoantigens in Man. *J. Immun.* **87**, 189.

Tovey G.H., Darke C.C. & Fraser I.D. (1970) Significance of HL-A cytotoxic antibodies in Rh haemolytic disease. *Lancet* **1**, 1234.

Voak D. (1969) The pathogenesis of ABO haemolytic disease of the newborn. *Vox Sang.* **17**, 481.

Walknowska J., Conte F.A. & Grumbach M.M. (1959) Practical and theoretical implications of fetal/maternal lymphocyte transfer. *Lancet* **1**, 1119.

Woodrow J.C. (1969) Paper presented at the meeting of the European Society of Human Genetics, Liverpool, April 1969.

Woodrow J.C. & Donohoe W.T.A. (1968) Rh-immunisation by pregnancy. Results of a survey and their relevance to prophylactic therapy. *Br. med. J.* **2**, 139–144.

Zipursky A., Pollack J., Chown B. & Israels L.G. (1965) Transplacental iso-immunisation by foetal red blood cells. *Orig. Article Series* **1**, 184.

Note added in proof: Since this went to press Bagshawe and his colleagues (*Lancet* 1971, 1, 353–357) report on 260 cases of choriocarcinoma. Whereas women of group A married to males of group O had the highest risk, women of group A married to males of group A (and therefore also compatible) had the lowest. It is clear, therefore, that some other mechanism is operative besides that of compatibility between the conceptus and the patient.

18 ❅ Drug Therapy as an Aspect of Ecological Genetics

D.A.P.EVANS

INTRODUCTION

Human beings have been subjected to many changes in their environment in the last century and most of these cannot be accurately defined and quantified. However, a fairly recent one which can be studied with some precision relates to the fact that about half the population now consumes some type of drug medication more or less regularly.

The work of Harris (1969) and his group has shown that, in healthy human subjects, common enzymes when examined by electrophoretic techniques exhibit polymorphisms and that these depend mainly on charge differences produced by amino-acid substitutions in the enzyme molecules. Many substitutions can also occur without changes in charge, and so these most impressive electrophoretic demonstrations provide only a modest estimate of the frequency and complexity of enzyme polymorphisms. Furthermore, some of the morphs have different activities when presented with the same substrate under the same conditions, a notable example of this being red cell acid phosphatase (see Harris 1966).

The actions of drugs in the body depend ultimately upon enzymic mechanisms. A drug molecule frequently undergoes enzymic biotransformation before elimination, and usually drug molecules influence enzyme systems in order to produce their pharmacological effects. Because of this, it is hardly surprising that there should exist considerable variability in the way populations of persons respond to drugs. This finds expression in the clinical sense in various ways such as lack of response ('drug resistance'), over response or unwanted response ('adverse reaction', 'idiosyncrasy') in individual patients when large numbers of people are given standard dosage regimes. Some of this variability is attributable to environmental factors and some to genetic differences between individuals.

345

The observed variability has in some instances revealed the occurrence of two or more discontinuous forms or 'phases' of humans, i.e. polymorphic systems, whereas in others it is unimodal and continuous. An example of each of these different types of variability which have been studied and which have medical relevance will now be described.

CORTICOSTEROIDS AND GLAUCOMA

AN ADVERSE REACTION

The introduction of corticosteroids into ophthalmological practice caused a considerable step forward in the treatment of a number of eye disorders, as for example iritis. However, it became apparent that the same corticosteroids could have deleterious effects, such as the production of subcapsular cataracts, activation of herpetic keratitis and other corneal infections.

Some cases of raised intraocular pressure caused by corticosteroids were described by Stern (1953) and further ones were reported by François (1961) and Goldman (1962). These papers indicated that raised intraocular pressure in patients on corticosteroids occurred more frequently than had been suspected; and that it could on occasion be severe enough to impair vision and a patient described by Armaly (1963a) demonstrates the clinical picture.

A 42-year-old man, with no previous or family history of glaucoma, had been diagnosed as having pars planitis (peripheral uveitis, see fig. 18.1) in the left eye. Treatment had been given thrice daily for three months with 0·1 per cent ophthalmic solution of dexamethasone* in the left conjunctiva only. At this time tonometric and tonographic findings revealed marked increase in tension in the left eye with a reduction in the tonographic estimate of outflow facility, but the filtration angles were open bilaterally (see fig. 18.1). Visual fields were normal with a 1/1000 test objective.

This uniocular hypertension disappeared seven weeks after stopping the eye drops, but could be made to reappear by giving them again.

---

* Dexamethasone and betamethasone are cortisone-like anti-inflammatory glucocortico-steroids.

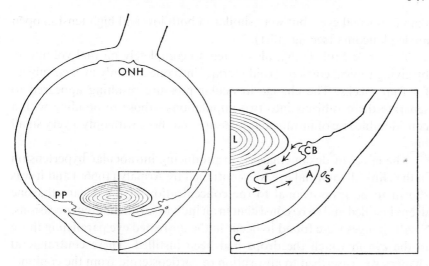

FIGURE 18.1. Diagram of the eye showing the principal features relating to glaucoma. L, lens; I, iris; CB, ciliary body; A, filtration angle; S, canal of Schlemm; C, cornea; ----→, path of the aqueous humour; ONH, optic nerve head; PP, site of pars plana. In glaucoma, the intra-ocular pressure is high and in the open angle variety there is no gross abnormality to be seen in the filtration angle (A) of the eye. The probable cause is some absorptive abnormality in the trabecular mesh-work through which the aqueous humour filters before discharging into the canal of Schlemm (S). In closed angle glaucoma, on the other hand, inflammatory or structural changes may result in obstruction to the filtration angle. The aqueous humour is secreted by the ciliary body (CB) and takes the path as indicated by the arrows. The uveal tract comprises the choroid, the ciliary body and the iris, 'pars planitis' is a peripheral uveitis, i.e. inflammation affecting the pars plana (PP) at the periphery of the retina. An applanation tonometer measures the pressure needed to flatten, rather than indent, the cornea over a specific area (Trevor-Roper 1962).

POPULATION SURVEYS

Detailed surveys of the effects of topical corticosteroids on intraocular pressure and fluid dynamics were undertaken by Armaly (1963a,b) and Becker & Mills (1963). These showed that application of dexamethasone eye drops in one normal eye consistently resulted in an increase in intra-ocular pressure and a diminished outflow facility in the population tested. The magnitude of the effects was greater in subjects over forty than in those under this age, and the effects were greater in glaucomatous than in normal eyes, but were similar in both low and high tension open angle glaucoma (see fig. 18.1).

Becker & Mills (1963) also tested 30 completely normal volunteers by giving topical corticosteroid therapy into one eye only in each subject for four weeks. The change in ocular pressure resulting appeared to separate these subjects into two populations—those responding with a considerable rise of intraocular pressure and those with only a very small one.

The effect of dexamethasone in producing intraocular hypertension in the clinically normal eye was continued by Armaly (1966a) and it was shown to be proportional to the concentration of the dexamethasone drops instilled and to be consistent in an individual on different occasions. Small changes were found to occur in the untreated eye paralleling those in the eye in which the drops had been instilled. These contralateral changes were ascribed to absorption of corticosteroid from the conjunctival sac into the blood-stream as well as directly into the eye by diffusion.

A GENETIC POLYMORPHISM

Armaly (1965, 1966b, 1967a,b, 1968) has clarified the genetics of this ocular pressure response to topical corticosteroids. It was shown that in normal people (i.e. healthy persons with eyes completely normal) there exist three types of responders and not two as had earlier been thought. The standard test applied was as follows: the change in applanation pressure (see fig. 18.1) in an eye was measured during four weeks in which 0·1 per cent dexamethasone drops were instilled into the conjunctivae thrice daily. The three groups were divided at the two points 6 and 15 mm mercury (see fig. 18.2). The hypothesis was made that these pressure responses were controlled by two allelic autosomal genes P^L

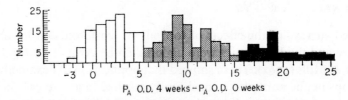

FIGURE 18.2. The change in applanation pressure over four weeks in eyes treated with 0·1 per cent dexamethasone eye drops three times daily (Armaly, personal communication).

TABLE 18.1. Evidence for intraocular pressure response to corticosteroids being controlled by two allelic genes P^L and P^H (compiled from Armaly, 1968).

Presumed genotypic mating	No. of matings	No. of offspring	Presumed genotypes of offspring		
			$P^L P^L$	$P^L P^H$	$P^H P^H$
$P^L P^L \times P^L P^L$	7	41	40 (41)	1	
$P^L P^L \times P^L P^H$	7	33	18 (16·5)	13 (16·5)	2
$P^L P^L \times P^H P^H$	3	14	1	11 (14)	2
$P^L P^H \times P^L P^H$	7	27	6 (6·75)	16 (13·5)	5 (6·75)
$P^L P^H \times P^H P^H$	4	18	1	8 (9)	9 (9)

and P^H, so that there were three genotypes (<6 $P^L P^L$; 6 to 15 $P^L P^H$; >15 $P^H P^H$) detected by the above procedure. The percentage frequencies in the random population were $P^L P^L$ 66; $P^L P^H$ 29 and $P^H P^H$ 5. The hypothesis was upheld by family studies (table 18.1).

POLYGENIC INHERITANCE

The approximately normal unimodal distribution curve of intraocular pressure in the normal eye before the instillation of steroid eye drops has also been investigated in families by Armaly (1967b). He found that there was a correlation of mean-offspring pressure with mid-parent pressure, and that there was an absence of significant correlation between the pressures of spouses. These findings are strongly suggestive of polygenic control.

It is particularly interesting that Armaly, Montstavicius & Sayegh (1968) have examined whether the P^L and P^H alleles make any significant contribution to the determination of the polygenically controlled resting intraocular pressure. They found that the applanation pressure corrected for age was significantly higher in the $P^H P^H$ homozygote than in the $P^L P^L$ homozygote and that the heterozygous $P^L P^H$ occupied an intermediate position. These results indicate, therefore, that these two alleles at one locus are involved in determining ocular pressure in the clinically normal eye. When the environment is changed by the instillation of corticosteroid eye drops the effects of these alleles become much more obvious, i.e. the variability due to this allele pair becomes very large relative to the variability due to other factors, and the phenomenon of polymorphism is observable.

AN ASSOCIATION BETWEEN
GENOTYPE AND DISEASE

Glaucoma* has long been suspected of having an hereditary basis and consequently the bimodal frequency distribution curve of the ocular responses found in normal volunteers by Becker & Mills (1963) prompted Becker & Hahn (1964) to attempt to obtain data on the genetics of both conditions. Seventy-five offspring of glaucomatous subjects were shown to respond very differently from a control population of volunteers. A rise in applanation pressure of over 6 mm mercury occurred in 80 per cent of those under 40 years of age compared with an incidence of 25 per cent of the same response in the control group. Thus it seemed possible that both glaucoma and the response to conjunctival steroids by production of intraocular hypertension were controlled by a common genetic mechanism.

Becker (1964) tested 28 subjects with various types of secondary glaucoma (some following cataract surgery or trauma and some with angle closure) and found that they resembled normal volunteers without a family history of glaucoma in that 32 per cent of both groups 'responded' to betamethasone with a raised intraocular pressure. This pharmacologic effect was thus different in the primary glaucoma population (92 per cent 'responders', Becker & Hahn 1964) as compared with normal subjects and those with secondary glaucoma.

Armaly (1968) presented evidence to show that the frequency of the allele P^H was much higher in open-angle hypertensive glaucoma, low tension glaucoma and recessed angle glaucoma (all varieties of 'primary' glaucoma, see fig. 18.1) than in a random sample of the population (see table 18.2).

Confirmation of the association between the pre-glaucomatous constitution and the P^H gene as shown by steroid-induced intraocular

---

* Any condition in which the intraocular pressure is high enough to present a threat to the optic nerve (Grant 1964).

TABLE 18.2. Dexamethasone hypertension in different clinical categories (Armaly 1968).

Category	Number of Subjects Tested	Percent frequency of genotypes		
		$P^L P^L \Delta P_A < 6$ mm Hg (%)	$P^L P^H \Delta P_A$ 6-15 mm Hg (%)	$P^H P^H \Delta P_A > 15$ mm Hg (%)
Random sample	80	66	29	5
Open-angle hypertensive glaucoma	33	6	48	44
Low tension glaucoma	15	7	53	40
Normal eye in recessed angle glaucoma	15	—	53	47
Normal eye in angle recession without glaucoma	4	75	25	—

hypertension was forthcoming in an elegant experiment by François *et al.* (1966). Two population samples were studied; firstly 480 normal subjects without a family history of glaucoma, and secondly 396 apparently normal individuals from families in whom an older member was known to suffer from glaucoma. After the instillation of 0·1 per cent betamethasone into one eye four times daily for three weeks, the pressure frequency distribution was trimodal. A very significantly greater incidence of a rise of pressure larger than 5 mm mercury was seen in the second group as compared with the first (table 18.3). Furthermore, and rather alarmingly, six previously normal subjects among the glaucoma

TABLE 18.3. Results of instilling 0·1 per cent betamethasone eye drops into one eye four times for three weeks.* Compiled from François *et al.* (1966).

Rise in intraocular pressure (mm mercury)	Normal subjects from families free of glaucoma	Normal subjects from glaucomatous families
> 5 mm	86	124
< 5 mm	226	226
No change	168	46

* If the increase in intraocular pressure was less than 5 mm mercury after three weeks the instillations were continued for a further three weeks.

relatives became frankly glaucomatous with demonstrable field defects as a result of this test procedure.

Thus an adverse reaction (ocular hypertension) to a drug (corticosteroid) has led to the discovery of a new polymorphism in Man. The genotypes in the polymorphism have different degrees of risk to develop this adverse reaction. The genotypes also have different liabilities to develop glaucoma in later life. Armaly (1967a) computes that if the risk for $P^L P^L$ developing glaucoma is 1 then the risk for $P^L P^H$ is 18 and for $P^H P^H$ is 101. Here then is one genetic component forming a disease prone 'diathesis', and knowledge of the genotype of an individual has predictive value.

TWO OTHER PHYSIOLOGICAL FUNCTIONS INFLUENCED BY THE ALLELIC GENES P^L AND P^H

1. GLUCOSE TOLERANCE

The finding that dexamethasone induced ocular hypertension was a genetically determined response encouraged the search for other glucocorticoid responses in extraocular systems that might be similarly determined. Thus Armaly (1967c) studied a population of 55 subjects (aged 20 to 41 years, median 25 years) with clinically normal eyes with no history of diabetes and who had their genotypes for P^L and P^H determined with a dexamethasone test. On another occasion they had an oral glucose tolerance test (75 g glucose; blood taken fasting and at 30-minute intervals for 2 hours). The results are shown in table 18.4 indicating that the P^H gene confers less glucose tolerance than P^L.

This finding has considerable clinical relevance since there is a known clinical association between glaucoma and diabetes mellitus, and a raised intraocular pressure has been demonstrated in diabetic patients by Armstrong *et al.* (1960), Christiansson (1961), Safir, Paulsen & Klayman (1964), and Becker *et al.* (1966). The last mentioned study reveals that the frequency of the P^H allele is in fact greater in diabetics than in non-diabetics.

2. PLASMA CORTISOL SUPPRESSION TEST

The plasma cortisol suppression test for the function of the pituitary-adrenal axis is performed as follows: a base line plasma cortisol concen-

TABLE 18.4. Results of oral glucose tolerance tests in relation to the polymorphism controlled by the alleles P^L and P^H compiled from Armaly (1967c).

Genotype	No. of subjects	Sum of glucose concentrations in mg per 100 ml at 30, 60 and 90 minutes after ingestion of 75 g glucose									Reducing substance in urine detected by Clinitest
		200– 249	250– 299	300– 349	350– 399	400– 449	450– 499	500– 549	550– 599	600– 649	
$P^L P^L$	14	5	5	2	2	—	—	—	—	—	1
$P^L P^H$	25	4	6	9	5	—	—	—	—	—	3
$P^H P^H$	16	—	3	4	2	2	1	2	1	1	4

tration value is determined; a single dose of 1 mg dexamethasone is then administered to suppress the pituitary production of ACTH and so secondarily suppress the adrenal production of cortisol assessed by a second determination of plasma concentration.

A negative correlation between the suppression of plasma cortisol by dexamethasone (y) and rise in intraocular pressure in response to topical corticosteroids (x) has been demonstrated by Levene & Schwartz (1968), Rosenbaum, Alton & Becker (1970) and Becker & Ramsey (1970). The results of Rosenbaum *et al.* (1970) can be taken as an illustration. They use a different nomenclature from Armaly and define their genotypes as follows: *nn* ocular pressure < 20 mmHg; *ng* 20–31; *gg* > 31 mmHg (measurements taken after 0·1 per cent dexamethasone drops four times daily in one eye for six weeks).

The suppression ratio is the ratio of plasma cortisol concentration after dexamethasone expressed as a percentage of the base line value. In 22 persons of genotype *nn* the mean value of the suppression ratio was 20 per cent ± SD 10 %; and therefore for this genotype the mean + 2SD = 40 per cent. Suppression ratio results greater than 40 per cent were given by 0 out of 22 *nn* subjects; 3 out of 17 *ng* subjects; and 4 out of 7 *gg* subjects.

CONCLUSION

This polymorphism controlled by the alleles P^L and P^H provides a striking and practically important validation of the prediction made by Ford (1965): 'Many cryptic human polymorphisms are likely to be of physiological significance ... while all of them are of relevance to medical studies.'

POLYGENIC CONTROL OF HUMAN
PHENYLBUTAZONE METABOLISM

Many drugs are metabolized in the body by the actions of enzymes located in the endoplasmic reticulum of the liver, and this after homogenization is transformed into the particulate 'microsomes' (Fouts 1961). A common biotransformation which many drugs undergo as a result of microsomal enzymic activity is hydroxylation (fig. 18.3), and clearly many enzymes could be involved in such a complex process.

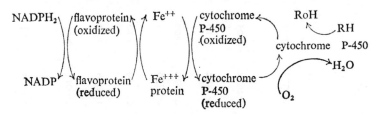

FIGURE 18.3. The electron transport system involved in the microsomal hydroxylation of the drug R (from Robson & Stacey 1968).

Microsomal enzymes including those concerned with hydroxylation of drugs are known to be influenced by chemical compounds administered *in vivo* to mammalia. Phenobarbitone is a typical example and causes 'induction', i.e. increased synthesis of liver hydroxylase which results in the more rapid metabolism of substrates such as other drugs. This increased enzymic activity can be demonstrated *in vivo* and also, using isolated liver microsomes, *in vitro*. Since drug metabolites are generally more polar and more water-soluble than their parent compound they are more readily excreted especially through the kidney. This process of induction can, therefore, be viewed as a protective mechanism in that it is a change which often enables an animal more efficiently to excrete foreign compounds which may be toxic. Genetic polymorphism is a protective mechanism for a population operating over many generations, whereas induction is a protection for an individual enabling him to respond in a manner which is likely to improve survival in a hostile environment.

There is a calculable phenotypic variance of any measurement of microsomal drug metabolism, and this is made up of genetic and environmental components. The environmental component receives a contribution from the fact that different individuals are exposed to different environments and so are induced to a varying degree.

One technique to study the metabolism of a drug is to measure its plasma half-life under appropriate conditions and the commonly used drug phenylbutazone* has been investigated in this way by two groups of authors. Levi, Sherlock & Walker (1968) measured the half-life of phenylbutazone in 36 persons who had not received any drug pre-treatment and in 19 who had received pre-treatment with various drugs.

*A substituted pyrazoline drug with analgesic, antipyretic and anti-inflammatory properties used to treat rheumatic conditions especially rheumatoid arthritis.

TABLE 18.5. Regressions calculated using plasma phenylbutazone half-lives after phenobarbitone (from Whittaker & Evans 1970).

Line No.	x	y	n	b	a	r	t	p
1	First estimate of $\log_{10}$ PBZ $T\frac{1}{2}$	Second estimate of $\log_{10}$ PBZ $T\frac{1}{2}$	16	0·6919	0·5555	0·7825*		<0·01
2	Steady state plasma concentration of phenylbutazone (μg per ml)	$\log_{10}$ PBZ $T\frac{1}{2}$	19	0·0080	1·5952	0·5802	2·9373	<0·01
3	Mid parent value (with no adjustment) of $\log_{10}$ PBZ $T\frac{1}{2}$	Mean offspring value (with no adjustment) of $\log_{10}$ PBZ $T\frac{1}{2}$	24	0·6636	0·6105	0·4701	2·4984	<0·05
4	Husband value (with no adjustment) of $\log_{10}$ PBZ $T\frac{1}{2}$	Wife value (with no adjustment of $\log_{10}$ PBZ $T\frac{1}{2}$	24	0·0632	1·7742	0·0751	0·3532	>0·10
5	Height (ins.)	$\log_{10}$ PBZ $T\frac{1}{2}$	142	0·0094	1·2873	0·3266	4·0880	<0·001
6	Weight (lbs)	$\log_{10}$ plasma PBZ $T\frac{1}{2}$ adjusted to a height of 66 inches	142	−0·0001	1·9191	0·0231	0·2738	>0·10
7	Age (years)	$\log_{10}$ plasma PBZ $T\frac{1}{2}$ adjusted to a height of 66 inches	142	−0·0001	1·9097	0·0153	−0·1809	>0·10
8	$\log_{10}$ plasma PBZ concentration at zero time	$\log_{10}$ plasma PBZ $T\frac{1}{2}$ adjusted to a height of 66 inches	142	−0·1687	2·1836	0·1542	−1·8468	>0·05
9	Husband value of $\log_{10}$ plasma PBZ $T\frac{1}{2}$ adjusted to a height of 66 inches	Wife value of $\log_{10}$ plasma PBZ $T\frac{1}{2}$ adjusted to a height of 66 inches	24	0·1110	1·7075	0·1271	0·6011	>0·10
10	Mid parent value of $\log_{10}$ plasma PBZ $T\frac{1}{2}$ adjusted to a height of 66 inches	Mean offspring value of $\log_{10}$ plasma PBZ $T\frac{1}{2}$ adjusted to a height of 66 inches	24	0·6529	0·6351	0·5588	3·1605	<0·01

n = number of pairs of observations.
b = regression coefficient (slope).
a = intercept, i.e. value of y when x = 0 (in units of y).
r = correlation coefficient
t = to test significance of r and b.
p = probability.
$T\frac{1}{2}$ = half-life.
PBZ = phenylbutazone.
* = the approximate 95 per cent confidence limits are: 0·445 and 0·925. Intraclass correlation 0·89.

The values had a skewed unimodal frequency distribution varying between 50 and 120 hours in the non-pre-treated subjects and between 30 and 100 hours in the pre-treated ones (many of whom had received known inducing drugs such as barbiturates).

On the genetic side, the contribution to the variability of phenyl-butazone metabolism between individuals was investigated by Vesell & Page (1968) in a study of twins who had not been subjected to pre-treatment with any other drugs. They found the plasma half-life of phenylbutazone was much more similar in identical than in fraternal twins.

In view of the foregoing, it seemed possible that phenylbutazone metabolism (as measured by its plasma half-life) was controlled by an unknown number of allelic genes (none individually identifiable), situated at an unknown number of loci which again cannot be individu-ally defined, i.e. polygenic inheritance. The most straightforward proof of this proposition was deemed to be the estimation of the 'heritability' i.e. V_A/V_p (where V_A = the variance due to the additive effects of the genes controlling the character, and V_p = phenotypic variance). This experi-ment has now been performed by Whittaker & Evans (1970).

One of the difficulties in the carrying out of the work was the existence of this phenomenon of 'induction'. It was found that giving all the subjects in the study a three-day course of phenobarbitone before their phenylbutazone test, caused an approximate normalization of the $\log_{10}$ phenylbutazone half-lives. When the same individuals had their plasma phenylbutazone half-lives determined without pre-treatment the distribu-tion of values had been markedly skewed. Persons who without pheno-barbitone pre-treatment had long half-lives had them reduced after the short course of phenobarbitone. This is interpreted as bringing all the persons in the experiment to a more standardized level of induction, i.e., reducing the environmental component of variance.

When a standard dose of a drug is given at regular intervals and the plasma concentration monitored it is found to reach a plateau level called the steady state plasma concentration (SSPC). This is regarded as a reliable index of the metabolic capacity of an individual since it represents an equilibrium between regular input and the processes of metabolism and elimination.

$T\frac{1}{2}$ (the post-phenobarbitone $\log_{10}$ plasma half-life of phenyl-butazone) was found to be repeatable (table 18.5, line 1) and to correlate with the phenylbutazone SSPC determined on another occasion (table 18.5, line 2). No significant correlation was found between husband and wife value (table 18.5, line 4), but there was a significant regression of mean offspring value on mid-parent value (table 18.5, line 3).

Some of the variability in $T\frac{1}{2}$ seemed to be due to variations in body build. There was a significant regression of $T\frac{1}{2}$ on height, and therefore $T\frac{1}{2}$ values were adjusted to a standard height of 66 inches using the calculated regression coefficient. After this adjustment had been performed the $T\frac{1}{2}$ values were again approximately normally distributed

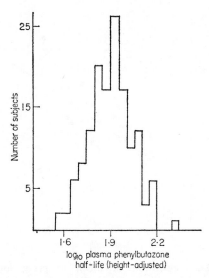

FIGURE 18.4. The frequency distribution of $\log_{10}$ post-phenobarbitone plasma phenylbutazone half-lives adjusted to a standard height (from Whittaker & Evans 1970).

(fig. 18.4) and there was no significant regression of half-life on weight (table 18.5, line 6), age (table 18.5, line 7), or $\log_{10}$ plasma phenylbutazone concentration at zero time as derived from the computation of $T\frac{1}{2}$ (table 18.5, line 8).

The $T\frac{1}{2}$ values adjusted for height showed no significant sex difference. When parents were compared with offspring the mean of 49 parents was $1\cdot898 \pm$ SE 0.018 and for 52 offspring the mean was $1\cdot870 \pm$ SE 0·018. Since there is a correlation, a special value of t needs to be derived (Tweedie, personal communication). This has been computed as $1\cdot23$ with about 25 degrees of freedom; indicating no significant difference between parents and offspring.

Using the height adjusted values of $T\frac{1}{2}$ for computation there was still no correlation between husband and wife values (table 18.5, line 9). The regression of mean offspring values upon mid-parent values retained a very similar slope, but the scatter of points about the regression line was reduced as a result of the height adjustment (fig. 18.5, table 18.5, line 10).

Thus the estimate of V_A/V_p i.e. the heritability for phenylbutazone metabolism is 0·65 ± SE 0.21 (approximate).

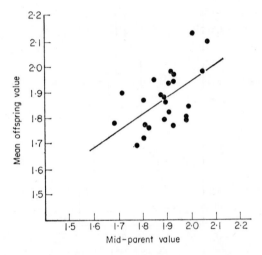

FIGURE 18.5. The regression of mean offspring $\log_{10}$ post-pheno-barbitone plasma phenylbutazone half-lives on mid-parent values. These half-lives have been adjusted to a standard height (from Whittaker & Evans 1970).

In order to serve as a comparison with the foregoing, the heights of the family members in the study of Whittaker & Evans (1970) were adjusted for age and sex. Following this adjustment the heritability V_A/V_p was computed as 0·62. The data of Galton (1889) give a heritability for height of approximately 0·66. It will be seen, therefore, that phenyl-butazone metabolism in Man is genetically controlled in a similar manner and to a similar degree as body height. The same genes do not control the two characters since the $T\frac{1}{2}$ values had been adjusted to a standard height.

Many drugs are known to show a correlation between plasma level and (1) their therapeutic effect and (2) the occurrence of toxic or adverse effects. Bruck *et al.* (1954) showed that various toxic effects were more common in patients receiving phenylbutazone whose blood levels exceeded 100 μg per ml than in those with concentrations below this level. On the other hand, subjective improvement in rheumatoid arthritics was more frequent above blood phenylbutazone concentrations of 50 μg per ml than below this level. Thus:

1   the variability in phenylbutazone half-life may well have a bearing on the consequences of phenylbutazone therapy;

2   for a pharmacological phenomenon in a population of patients, V_p may be increased by additions to V_E (the environmental component of variance) due for example to disease, differing modes of administration and other factors; on the other hand, V_G cannot be diminished;

3   it is possible that in the future, attention to the variability between persons may result in the development of improved therapeutics.

REFERENCES

ARMALY M.F. (1963a) Effect of corticosteroids on intraocular pressure and fluid dynamics I the effect of dexamethasone in the normal eye. *Archs. Ophthal. N.Y.* **70**, 482–491.

ARMALY M.F. (1963b) Effect of corticosteroids on intraocular pressure and fluid dynamics. II the effect of dexamethasone in the glaucomatous eye. *Archs Ophthal. N.Y.* **70**, 492–499.

ARMALY M.F. (1965) Statistical attributes of the steroid hypertensive response in the clinically normal eye. I the demonstration of three levels of response. *Invest. Ophthal.* **4**, 187–197.

ARMALY M.F. (1966a) Dexamethasone ocular hypertension in the clinically normal eye. II The untreated eye outflow facility and concentration. *Archs. Ophthal. N.Y.* **75**, 776–782.

ARMALY M.F. (1966b) The heritable nature of dexamethasone-induced ocular hypertension. *Archs. Ophthal. N.Y.* **75**, 32–35.

ARMALY M.F. (1967a) Inheritance of dexamethasone hypertension and glaucoma. *Archs Ophthal. N.Y.* **77**, 747–751.

ARMALY M.F. (1967b) The genetic determination of ocular pressure in the normal eye. *Archs Ophthal. N.Y.* **78**, 187–192.

ARMALY M.F. (1967c) Dexamethasone ocular hypertension and eosinopenia and glucose tolerance test. *Archs Ophthal. N.Y.* **78**, 193–197.

ARMALY M.F. (1968) Genetic factors related to glaucoma. *Ann. N.Y. Acad. Sci.* **151**, 861–875.

ARMALY M.F., MONTSTAVICIUS B.F. & SAYEGH R.E. (1968) Ocular pressure and aqueous outflow facility in siblings. *Archs Ophthal. N.Y.* **80**, 350–360.

ARMSTRONG J.R., DAILY R.K., DOBSON H.H. & GIRARD L.J. (1960) The incidence of glaucoma in diabetes mellitus. *Am. J. Ophthal.* **50**, 55–63.

BECKER B. & MILLS D.W. (1963) Corticosteroids and intraocular pressure. *Archs Ophthal. N.Y.* **70**, 500–507.

BECKER B. (1964) The effect of topical corticosteroids in secondary glaucomas. *Archs Ophthal. N.Y.* **72**, 769–771.

BECKER B. & HAHN K.A. (1964) Topical corticosteroids and heredity in primary open angle glaucoma. *Am. J. Ophthal.* **57**, 543–551.

BECKER B., BRESNICK G., CHEVRETTE L., KOLKER A.E., OAKS M. & CIBIS M. (1966) Intraocular pressure and its response to topical corticosteroids in diabetics. *Archs Ophthal. N.Y.* **76**, 477–483.

BECKER B. & RAMSEY C.K. (1970) Plasma cortisol and the intraocular pressure response to topical corticosteroids. *Am. J. Ophthal.* **69**, 999–1002.

BRUCK E., FEARNLEY M.E., MEANOCK I. & PATLEY H. (1954) Phenylbutazone therapy: relation between the toxic and therapeutic effects and the blood level. *Lancet* **1**, 225–228.

CHRISTIANSSON J. (1961) Intraocular pressure in diabetes mellitus. *Acta ophthal.* **39**, 155.

FORD E.B. (1965) *Genetic Polymorphism*, p. 87. Faber and Faber, London.

FOUTS J.R. (1961) The metabolism of drugs by subfractions of hepatic microsomes. *Biochem. biophys. Res. Commun.* **6**, 373–378.

FRANÇOIS J. (1961) Glaucome apparemment simple secondaire à la cortisono-theràpie locale. *Ophthalmologica, Basel* **142**, 517–523.

GALTON F. (1889) *Natural Inheritance*. Macmillan, London.

GOLDMANN H. (1962) Cortisone glaucoma. *Archs Ophthal. N.Y.* **68**, 621–626.

GRANT W.M. (1964) Glaucoma Research Conference. *Am. J. Ophthal.* **58**, 1065.

HARRIS H. (1966) Enzyme polymorphisms in Man. *Proc. R. Soc.* B **164**, 298–310.

HARRIS H. (1969) Genes and Isozymes. *Proc. R. Soc.* B **174**, 1–31.

LEVENE R.Z. & SCHWARTZ B. (1968) Depression of plasma cortisol and the steroid ocular pressure response. *Archs Ophthal. N.Y.* **80**, 461–466.

LEVI A.J., SHERLOCK S. & WALKER D. (1968) Phenylbutazone and isoniazid metabolism in patients with liver disease in relation to previous drug therapy. *Lancet* **1**, 1275–1279.

ROBSON J.M. & STACEY R.S. (1968) *Recent advances in pharmacology*, 4th edn., p. 34. Churchill, London.

ROSENBAUM L.J., ALTON E. & BECKER B. (1970) Dexamethasone testing in South-western Indians. *Invest. Ophthal.* **9**, 325–330.

SAFIR A., PAULSEN E.P. & KLAYMAN J. (1964) Elevated intraocular pressure in diabetic children. *Diabetes* **13**, 161–163.

STERN J.J. (1953) Acute glaucoma during cortisone therapy. *Am. J. Ophthal.* **36**, 389–390.

TREVOR-ROPER P.D. (1962) *Ophthalmology*. A Textbook for Diploma Students, p. 512. Lloyd-Luke, London.

VESELL E.S. & PAGE J.G. (1968) Genetic control of drug levels in Man: phenylbutazone. *Science* **159**, 1479–1480.

WHITTAKER J.W. & EVANS D.A.P. (1970) Genetic control of phenylbutazone metabolism in Man. *Br. med. J.* **4**, 323–328.

19 ❋ Ecological Genetics and Biology Teaching

W.H.DOWDESWELL

Traditionally, the teaching of biology has been characterized by abrupt discontinuities with one branch of the subject largely isolated from another; thus botany has been distinct from zoology intellectually and sometimes in geographical location as well. During recent years welcome changes have been initiated in the direction of a more interdisciplinary approach, beginning in the schools under the influence of the Nuffield Projects and spreading more slowly within the universities. In the midst of these changes in biological thought and attitudes, one of the great unifying themes has been heredity. It is a great pleasure to contribute this short paper in honour of Professor Ford, since it provides me with an opportunity to pay tribute to him on behalf of the many students of biology both at Oxford and elsewhere, who have been stimulated and inspired by his teaching of genetics, particularly that aspect of the subject which has become so peculiarly his own—ecological genetics.

What is ecological genetics? The answer can best be given in Ford's own words (1969) used when discussing the need for setting up evolutionary studies on a strictly scientific basis as envisaged by Darwin himself. 'It seemed to me that this object could be attained by a joint programme of laboratory genetics combined with observation and experiment in the field, using the statistics appropriate to such studies ... This twofold technique is today known as ecological genetics. It is essentially evolutionary in aim, for it deals with the adjustments and adaptations of wild populations to their environment.' As defined by Ford the subject stands, as it were, at the meeting point of ecology, genetics and evolution, and hence demands serious consideration in the planning and teaching of any integrated course of biology. Moreover, it serves to enhance our understanding of important areas of ecology which are conventionally regarded as being outside the scope of ordinary ecological research and teaching.

This paper makes no claim to be a survey or to be exhaustive in its coverage. Rather, its aim is to highlight a few areas of biology teaching which are widely regarded as important and to draw attention to the contributions that ecological genetics has made towards a broader and more profound understanding. Some of the examples chosen are essentially of a theoretical kind while others lend themselves well to a more practical treatment.

NUMBERS OF ORGANISMS

One of the fundamental changes that has taken place in ecological thinking during the last forty years or so has been a transition from a largely descriptive approach to an increasingly quantitative one. The need for this was foreshadowed by Elton as long ago as 1924 when he pointed out that the numbers of animals such as rabbits, hares and jack-rabbits in North America and elsewhere are subject to violent fluctuations. Moreover, these fluctuations tend to occur with a regular periodicity. As a result of the work of Chitty and others we now know that the same is true of many of our native species, for instance the Short-tailed Vole, *Microtus agrestis*, whose populations fluctuate with a frequency of three to four years. Like mammals, insect populations too, are subject to considerable variations in density although these tend to be less regular than in mammals. The work of Ford & Ford during the period 1920–24 showed that such changes in density provide a unique opportunity for the application of the principles of ecological genetics. From studies of a colony of the Marsh Fritillary butterfly, *Euphydryas* (*Melitaea*) *aurinia*, they showed that while numbers remained low a stable type prevailed. But a rapid outburst over a period of about five years was accompanied by a great increase in variation including a high incidence of abnormalities. With a subsequent stabilization of numbers variability declined and a comparatively uniform type prevailed once more—one which was recognizably different in appearance from the form which had prevailed at the outset (Ford & Ford 1930).

The evolutionary implications of this work are of such fundamental importance that one might have expected similar observations to be made with the same or other variable species. Yet, to the best of my knowledge, no such follow-up has taken place, either under natural conditions or as a model in the laboratory. That such situations are probably quite common is suggested by largely qualitative observations of mine on the same species in a restricted locality near Winchester. Here, over a period of three years, numbers rose from a few dozen to several hundred, the increase being accompanied by a noticeable outburst of deformities which rose to a maximum of about 10 per cent, before declining again. Unfortunately, no evidence is available regarding the type which prevailed before the fluctuation took place.

Ford sums up the situation succinctly when he says 'a population can increase in numbers only if certain aspects of selection bearing upon it are relaxed. In such circumstances it will therefore become more variable. On the other hand, a numerical decline indicates more rigorous selection, and this must tend to reduce diversity'.

Paradoxically, one of the animals which has contributed most to our knowledge of ecological genetics, the Meadow Brown butterfly (*Maniola jurtina*), was originally selected for study in the Isles of Scilly, not on account of its spot-variation for which it has become so well known, but because of its suitability for marking, release and recapture (Dowdeswell, Fisher & Ford 1949). The methods now available for estimating animal numbers which owe much to these early experiments, have added greatly to the validity and precision of subsequent studies of population density. Used in conjunction with the kind of observations outlined above, such methods could contribute greatly to our knowledge of the fundamental principles underlying population dynamics. Moreover, the marking procedure involved is so simple and reliable that it is well suited to classwork, particularly if this is conducted on a project basis. Colonies of butterflies which are often surprisingly localized, aquatic Hemiptera such as *Corixa* or *Notonecta* in a small pond, and numerous other mobile invertebrates are well suited to this form of study. Conducted on an annual basis, such work can also provide valuable long-term information on changes in density.

STABILITY IN POPULATIONS

In teaching biology there is a great temptation to place a disproportionate emphasis on variation and change, and to disregard the equally striking and fundamental fact of stability. It is an axiom of ecology that under natural conditions the larger and more diverse a community, the more stable does it tend to become. Clearly, the greater the number of ecological niches available, the more flexible will the situation be, so that if, for example, the food plant of a particular herbivore should happen to be in short supply the chances are that an alternative may be forthcoming. It is not easy to provide an omnibus definition of 'stability' in a species. As we have seen, its numbers may fluctuate and frequently do so, and this, in time, may have an effect on variability. In a polymorphic situation, there may well be a balance of some antiquity between the proportions of two or more phenotypic forms. However, under changing environmental conditions, particularly those resulting from the activities of man, a state of instability may be set up in which a transient polymorphism occurs with one form at an increasing advantage over the other in the new circumstances. Such a situation must have occurred in the Peppered moth, *Biston betularia*, during the early years of this century when, in the industrial Midlands, the black form *carbonaria* was in the process of replacing the typical one.

Relatively few plant and animal populations have been studied for long enough and with sufficient precision for a quantitative assessment of their degree of stability to be gauged. Fortunately, there are some notable exceptions in Britain most of which owe their initiation in varying degree to the foresight of Professor Ford. Mention has already been made of the Meadow Brown butterfly, *Maniola jurtina*, in which the number of spots on the underside of the hind-wings provides a useful index of evolutionary divergence and the varying effects of selection. Moreover, they are now known to have a high female heritability (McWhirter 1969). Our work with this species dates from 1946 but the British Museum collections, some of which are substantial, extend as far back as 1890. Many of them are due to the influence of Walter, Second Baron Rothschild, who was far ahead of his time in believing that species should be defined on a wide basis involving extensive sampling of populations rather than on a few 'type' specimens.

Extensive sampling of Southern English populations from 1950 onwards has established that spot-values are stabilized throughout most of the region from Kent westwards (*Southern English* stabilization) with distinct spot-frequencies in the two sexes, the males being unimodal at 2 spots and the females unimodal at 0 (fig. 19.1). European data for the years 1890–1935 show that this situation extends north to Finland, eastwards to Bulgaria and into the South of France (Dowdeswell & McWhirter 1967). Evidently, the Southern English stabilization is but a small part of a far larger *General European* one. Museum collections are

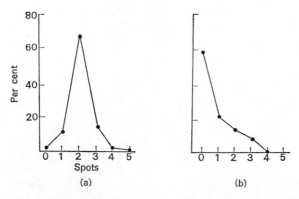

FIGURE 19.1. 'Southern English' spot-stabilization in *Maniola jurtina* (a) males; (b) females. (From *Heredity* 6, p. 104, 1952.)

inevitably somewhat fragmentary, but even so they show clearly that round the periphery of the area a diversity of other stabilizations is to to be found, indicating that the species achieves widely differing adjustments to the diverse environments existing on the edge of its range (fig. 19.2). One such situation we have been able to study annually since 1956—the so called 'Boundary Phenomenon', occurring in the neighbourhood of the Devon–Cornwall border. Here the Southern English stabilization breaks down to be replaced by an *East Cornish* pattern, still unimodal at 2 spots in the male but bimodal in the female with the greater mode at 0 spots and the lesser at 2 (fig. 19.3) (Ford 1965). The circumstances in which the change occurs from one to the other are of some interest:

(a)   no physical barrier is involved;
(b)   the transition takes place with extraordinary abruptness, sometimes within only a few yards;
(c)   the location of the boundary has fluctuated, moving eastwards and then westwards again over a distance of approximately 40 miles during a period of 13 years (Creed, Dowdeswell, Ford & McWhirter 1970).

These results accord well with evolutionary theory and throw an interesting light on the ecology of *Maniola jurtina*. For it must be remembered that the species in Cornwall is nearing the westerly extremity of its range—the Scillies to the southwest and Ireland farther north. It may seem surprising that in spite of the gene-flow that must take place at the Boundary, variation in *jurtina* can exhibit such an abrupt dis-

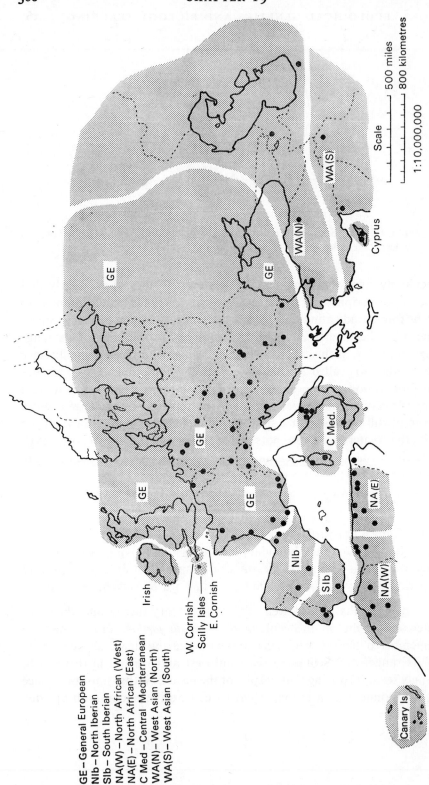

GE – General European
NIb – North Iberian
SIb – South Iberian
NA(W) – North African (West)
NA(E) – North African (East)
C Med – Central Mediterranean
WA(N) – West Asian (North)
WA(S) – West Asian (South)

FIGURE 19.2. Spot-stabilizations in *Maniola jurtina* throughout its range. Spots denote sampling areas. (Redrawn from Dowdeswell & McWhirter 1967.)

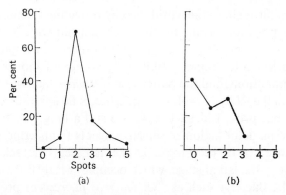

FIGURE 19.3. 'East Cornish' spot-stabilization in *Maniola jurtina* (a) males; (b) females. (From *Heredity* **6**, p. 105, 1952.)

continuity. However, the situation conforms closely to a laboratory model set up by Millicent & Thoday with *Drosophila melanogaster*, to investigate the effects of disruptive selection using the character, sterno-pleural chaeta number. They found that with a 25 per cent gene-flow, the equivalent of random mating, divergence of the two selected lines, although slower, eventually achieved the same magnitude as complete isolation (i.e. with no gene-flow).

Considerations such as those outlined above emphasize the urgent need for setting up model situations under field conditions which can be studied continuously for a number of years. Collecting and handling the resulting data could provide valuable experience for students as well as yielding the kind of long-term information which is still badly needed.

SELECTION AND CHANGE

One of the difficulties that arises in studying evolutionary change is that precise evidence for the action of selective agents is so frequently lacking. Selection by predation on the adult, as Kettlewell has demonstrated so elegantly in the Peppered Moth, *B. betularia*, may well turn out to be the exceptional situation rather than the rule. It does not even apply to melanic moths in general, for in species such as the Mottled Beauty, *Cleora repandata*, the genotype producing the adult form *nigra* is also known to confer an advantage of superior physiological hardiness in the larva. The human ABO blood groups exhibit characteristic regional distributions throughout the world thereby providing splendid examples of balanced polymorphism. But their relationship with human bodily deficiencies or advantages is at present no more than statistical. All organisms selected for experimental study are found sooner or later to have their limitations, and the butterfly *M. jurtina*, which we have used so extensively in the study of ecological genetics, is no exception. For some purposes it has proved ideal, but we are still a long way from understanding the mode of action of selective agents in affecting spot-distribution. From extensive breeding experiments it seems that selection takes place largely in the larval stage, which occupies nearly three-quarters of the life cycle. Bacteria such as *Pseudomonas fluorescens* probably play some part, but in a herbivorous animal, elucidating the roles of the various members of the internal bacterial community is a complex undertaking (McWhirter & Scali 1966). The relationship between larval susceptibility to bacteria and adult spotting still remains obscure. The main reason why it has taken so long merely to establish heritability of spotting in *Maniola*, has been that bacterial selection may well have been more powerful inside our breeding cages than in nature—a salutory reminder to those who conduct genetic experiments on the assumption that mortality in the young stages is at random!

An important contribution of recent studies in ecological genetics has been to highlight the fact that much of selection is endocyclic. This means that the direction of selection is reversed in the young stages of plants and animals, for instance in seeds or larvae, compared with later phases. Certain species lend themselves particularly well to such investigations, for instance molluscs of the genus *Cepaea*. The work of Cain and others has established that selection of adult snails by Song Thrushes for shell background colour and degree of banding, can fluctuate both with different kinds of vegetation and in the same locality at varying times of the year (Cain & Sheppard 1950). Animals whose shells achieve the best matching with their background are those most likely to survive. In areas where birds tend to congregate temporarily such as Portland Bill, predation may reach astonishing proportions. For instance, along a small grassy bank with an estimated population of 1,500 *C. hortensis*, no less than 580 were eaten by birds (38·7 per cent) during the spring of 1962, most of this decimation taking place in the course of a few days. Thereafter the size of the colony rapidly declined until three years later it was only about 100; but whether this could be attributed to the onslaught of three years earlier is uncertain. In snails, little attention seems to have been paid to endocyclic selection partly because of the difficulty of separating *C.nemoralis* and *C.hortensis* in the young stages. However, colonies exclusively of one species are not uncommon and occur on Portland Bill and at Atworth (Wiltshire). The shells can be separated into size groups using a series of sieves of varying mesh. These are constructed to fit one on top of another with a lid above and a collecting tray below the smallest mesh. With such an apparatus, a sample of snails can be sorted for size accurately and quickly. In constructing the sieves, the strong corrosive action of snail slime must be borne in mind and nylon should be used in preference to wire. Young snails are to be found during spring and summer, and comparisons with adults in respect of ground colour and banding is a field exercise which can be well worth while. Some typical results from a locality on Portland Bill are shown in table 19.1 (all shells in this area are yellow).

TABLE 19.1. A comparison of different ages in *Cepaea hortensis* from Portland Bill with respect to banding.

Date	Shells over 10 mm diam			Shells 10 mm diam and less		
	Unbanded	Banded	Total	Unbanded	Banded	Total
12–13.8.67	57	12	69	19	4	23
7–8.10.67	124	27	151	18	—	18
16–17.8.68	100	31	131	35	4	39
Total	281	70	351 (19·9 per cent banded)	72	8	80 (10·0 per cent banded)

Comparison of these data gives $\chi^2_{(1)} = 4\cdot35$; $0\cdot05 > P > 0\cdot02$, indicating a significant deficiency of banded individuals among the young.

In the Atworth population of *C.hortensis* which occupies a very different type of locality from that on Portland, selection seems to be tending in the same direction against banded, but the divergence has not yet reached the level of statistical significance.

In evaluating results such as these, it must be remembered that birds probably seldom eat young snails. Evidence for this is provided by snail colonies where significant differences in banding occur between young and adults, but bird predation on adult snails is absent. In such circumstances selection must be of some other kind—perhaps physiological. This is no doubt true to some extent of adults as well, as has been pointed out by Lamotte (1959). In our work on Portland Bill, Day and I have studied two populations of *C. hortensis* occurring on grassy banks sloping from west to east, and about 300 yards long, both of which exhibit a steep cline of banding ranging from approximately 19 per cent banded at one end to over 50 per cent at the other (Day & Dowdeswell 1968). Extensive data on predation by birds and mammals suggest that both the intensity of selection and its pattern are inadequate to account for the establishment and maintenance of such a cline over a number of years. Here again, it seems that the gene systems controlling shell banding may also exert physiological effects which somehow affect adaptation to a varying environment.

An important outcome of the kind of investigations outlined above is the realization that selection pressures in nature are far higher than was originally supposed. In our work with *M. jurtina*, we have frequently encountered selective elimination of the order of 60 per cent and on occasions it has reached 80 per cent. The data for survival in young *C. hortensis* on Portland Bill mentioned earlier suggest a selection pressure against banded morphs of about 55 per cent.

If pressures of this magnitude extend to other organisms, both plant and animal, and there is every indication that they do, it should be possible to detect evolutionary changes in the field quite easily and to establish situations suitable for teaching along the lines suggested earlier. The process can, of course, be enormously enhanced by the interference of man. Thus on the small island of Tean (Scilly Isles) a herd of cattle had kept the vegetation consistently grazed, and in places, almost lawn-

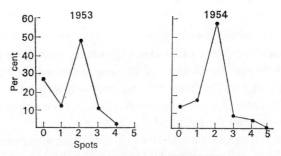

FIGURE 19.4. Spot-distribution in female *Maniola jurtina* on part of Tean, Isles of Scilly, showing the change associated with altered ecology in 1953 and 1954. (From Ford: *Mendelism and Evolution*. 7th ed. 1960.)

like. Their removal in 1950 resulted three years later in a growth of rank vegetation, the 'lawns' resembling hay fields. In one locality, this enabled a colony of *Maniola* to spread from a small restricted locality into an adjoining one which now afforded the necessary protection from the wind whereas it had not previously done so. The resulting adjustment in spotting of the females over a period of a year is shown in fig. 19.4 ($\chi^2_{(3)} = 8.8$; $0.05 > P > 0.02$). Evidently, a significant evolutionary change had taken place in a single generation (Ford 1965).

A LABORATORY MODEL

There are numerous instances in which the process of selection has been simulated and studied under laboratory conditions. Mention has already been made of one example, notably the work of Thoday and his associates on the effects of disruptive selection in *Drosophila*. However, this and similar models provide only a limited scope in teaching. Ideally, students need to be placed initially in a situation in which they can make an observation in the field, set up a hypothesis to explain it and then test that hypothesis under laboratory conditions. Finally, it may be necessary to refer the laboratory findings back to the field situation. Devising a laboratory study that meets these requirements is difficult but well worth the trouble involved, and the following suggestion is intended to emphasize, not so much practical details (which are readily obtainable from the sources indicated) as the important biological principles that can be brought out from such an exercise.

White Clover, *Trifolium repens*, occurs both as wild and cultivated populations throughout the temperate zones and exhibits an interesting dimorphism for cyanogenesis. This involves the production of easily detectable quantities of hydrogen cyanide should the cells of the leaves be damaged in any way. Two cyanogenic glucosides, linamarin and lotaustralin, provide the source of cyanide and are broken down very slowly

$$H_3C \diagdown \diagup C_2H_5$$
$$C\!-\!O\!-\!Glucose$$
$$|$$
$$CN$$

Lotaustralin

in vitro at normal temperature to ketone and hydrogen cyanide. But in

the presence of the enzyme linamarase the breakdown is greatly acceler-
ated and the resulting cyanide can be detected in a matter of a few hours.
In a normal leaf, glucoside and enzyme must somehow be kept apart but
it is not known how this is achieved. Conveniently, sodium picrate paper
provides a simple and rapid test for cyanogenesis, turning a reddish
brown colour in a matter of a few hours in the presence of HCN. The
reaction can be carried out easily in a small stoppered specimen tube or a
combustion tube sealed at the mouth with sellotape. A standard amount
of clover (say two leaves) is pushed to the bottom of the tube and two
drops of toluene added with a glass rod. This has the effect of breaking
down the cells so bringing together any glucoside and enzyme that may
be present. A strip of picrate paper is inserted into the tube and adjusted
so that it just does not touch the leaves. One advantage of this simple
procedure is that the tubes can be inserted in the soil beside each plant
under test, so making visual comparison easy.

The genetics of the situation involves two pairs of alleles. The presence
of glucoside is controlled by a dominant *Ac*, the homozygous recessive
ac ac causing its absence. Similarly, the dominant *Li* determines the
presence of enzyme and the recessive homozygote *li li* its deficiency. The
reactions of the four possible combinations to the sodium picrate test
are therefore as shown in table 19.2.

TABLE 19.2. The reactions of the four genotypic combinations of *Trifolium repens* with sodium picrate paper.

Genotype	Reaction with sodium picrate paper
Ac Li	Rapid and strong
Ac lili	Slow and weak
acac Li	None, unless glucoside added
acac lili	None, even when mixed with either enzyme or glucoside

Apart from its significance in the study of ecological genetics, the
situation is of some interest from the purely genetical standpoint in that
it provides a good example of a one gene—one enzyme relationship.

The adaptive aspects of the cyanogenesis polymorphism have been
studied in some detail by Daday (1954) not only in *Trifolium*, but also in
the closely related Bird's-Foot Trefoil, *Lotus corniculatus* (Family

Papilionaceae), and offer considerable possibilities for teaching, particularly for student project work. Jones (1962) has shown that cyanogenic plants of *Lotus* are avoided by a variety of herbivores, particularly certain insect larvae and molluscs. The existence of preferential feeding on *Trifolium* is more debatable and is badly in need of further study, particularly in the field. In my experience and that of several other teachers of biology, the snail *Helix aspersa* and slug *Agriolimax reticulatus* both tend to prefer acyanogenic forms of clover where these are available as an alternative to cyanogenic. However, recent work of Bishop and Korn (1969) under carefully controlled conditions indicated no significant difference in the amount of cyanogenic and acyanogenic plants eaten by *Helix* and *Agriolimax*. Clearly, there is a promising field here for further study not only in the direction already outlined, but also from the point of view of the feeding habits of invertebrates like snails. Could a tendency to eat acyanogenic plants depend to some extent upon the kind of diet to which a herbivore has been most used; does conditioning play a part in selection? Alternatively, do genetic strains of animals exist with different degrees of preference?

Daday has investigated another aspect of the cyanogenesis polymorphism in *Trifolium*—that concerned with the effect of temperature, and has shown that the occurrence of *Ac* and *Li* is closely correlated with the January isotherms, but not with those of July or with annual rainfall. The evidence that mean temperature has played a part in the evolutionary adjustment of *Trifolium* is supported by a study of the distribution of the two forms in the Alps (fig. 19.5), where the incidence of cyanogenesis decreases with altitude and increasing cold. It would be interesting to study distribution in this country both on mountains and in frost pockets. The Alpine situation can be simulated for teaching purposes by placing clover or trefoil leaves in a refrigerator at 4°C or below for about half an hour. The production of cyanide can then be tested in the usual way. Evidently, frost has the same effect as bruising in causing damage to the cells and enabling the glucoside and enzyme (if present) to come together.

In *L. corniculatus* at least, we thus have a double balanced polymorphism, the action of herbivores tending to favour cyanogenic phenotypes, the action of frost tending to decrease them. The situation in *T. repens* as far as animals are concerned is less clear, possibly because a smaller amount of cyanide is emitted from cyanogenic tissues than in *Lotus*. Some invertebrates such as the larvae of the Burnet moths, *Zygaena filipendulae* and *Z. lonicerae*, and of the Common Blue butterfly,

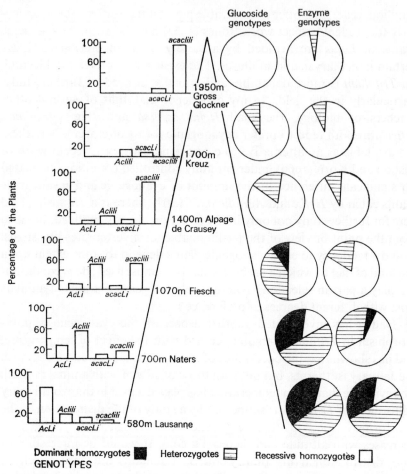

FIGURE 19.5. Cyanogenesis of *Trifolium repens* in relation to altitude (after Daday).

Polyommatus icarus, habitually feed on vetches and trefoils as their principal food plants. These animals have been shown to be quite un-selective when cyanogenic forms are present. However, while *Zygaena* builds up HCN in the body of the larva (detectable with sodium picrate) no doubt using it as some kind of defence mechanism, *Polyommatus* does not. Significantly, both are now known to contain the enzyme rhodanese which probably provides protection against the harmful effects of cyanide by breaking it down into the harmless thiocyanate ion. As might be expected, this adaptive mechanism also extends to the parasites of *Zygaena* larvae, notably the Braconid, *Apanteles zygaenarum*, which also possesses rhodanese. On the other hand, the nearly related *A. tetricus*, a parasite of the Meadow Brown butterfly, *M. jurtina*, whose larva is a grass feeder, contains no rhodanese—an interesting commentary on the biochemical nature of parasitic adaptation in the Genus.

TRENDS IN BIOLOGY TEACHING

Street has recently drawn attention to the implications for teaching of the escalation in the growth of biological knowledge during the last 20 years or so. Recent discoveries ranging over a wide field of biological science are inevitably promoting new syntheses in our knowledge and a gradual breakdown of some of the traditional interdisciplinary boundaries. For instance, the relationship between comparative anatomy, physiology and biochemistry has undergone a radical transformation as a result of the expansion of knowledge at the tissue, cellular and molecular levels. There is evidence that in botany, at least, the pendulum may already have swung too far and that a reduced knowledge of plant anatomy can have effects detrimental to a proper understanding of physiology. As Street points out, for those involved in teaching at all levels this is a period of experiment and debate. Just as major changes are taking place in attitudes to biology in our Secondary Schools, so too is a reorientation occurring in University Departments, both organizational and intellectual. Running parallel with these changes, are great advances in the broad field of educational technology concerned with the diversity of methods now available for presenting educational materials of all kinds. This is the context in which biology teaching is changing and in which new fields of knowledge such as ecological genetics can make an important contribution, both as subjects in their own right and by linking areas such as genetics and ecology that have tended to be unduly isolated from each other in the past.

REFERENCES

BISHOP J.A. & KORN M.E. (1969) Natural Selection and Cyanogenesis in White Clover, *Trifolium repens*. *Heredity* 24, 423–430.
CAIN A.J. & SHEPPARD P.M. (1950) Selection in the Polymorphic Land Snail *Cepaea nemoralis* (L). *Heredity* 4, 275–294.

CHITTY D. (1958) Self-regulation of numbers through changes in viability. *Cold Spring Harb. Symp. quant. Biol.* **22,** 277–280.

CREED E.R., DOWDESWELL W.H., FORD E.B. & MCWHIRTER K.G. (1970) Evolutionary Studies on *Maniola jurtina* (Lepidoptera, Satyridae). The 'Boundary Phenomenon' in Southern England 1961 to 1968. In Hecht & Steere (eds.), *Essays in Evolution and Genetics in Honour of Theodosius Dobzhansky.* Appleton-Century-Crofts, New York.

DADAY H. (1954) Gene frequencies in wild populations of *Trifolium repens* L. II. Distribution by Altitude. *Heredity* **8,** 377–384.

DAY J.C.L. & DOWDESWELL W.H. (1968) Natural Selection in *Cepaea* on Portland Bill. *Heredity* **23,** 169–188.

DOWDESWELL W.H., FISHER R.A. & FORD E.B. (1949) The Quantitative Study of Populations of the Lepidoptera. 2 *Maniola jurtina. Heredity* **3,** 67–84.

DOWDESWELL W.H. & MCWHIRTER K.G. (1967) Stability of Spot Distribution in *Maniola jurtina* throughout its Range. *Heredity* **22,** 187–210.

ELTON C.S. (1924) Periodic fluctuations in the numbers of animals: their causes and effects. *Br. J. exp. Biol.* **2,** 119.

FORD H.D. & FORD E.B. (1930) Fluctuation in Numbers and its Influence on Variation in *Melitaea aurinia. Trans. R. ent. Soc.* **78,** 345–351.

FORD E.B. (1965) *Ecological Genetics,* 2nd edn. Methuen, London.

FORD E.B. (1969) Ecological Genetics. In R. Harre (ed.), *Scientific Thought, 1900–1960,* Chap. 8, pp. 173–195. Oxford University Press, London.

GRIME J.P., BLYTHE G.M. & THORNTON J.D. (1970) Food Selection by the Snail *Cepaea nemoralis* L. In A.Watson (ed.), *British Ecological Society Symposium Number Ten.* Blackwell Scientific Publications, Oxford.

JONES D.A. (1962) Selective Eating of the Acyanogenic Form of the Plant *Lotus corniculatus* L. by Various Animals. *Nature* **193,** 1109.

JONES D.A., PARSONS J. & ROTHSCHILD M. (1962) Release of Hydrocyanic Acid from crushed tissues of all stages in the life-cycle of species of the Zygaeninae (Lepidoptera). *Nature* **193,** 52–53.

LAMBERT J.M. (ed.) (1967) The Teaching of Ecology. In *British Ecological Society Symposium Number Seven.* Blackwell Scientific Publications, Oxford.

LAMOTTE M. (1959) Polymorphism of natural populations of *Cepaea nemoralis. Cold Spring Harb. Symp. quant. Biol.* **24,** 65–84.

MCWHIRTER K.G. (1969) Heritability of spot-number in Scillonian strains of the Meadow Brown butterfly (*Maniola jurtina*). *Heredity* **24,** 313–318.

MCWHIRTER K.G. & SCALI V. (1966) Ecological Bacteriology of the Meadow Brown Butterfly. *Heredity* **21,** 517–521.

MILLICENT E. & THODAY J.M. (1961) Effects of disruptive selection. IV Gene-flow and divergence. *Heredity* **13,** 205–218.

PARSONS J. & ROTHSCHILD M. (1964) Rhodanese in the larva and pupa of the Common Blue Butterfly, *Polyommatus icarus* Rott. (Lepidoptera). *Entomologist's Gaz.* **15,** 58–59.

PUSEY J.G. (1966) Cyanogenesis in *Trifolium repens.* In C.D.Darlington & A.D. Bradshaw (eds.), *Teaching Genetics.* Oliver and Boyd, Edinburgh.

STREET H.E. (1970) A Degree in Biological Sciences. *Biologist* **17,** 92–96.

Index

379